The symbiotic city

THE SYMBIOTIC CITY

Voices of nature in urban transformations

edited by:

Marian Stuiver

Wageningen Academic
Publishers

EAN: 9789086863839
e-EAN: 9789086869350
ISBN: 978-90-8686-383-9
eISBN: 978-90-8686-935-0
DOI: 10.3920/978-90-8686-935-0

First published, 2022

Wageningen Academic Publishers,
P.O. Box 220,
NL-6700 AE Wageningen,
The Netherlands.
www.WageningenAcademic.com
copyright@WageningenAcademic.com

Table of contents

About the editor

Marian Stuiver (PhD) is head of the Green Cities programme at Wageningen Environmental Research, the Netherlands. Marian is a speaker, senior scientific researcher and advisor in multi-stakeholder programmes for green cities, food security and inclusive urban development. The work focuses on strategic as well as practical solutions to combat biodiversity decline in cities with a range of stakeholders from industry, civic society and government. Marian is a member of the Programme Panel of Wageningen Dialogues, where people are connected from different backgrounds and cultures for a common goal. Marian is also active in the LGBTIQA+ movement to develop more distributional, recognitional and participative equity for all people of all ages.

About the authors

Jan Broeze (PhD) works at Wageningen Food and Biobased Research, the Netherlands, as a senior scientist in food chain analysis and applications. He applies his analytical skills for analysing inefficiencies and impacts in food supply chains and for identifying the effects of interventions at food system level. He analyses and optimises food production systems, including circular utilisation of residue and waste streams. In this way he contributes to the sustainable development of agri-food chains through the valorisation of by-products.

Ingrid Coninx (PhD) is a social scientist working at Wageningen Environmental Research, the Netherlands. Her research focuses on impact and policy aspects related to climate change adaptation. She translates complex climate science into practical policy recommendations for climate action. Climate services are one way of easing the communication about climate change and actions. Her projects often take a systemic approach and are built upon theories and concepts of behavioural and organisational sciences.

Karianne de Bruin (PhD) works at Climate Adaptation Services (CAS), the Netherlands. Trained as an environmental economist, she has over 15 years of experience in research, project management and consulting in the areas of adaptation economics, decision-making under uncertainty, and climate services. She worked at the Center for International Climate Research (CICERO) in Oslo and at Wageningen Environmental Research (WENR) in Wageningen, where she led, among other things, the Dutch part of the European project ClimINVEST and worked on the economic assessment of urban blue-green infrastructures.

Bertram de Rooij (MSC) is senior researcher and landscape architect at Wageningen Environmental Research, the Netherlands. His focus is on strategic integrative spatial planning, concept development and policy support. Key topics are food-secure cities, climate change adaptation and green cities. He has developed research by design and participatory approaches. He is the creator of many of the illustrations in Chapter 2, 3 and 7. He is also coordinator of the Wageningen Fire Centre: integrating knowledge about wild (land) fire management and spatial planning/adaptation especially in terms of ecological management and safety of inhabitants.

Sjerp de Vries (PhD) is a senior scientific researcher at Wageningen Environment Research, the Netherlands. As an environmental psychologist he conducts research into the effect of contact with nature on the health and well-being of people. This concerns issues such as: which greenery is particularly important for the well-being of city residents? Does the biodiversity of that greenery play a role in this? Is green in the living environment equally important for everyone, or do certain population groups benefit more from it? And exactly how big are the benefits?

Ivo Demmers (MSc) is the director of the Food Security and Valuing Water programme of Wageningen University and Research, the Netherlands, which is a multi-disciplinary, multi-annual and organisation-wide programme on research, impact and valorisation in food security and the value of water. Ivo's expertise includes organisational development, entrepreneurship, environment, water policy, food security and international cooperation.

Luis Filipe Dias (PhD) received a PhD in Urbanism from the Faculty of Architecture, University of Lisbon, Portugal in 2016 entitled 'City, climate change and floods. A contribution to the urban resilience study'. He is currently working in the Centre of Ecology, Evolution and Environmental Change (cE3c) at the Faculty of Science, University of Lisbon, Portugal. His research focuses on climate change extremes, flood risk, urbanism, and socio-ecological resilience.

Anna Franz (PhD, FAIA, NAC) is an Associate Professor, Chair of the Department of Architecture and Urban Sustainability in the College of Agriculture, Urban Sustainability and Environmental Sciences at the University of the District of Columbia, and Program Director for Architecture and Community Planning. As a registered architect with degrees in nursing, architecture and engineering management, she specializes in whole system design and architectural research promoting healthy urbanism for people, society and culture. She is a Fellow in the American Institute of Architects and member of the National Academy of Construction.

Jim Groot (MSc) works as Scientist Supply Chain Management at Wageningen Food and Biobased Research, the Netherlands, and is a PhD student in Adaptive Value Chains. He is a project leader and researcher and focuses on design, development and optimisation of supply chain logistics for perishable products. More specifically his work entails topics such as redesigning food system networks dealing with indicators like cost, time, circularity, food quality, waste, protein and nutrients.

Sabine O'Hara (PhD) is an Agricultural Economist and Ecological Economist known for her expertise in sustainable community development and participatory valuation. Her work has focused consistently on the quality of life and economic opportunity of local communities through multi-dimensional intellectual, social, and physical capacity development. She currently leads an innovative PhD program in Urban Leadership & Entrepreneurship at the University of the District of Columbia (UDC), Washington DC, United States of America. Prior to her current appointment she served as founding Dean of the College of Agriculture Urban Sustainability and Environmental Sciences (CAUSES) and led UDC's efforts to build a cutting-edge model for urban agriculture and urban sustainability.

Eric Harris (Associate AIA, AIAS, NOMA) is a PhD student in Urban Leadership and Entrepreneurship at the University of the District of Columbia (UDC) in the College of Agriculture Urban Sustainability and Environmental Sciences (CAUSES), Washington DC, United States of America. His research focuses on expanding architectural education and practice through value-inclusive design. An advocate for inclusive design practices, Mr. Harris is the CAUSES Aspiration

Award recipient for Architecture and Urban Sustainability, the Clarence Pearson Award recipient in Recognition of Overall leadership and academic performance, and the 2020 Jason Pettigrew Memorial ARE Scholar. Eric currently works for Samaha Associates, P.C., a multi-disciplinary architecture, interior design, and planning firm. In addition, he is a coach/expert for Wageningen University and Research's Urban Greenhouse Challenge #3, the social edition at East Capital Urban Farm, Ward 7, in Washington, D.C.

Menno Houtstra (MSc) is the founder and manager of Urban Farm Cafe 'The KasKantine' in Amsterdam, the Netherlands, together with his partner Betty Gebhard. The KasKantine started in 2014 as an off-grid restaurant designed to temporarily occupy unused plots of land in the city, and uses locally sourced materials and ingredients. Currently, The KasKantine is being built up for the fourth time and consists of 13 refurbished shipping containers, three greenhouses, and a large garden. The project is supported by more than 30 volunteers and regularly hosts climate discussions and garden workshops. Menno Houtstra also co-manages a project to introduce the Doughnut Economy on a neighbourhood level, with a Doughnut Fund that connects grassroots organisations with local businesses and rebuilds community wealth.

Jan Eelco Jansma (MSc) works at Wageningen Plant Research, the Netherlands. He has been operating as action researcher in the field of sustainable urban food systems. He coordinated the Agromere research team (2006-2011) that inspired the city of Almere to include urban agriculture in the planning of the new area Oosterwold. He initiated and coordinated the Dutch city network of urban agriculture (2010-2017), which was able to put urban agriculture on local as well as national policy agendas, and he co-founded the Flevo Campus research programme (2017-), which studies new pathways for sustainable food consumption in the urban context of Almere. In addition, he is a PhD student focusing on urban planning of peri-urban agriculture.

Katarzyna Kujawa-Roeleveld (PhD) works at Wageningen University and Research as a Senior Lecturer and Researcher in Urban System Engineering. She lectures on water technologies, sustainable water management, closing urban and industrial water cycles, resource-oriented sanitation, removal of micropollutants from water, baseline assessment and design. She is the coordinator of MSc courses and supervises students. She coordinates international projects on sustainable water management and reuse in South East Asia (India, Bangladesh, Vietnam and Indonesia). She participates in sanitation projects for resource recovery in African countries (Ghana, Tanzania, South Africa).

Clara McDonnell (MSc) is a PhD student at the University of Amsterdam, the Netherlands in the Governance and Inclusive Development (GID) programme group. Her research focuses on the role of large investors, including pension funds, in leaving fossil fuels underground. She is also a volunteer in the 'De KasKantine' in Amsterdam and has previously worked as a consultant for Both Ends on fossil-free finance campaigning.

Lotte Roosendaal (MSc) works as Urban Food Systems advisor at Wageningen Centre for Development Innovation, the Netherlands. She has a background in global health and international development studies and her work focuses on urban food systems governance and food and nutrition from a spatial perspective as well as capacity strengthening on these topics. Landscape approaches, multi-stakeholder collaboration, transdisciplinary action research and facilitating learning processes are key areas of her expertise.

Shahryar Sarabi (MSc) is a PhD student at Eindhoven University of Technology, the Netherlands. He is a planner interested in bridging scientific and professional disciplines to help cities in their efforts toward a more sustainable and resilient future. His main research interests include adopting systemic views to study urban sustainability innovations. He has authored several articles, handbooks, and reports, including his recent publications investigating the challenges of implementing Nature-Based Solutions (NBS). He has developed a number of intelligent information systems facilitating the uptake of NBS.

Robert Snep (PhD) is a senior scientific researcher at Wageningen Environmental Research, the Netherlands. He has over 20 years of experience in Green City projects on the topics of biodiversity and urban ecosystems, climate adaptation, green business sites, green architecture and urban wildlife research. He is an author and speaker in scientific and scholar publications for diverse audiences. His focus is on having an impact in making cities greener with his wide (inter)national network of colleagues who all work on nature-based city planning & design.

Joop Spijker (MSc) works at Wageningen University and Research, the Netherlands. His research focuses on how urban greenery and forests and nature can contribute to the challenges of the 21st century, such as rising population pressure, the desire for a green and healthy living environment and adaptation to climate change (adaptation) and reduction of greenhouse gas emissions (mitigation). He works together with municipalities, companies in the green sector and NGOs, such as the Steenbreek Foundation in the Netherlands.

Rob Swart (PhD) is currently retired and works as an independent advisor on international climate change adaptation and mitigation in a broader sustainability context. Rob headed the European Topic Centre on Air and Climate Change of the European Environment Agency (EEA). Before that, as lead author of various IPCC reports and head of the IPCC WGIII Technical Support Unit he contributed in particular to the work on emissions scenarios and mitigation. In the late 1990s, he co-introduced the proposal to establish a 2° target for climate change. For most of his career he worked at the interface of science and policy for sustainable development at the Netherlands Environmental Assessment Agency (PBL).

Wim Timmermans (PhD) works on sustainable urban planning for a wide range of Dutch cities. At Wageningen Environmental Research, the Netherlands, he leads the topic Climate Resilient Cities in the Green Climate Solutions research programme. His background is in urban green planning and his specialties are interdisciplinary work, sustainability and complex problems in urban environments. He is lead author on 'The Vision of Arnhem, the Netherlands in 2120: a nature-inclusive, circular, social and healthy city', which was produced by a team of experts from various disciplines.

Wim van Dijk (MSc) is a researcher on nutrient and soil management on arable as well as dairy farms at Wageningen Plant Research, the Netherlands. Currently, a main issue is improving the circularity of nutrients in the food system. His topics are decreasing emissions, optimisation of manure use and re-use of side streams in the food system. He is also involved in circularity aspects of the 'Farm of the Future' in Lelystad in which future arable farming systems are designed and tested.

Maarten van Ginkel (MSc) is senior advisor climate adaptation and urban development for the city of Nijmegen, the Netherlands. Nijmegen is a Dutch city on the river Waal and combines large-scale climate adaptation measures with a strong emphasis on spatial quality. A new bypass of the Waal, called the Mirror Waal, is one of the major urban examples of the Dutch 'Room for the River' project.

Tim van Hattum (MSc) works at Wageningen Environmental Research, the Netherlands, as a consultant, policy advisor, project manager and researcher in the field of integrated water management and climate change. He has led several (inter)national projects and public private partnerships focusing on innovative and nature-based climate solutions. He is leading the Green Climate Solutions Programme at Wageningen Environmental Research, which aims to contribute to climate action by providing tailor-made climate impact information and an evidence base for nature-based solutions for mitigation and adaptation.

Henk van Ruitenbeek is a visual and narrative artist and expresses himself in drawings and in making manuscripts for theatre and plays. He is also an actor himself and has directed numerous theatre performances. One of his highlights was the performance Halte Lingezegen, organised in the context of the Landscape Triennale, the summer festival that took place in Park Lingezegen near Nijmegen in 2014. During the performance, the audience boarded a bus at the forgotten bus station Halte Lingezegen and got to know Park Lingezegen in a completely new way. Along the way, the Betuwe landscape revealed itself with its inhabitants through theatre performances and music. Henk received a prestigious award in 2022 from the Cultural Committee of Wageningen for the complete works he has made.

Wolfgang Weisser (PhD) works as a Professor in the Department of Ecology and Ecosystem Management, at the Technical University of Munich. He is interested in how human land use affects biodiversity and ecosystem processes, and how to use this knowledge to derive solutions for biodiversity conservation. Most of his work involves insects, and work in grasslands, forests, and more recently also in cities. He is very interested in species interactions, how they work, and how they are affected by human action.

Acknowledgements of the editor

This publication is the result of cooperation between many different stakeholders who build cities that are biodiverse, climate adaptive and healthy. However, this book is also contextual and based on scientific practices performed mostly in European and North American countries. Although it might not cover all the challenges of greening cities, climate adaptation and inclusion in all cities worldwide, it may be inspiring for all those who want to transform their urban environments into nature-based cities for everyone. Chapter 1, 2 and 13 are the sole responsibility of the editor in that respect.

The 13 illustrations, one for each chapter, are drawn by the artist Henk van Ruitenbeek from the Netherlands. The illustrations are the result of a series of conversations between the editor and the artist, in which we explored ways in which to visualise, confront and imagine the possibilities of the symbiotic city, a city that is not a reality yet. Our dialogue was a meeting between the social sciences and the arts in order to create new future projections that can help to inspire and encourage open conversations and boost creativity among all kinds of people to build their dream future.

I first want to thank Dr Sabine O'Hara. You have helped me tremendously in building a methodological approach that would do justice to the many local initiatives that I have encountered in so many cities over the last two decades. It was great to understand and shift from abstract notions of biodiversity, equity, systems thinking, transformations and transition, to the everyday practical steps people make in their surroundings, be it streets, gardens, shops, offices, houses, farms or neighbourhoods. I also want to thank all the authors for their dedication and commitment to working on both the scientific as well as the practical steps needed to understand how to give nature a voice in urban transformations. Every chapter is a unique contribution to this complex endeavour.

I want to acknowledge all colleagues and co-workers that I have met over the last few years for the conversations, (international) workshops, tours and field visits that all contributed to this book. Thank you for giving me the insights into issues of biodiversity conservation, social equity and climate justice. I am thankful for gaining knowledge on critical theories of the embedded economies, radical transformation and socio-technological change. I admire your dedication to making this world greener and more just. I am especially grateful for the years of mutual cooperation with my esteemed colleagues of the programme Green Cities of Wageningen University and Research. If I have forgotten to acknowledge you, please feel free to let me know; then I will do so in person and with great pleasure.

I explicitly want to mention the following people for their inspiring work, contributions and healthy criticisms throughout the process: Margaretha Bakker of Impunity Watch Utrecht, Claudia Bouwens of the Dutch Network Climate Adaptive Building with Nature (KAN), Onno Dwars of Ballast Nedam, Egbert Fransen of Pakhuis de Zwijger, Amsterdam, Linneke Heemskerck Duker of Green Circles, Province of South Holland, Lodewijk Hoekstra of Greenlabel NL, Hillebrand

Koning of the province of Flevoland, emeritus Professor Michiel Korthals, Professor Eveline van Leeuwen of Amsterdam Institute of Metropolitan Solutions, Professor Liesje Mommer of the Wageningen Biodiversity Initiative, Yvon van der Pijl of Utrecht University, Gaby Schmelzer of the Bee Association Wageningen, Professor Annemieke Roobeek of Nijenrode University and owner of GrwNxt, Line Rondard of the Institute Agro Rennes-Angers in France, Haniyeh Shariatmadary of CAUSES Washington DC, Margreet Takken of Thuis Wageningen, Daniel Silverstein of Ag Salon Washington DC, Rianne Valkenburg of TU Eindhoven, Frank Verhoeven of Boerenverstand Utrecht, Stefan Vreugdehil of Birds Association, the Netherlands, Maaike Wijngaard of the Dutch Ministry of Agriculture, Nature and Food Quality and Tutku Yuksel of Thuis Wageningen.

This publication is funded by the Wageningen University and Research Knowledge Base Programme Metropolitan Solutions (KB-25-007-001) and Knowledge Base Programme Circular and Climate Neutral (KB-34-004-012), supported by finance from the Dutch Ministry of Agriculture, Nature and Food Quality. Therefore, I thank Imke de Boer and Saskia Visser of WUR and Niek Hazendonk of the Dutch Ministry of Agriculture, Nature and Food Quality for their contributions.

The publication also builds upon work performed for the Rockefeller Foundation in 2020 and the Circular and Inclusive Cities Programme of the Alliance of University of Eindhoven, Wageningen University and Research, Utrecht University and UMC Utrecht, the Netherlands in 2022. Therefore, I thank Professor Huub Rijnaards of WUR, Professor Richard van de Sanden of TU Eindhoven, Professor Chris Backes of Utrecht University and Professor Peter Blankestijn of UMC Utrecht.

PART 1
THE SYMBIOTIC CITY

AALSMEER

Chapter 1
Bringing nature back on stage

Marian Stuiver

1.1 Introduction

In the coming decades, the urgency of revising the human relationship with nature will put great demands on everyone's responsibility. One way to address this responsibility is hidden in the argument that we live in the Anthropocene (Crutzen and Stoermer, 2013), a new geological epoch, marked by the extreme consequences of human interference in the planet's resources, climate change, environmental degradation and loss of biodiversity (Yusoff, 2018). Many urban practices, especially those that have high social and economic capital (Bourdieu, 1979), affect crucial aspects of our future existence.

The book introduces symbiotic cities as possible future urban landscapes of a revived human relationship with nature, reflected in embedded social-ecological, circular, institutional and technical networks that embrace nature (Arnaiz-Schmitz *et al.*, 2018). Symbiotic cities are embedded in and restorative for nature (Aliste *et al.*, 2018; Soper, 1998), and demand that we move into a reflexive Anthropocene where humans become aware of the consequences of their actions and decide to reconnect with the abiotic and biotic entities of soil, water, wind and organisms in urban practices. The urgency of climate adaptation, better health-supporting urban environments and the economic status of cities are additional arguments that will have a large impact in mainstreaming this nature-based method of urban transformation (FAO, 2021; UN-Habitat III, 2017; UN, 2015).

Urban life is the most dominant form of living for human beings (UN-Habitat III, 2017). In 1900, 12% of the population lived in cities, while the year 2007 witnessed the turning point when more than 50% of the world's population were living in cities. With the expected increase in the global population from 7.7 billion in 2019 to 9.7 billion in 2050, it is estimated that more than two thirds of us are expected to be living in cities (UN DESA, 2019). Urban areas over the last century have become a prime location for industrialisation, population growth, well-being and increases in per capita income.

The Industrial Revolution marked a fundamental turning point in many ways, including the rise of welfare for a part of the global population, the rise of capitalism, and increased and continued social inequity across the globe. It also drastically changed human relationships with nature. Technological developments put nature in the background and detached human-made infrastructures from their biotic and abiotic foundation (White, 1967). These changes in the modern era (Jacobs, 1961) were accompanied by modernistic urban planning scenarios. Architects began to build dense concentrations of buildings in the urban core with high-rise buildings

10.3920/978-90-8686-935-0_1, © Wageningen Academic Publishers 2022

and skyscrapers (Figure 1.1). Large suburbs arose where people lived in blocks of apartments and houses. These new buildings were destined for housing and citizens became connected by technical networks of underground trains, highways and streets to commute to business parks and city centres.

> Examples of such cities are New York, London and Paris where skyscrapers are scattered around the city centre, and space for green is often limited to human leisure purposes. Other examples are Amsterdam and Rotterdam in the Netherlands where in the sixties new districts full of high-rise buildings were built. For instance, in Amsterdam Zuid Oost, where Le Corbusier (1935), an advocate of modern urban planning, built the Bijlmermeer, a separate city in the sky, with high-rise buildings surrounded by parks and easily accessible by roads and underground railways. Other examples of such modernistic urban planning schemes are the geometric patterned socialist suburbs such as Nova Huta (Figure 1.2) in Krakau. Poland or Drumul Taberei in Bucharest, Rumania, where agricultural landowners were forced to leave and the governments developed a dense pattern of apartments up to ten stories high, made of concrete panels, surrounded by green spaces of lanes and pathways, and with direct access to the city centre.

These modernistic planning principles were meant to replace the slums that were so dominant before. Their overcrowding, dirt and unhealthy conditions were replaced by healthier homes, with low-rent housing that was surrounded by parks and lanes for health and fresh air. The greenery, such as parks and sport facilities or swimming areas, contrasted with the modest way of living in the apartment blocks. Nevertheless, this modernistic approach has been criticised for destroying inner city communities, and also because the parks and other public green spaces were too anonymous, unsafe and gave rise to criminality (Jacobs, 1961).

Figure 1.1. High-rise buildings in Amsterdam, the Netherlands.

Figure 1.2. Concrete in Nova Huta, Krakow, Poland.

In the last decades there has been ample evidence that modernity and its allies, agro-industrialisation and urbanisation, have a huge effect on the planet's resources and its resilience and vulnerability as a whole (Lenton *et al.*, 2019; Mazzucato, 2018; Raworth, 2012). There has been an upsurge of literature describing the challenges of climate change and ecological degradation (Biermann *et al.*, 2012). Climate change and ecological degradation are highlighted by the World Economic Forum as the largest risks for the global economy in their Global Risk Report (WEF, 2020). In general, many indicators of our planet's biodiversity are showing a decline. Landscapes and landscape diversity, which provide habitats for all species, are disappearing at an alarming rate (IPCC, 2022). These developments necessitate urgent changes and a rethink of the relationships between nature and people.

In the twentieth century, the compact city became an important paradigm with which to counter the externalities and environmental trade-offs of building these modernistic metropolitan areas at the expense of land-take and urban sprawl surrounding the area (Artmann *et al.*, 2019; Lehmann, 2010; Westerink *et al.*, 2013). The compact city paradigm aims to keep the spaces on the outskirts of the city for the purposes of natural lands and forests in order to form a carbon sink and provide natural ecosystems as a buffer for the city. In the compact city paradigm, social and economic concentrations of related forms of activity in the city are central to the city's design. Suppliers, production companies, logistics companies, all kinds of service providers, etc., are closely linked in their location in the urban areas. In this way a more dense city is effective in reducing transport costs (Clos, 2014), energy consumption and environmental pollution such as greenhouse gas emissions.

The compact city paradigm was a first answer to the problem of biodiversity decline and climate change and included other advantages in terms of public transport, accessibility to the city centre and land-use mix (Mouratidis, 2018; Westerink *et al.* 2011 referring to Calthorpe (1993) and Duany *et al.* (2000)).

At present, the densification of people, infrastructures and businesses makes cities vulnerable to climate risks, as they have in the past rarely been designed with climate change in mind. Climate threats such as floods, heat and drought have a huge impact on the urban economy, human health and well-being. One other result of the ongoing urbanisation together with continuing densification policies all over the world can be observed in the functioning of urban ecosystems (Breheny, 1997; Morrison and Weckroth, 2018; Tzoulas *et al.*, 2007).

This book takes urban areas as a focal point where humans can respond to biodiversity decline and climate change adaptation. It acknowledges the disastrous impact cities have had on nature, but shows that urban environments offer unexplored opportunities for their inhabitants to achieve sustainability goals. The ecological system in urban areas can potentially offer biodiversity hotspots that increase the resilience of cities, for instance in their adaptivity to meteorological extremes and flood or drought hazards (see for an example Figure 1.3). However, biodiversity hotspots in the urban environment need optimal connections between the biotic and abiotic foundation and the grey and green networks that the city is built up from. And in order to have a sustainable functioning of social-ecological networks, one needs specific biotic and abiotic conditions, which need to be managed properly. This complexity poses a challenge to urban planning for biodiversity that goes beyond sustainable urban planning from the compact city paradigm.

The book proposes to develop symbiotic cities in which nature is perceived as a partner of all citizens in urban transformations (Haase *et al.*, 2017). We need to transform our cities irreversibly and build social-ecological networks, re-rooted in the abiotic and biotic foundations of soil, water, wind and organisms as well as in the socio-diversity of cities (Hölscher *et al.*, 2021). Participants within these networks must lead these re-rooting practices that promote urban health for both

Figure 1.3. Green wadi in Pijnacker, the Netherlands (courtesy of Nanda Sluysmans).

humans and non-humans (Latour, 1987). The symbiotic city makes all the voices of nature heard again in the design of cities (Pascual *et al.*, 2021). And because these voices of nature are diverse, they will contribute not only to biodiversity but also to social diversity in cities (Kruize *et al.*, 2019; Raworth, 2012). By working together with nature, urban actors and their practices of urban planning and design, real estate development, water management, and food production can transform their related practices into true social-ecological ones (Gunderson and Holling, 2002; Walker *et al.*, 2004).

This transformation, which is so essential to creating symbiotic cities, is already taking place in many places (Hölscher *et al.*, 2021) The absence of nature in the compact design of urban areas is increasingly being scrutinised by citizens and scholars who embrace concepts such as the organic city (Lehmann, 2010), rewilding (Arts *et al.*, 2022), urban landscaping (Van Stiphout, 2019), nature-based solutions (Roggema *et al.*, 2021), agro-ecology (Anderson *et al.*, 2021) and food localism (WUR, 2021). In recent decades, social and environmental factors have been introduced in urban planning theories to improve both the environmental and sustainability standards, and the health and well-being of people (Andersson *et al.*, 2014; Haase *et al.*, 2014). All these concepts are linked to wider narratives about designing the city of the future (Brenner and Schmid, 2015).

All these initiatives become connected in an age of the reflexive Anthropocene, where more and more urban practices occur, with growing coalitions of citizens critically rethinking their assumptions and making use of self-criticism to accompany the desired social and political progress (Smith, 2016). Related to this is the concept of 'ecological reflexivity (Faber, 2018)': recognition of the non-human world and its actors (Latour, 1996), reflection upon what functions and what we need to get rid of (Loorbach *et al.*, 2020), and developing new practices and experiments for regime change (Kemp *et al.*, 2001). In the reflexive Anthropocene, we know that the urban dweller and their economic and social activities are embedded (Polanyi, 1944) and connected with the flora and fauna in meadows and oceans through, for instance, global greenhouse gas emissions, plastic waste and other types of connections due to our production and consumption practices (Raworth, 2017; Schumacher, 1973; Spash, 1999). In other words, we are inherently connected with and have an influence on social-ecological systems (Carson *et al.*, 2016). In the reflexive Anthropocene we know that we can only support transformations if we become reflexive citizens and acknowledge our agency (Giddens, 1990; 1991) and moral responsibilities (Bauman, 1995; Beck, 1992) in shaping socio-ecological systems (Gunderson and Holling, 2002). In the reflexive Anthropocene, human agency (Westley *et al.*, 2013) can become the guiding principle in shaping socio-ecological systems (Figure 1.4).

The concept of reflexive citizenship (Dietz, 1991) goes beyond the dichotomy of the human being as a free (enlightened) being who makes rational choices and the human being who is controlled by his environment, the rulers, the advertising, the regime. In the first view, the human being is the rational actor who knows what he is doing, what he is deciding, what he is eating, the homo economicus who makes decisions based on his rational understanding. In the second view, people are the product of their environment and therefore need to be nudged and tempted to undertake certain behaviour, because the human is a routine animal and does not intend to change so

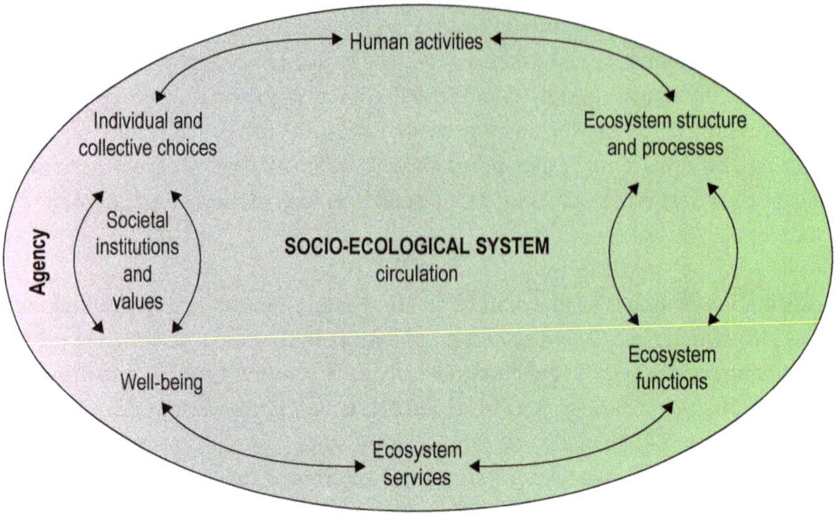

Figure 1.4. Agency in a social-ecological system (Carson *et al.*, 2016).

quickly. We take a third position and argue that; both the rationality of the first position and the routines of the second position are supported by agency and morality: it is in the hierarchy of values a person has that they make a choice. This hierarchy is not necessarily rational or routine, but it does presuppose a degree of self-reflection. The human being will have to become a learning human being if she does not want to be a robot of her own rationality, or of her made environment (Bauman, 1997, 2009; Bauman and Tester, 2001).

1.2 From local practices to urban transformations

Cities are composed of urbanites that are united in the double triple helix of Figure 1.5: municipal governments, large corporates, knowledge institutes, citizens and neighbourhood communities, NGOs and businesses. These actors need to work together and reflect on the meaning of nature in the city, as nature still has two dominant meanings and places in the city: natural entities are often perceived as either capital or consumer good. These perceptions of nature are among the root causes of the world's environmental crises. For example, when we talk of 'ecosystem services' and natural production values, we are suggesting that the Earth's fabric holds no value apart from what it provides to us (Routley and Routley, 1980; White, 1967) Urban dwellers have become separated from nature in their cities because of these dominant perceptions that frame humans as separate from and superior to the non-human world. Also, Actor Network Theory and other disciplines in Philosophy (Latour, 1996; Morton, 2010) have argued that the natural world is much closer and connected to us than the former belief systems tell us. Think of the microbes in our body for a start, but also our body itself.

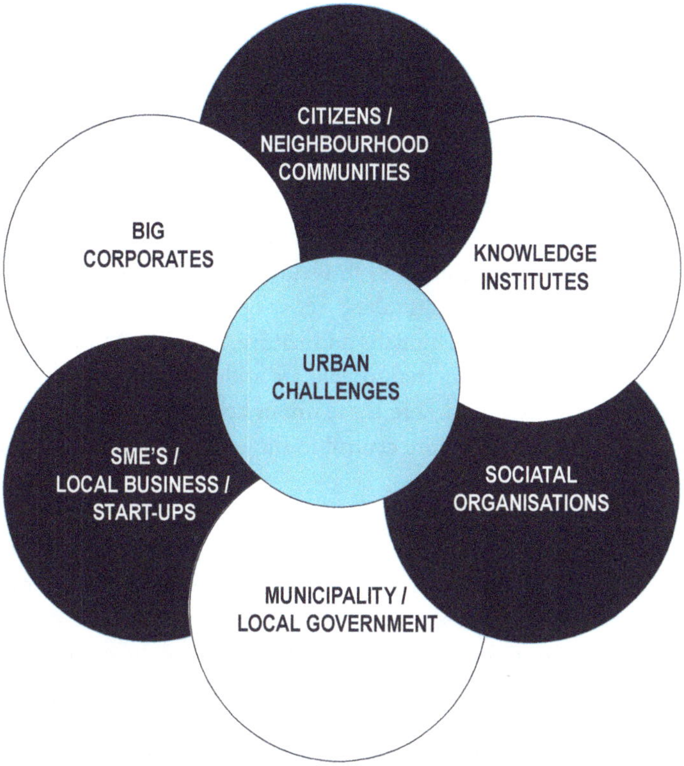

Figure 1.5. The double triple helix of the urban innovation ecosystem (source: Pakhuis de Zwijger).

When nature is perceived as a capital good, it is a means of production that indirectly, through the goods and services it produces, contributes to the satisfaction of people's needs. Traditional capital goods that can be found in the urban (and rural) areas are roads, railroads, machines, trucks, tankers, buildings, support services such as biomass production or soil fertility, supply services such as food, water, fibre, energy, medicine, and regulating services such as water retention, climate adaptation, disease prevention, and flood prevention. When nature is perceived as a consumer good, it is a source of inspiration, awe, mysticism, surprise, wonder, relaxation and beauty. In terms of ecosystem services, these are so-called cultural services that ecosystems provide. A similarity with nature as a capital good is that nature should not be totally consumed when we interact with it. For instance, if overfishing is avoided, the ponds and lakes from which people fish will also supply fish in the future. Or after an invigorating walk in the park, the urban forest should be no less attractive for the next user. With nature as a consumer good, there are opportunities in nature for recreation, spiritual enrichment and education.

We propose that we acknowledge the voices of urban actors that work on a third perception, of nature as heritage. In nature as heritage, nature has an 'intrinsic value', a value of itself, outside of man. In a nutshell, the Earth we inherited, including its species' richness, must be passed on to the generations after us in as complete a form as possible. This perception of nature fits with Brundtland's definition of Sustainable Development (1987) and the SDGs of the United Nations that focus on taking good care of the Earth to ensure that we pass it on to future generations in as good a condition as possible (UN, 2015). In recent decades many urbanites have started cooperating in order to restore their perception of nature as heritage. These urbanites are a source of inspiration and act as central figures in this book. They stress the symbiotic relationships between humans and nature, and focus on transformative action, beyond the Anthropocentric views on nature that are still so dominant. Throughout the book all types of practices are presented where these urban actors develop the necessary stepping stones (Figure 1.6) to make nature more central to the issues of urbanisation, rather than the buildings (Van Stiphout, 2019).

If one radically wants to transform present urban regimes, one needs to rethink the rules on how to act and not to act in these urban areas. In this book we describe several transformative practices that already question these rules and therefore confront the power vested in the present urban regimes. If our goal is 'ultimately influencing – the patterns and pathways of contemporary urbanization' (Brenner and Schmid, 2015), we need to be not just epistemological, but methodological, political, and action-oriented (Issar, 2018) with a new understanding of 'what to know' and 'how to know' (Rip, 2002; Stuiver, 2008). Crucial in this respect is the celebration of cultural diversity that embraces social diversity. If the transformation is only accessible for some groups of actors with their knowledge and information, this is bad for the legitimacy and diversity

Figure 1.6. Stepping stones in Utrecht, the Netherlands, courtesy of Nanda Sluysmans.

of perspectives (Berkes and Folke, 1998; Folke *et al.*, 2005; Olsson *et al.*, 2004). Transformations occurs when they establish direction, align, motivate and inspire a large diversity of people with the ultimate goal of producing movement and irreversible change (Hölscher *et al.*, 2018; Pascual *et al.*, 2021).

1.3 Outline of the book

Chapter 2 gives an in-depth account of the symbiotic city. The symbiotic city departs from the worldview that not humans but ecosystems are at the centre of things. According to this ecocentric viewpoint, nature has intrinsic value, a value in itself. Symbiotic cities are dynamic urban landscapes of beneficial relationships between people and nature. Urban actors can work on these beneficial relationships with nature in many different forms. Assumptions of the symbiotic city are therefore optimism, ecocentrism, inclusion, nature based thinking and the importance of embedded networks. The chapter provides the reader with the building blocks of the symbiotic city and guides the reader to 'symbiotic thinking' in the future design of cities. The chapter will set out a three-pronged framework which differentiates between the foundations, the networks and the human practices that build the symbiotic city. The framework in that way aims to re-root the practices of humans and the networks of the city in the biotic and abiotic foundations of the Earth.

In Chapter 3 future visions of the symbiotic city in Washington DC and a region central in the Netherlands in 2050 are presented. First, a vision for a City Region Food System (CRFS) in Washington DC in 2050 that focuses on eight different storylines to link food production with a nature-based and socially inclusive urban economy embedded in the wider metropolitan region. Second, a vision of a symbiotic city in the middle of the Netherlands in 2050 where nature, biodiversity, water, soils, climate and biodiversity go hand in hand with eight different social-ecological practices in which people express their values of nature. Together with stakeholders the eight storylines were co-created and translated into eight perspectives with transformative potential to make the city a reality in 2050.

The first set of practices focuses on the value of adaptation: in the city there are symbiotic connections between soil, water, biodiversity and technical networks that improve the adaptivity and resilience of cities toward future climate challenges. The second set of practices focuses on the cultural and communal values of nature. Social diversity stabilises communities in the same way that biodiversity stabilizes ecosystems. The third set of practices focuses on green neighbourhoods and districts that adopt nature-based thinking in urban practices to replenish resources and restore the foundations of soil, water and sustain ecosystem health. The fourth set of practices gives space to the intrinsic value of nature: urban forestry, restoration and regeneration, creating biodiversity hotspots. The fifth set of practices values water and gives space to water smart cities, smart water management to re-root the foundations of the city. The sixth set of practices focuses on circularity and nature-based technologies and innovation that enhance natural resources and strengthen sink capacities rather than depleting them. The seventh set of practices focuses on individual values of nature, personal lifestyles and health, ensuring lifestyles that integrate nature in their experience. The eighth set of practices focus on restorative urban food production and consumption: creating food production systems in the urban scape that are diverse, sustainable, and able to link urban, peri-urban, and rural communities.

After this, the chapters are presented in three parts with an emphasis on Nature, Food and People. The chapters present a diversity of actors that already work on the symbiotic city. First, through new projects to integrate cities with the abiotic and biotic foundations of nature, e.g. wildlife and nature restoration in cities, the use and necessity of water management, and the role of climate services to make cities more resilient. This includes Chapter 4 on the wildlife city, Chapter 5 on the water smart city and Chapter 6 on climate services. Second, through the development of nature-based City Regional Food Systems (CRFS) to restore the natural resources (water, soil, biodiversity) of the city and improve local food security issues (Wiskerke, 2015). This includes Chapter 7 on the importance of a nature-based CRFS, Chapter 8 on circularity and Chapter 9 on the importance of restorative economics. Third, through the importance of nature for health and well-being of citizens, the role of local communities in greening the city and the much-needed inclusion of all urban human actors in urban design. This includes Chapter 10 on nature and health, Chapter 11 on neighbourhoods and inclusive civic engagement and Chapter 12 on a design canvas for inclusion to build the symbiotic city.

Chapter 4 departs from the view that cities are landscapes accommodating a variety of different ecosystems that harbour wildlife. Creating wildlife habitats in an urban context provides opportunities to conserve biodiversity and reconnect humans with nature. Metropolitan regions are, however, not yet designed and managed for the needs of wildlife. Addressing wildlife habitat development in an urban context first means raising awareness among the involved city stakeholders, getting them enthusiastic about the benefits of a greener and more natural city environment, and providing clues on how to implement habitat development in the real-life case of a built-up location. In this chapter we discuss the overall phenomenon of green, biodiverse residential and business sites. The chapter describes which characteristics of such locations provide opportunities for which type of wildlife, identifies relevant stakeholder groups and accompanying arguments, and demonstrates the added value of biodiverse residential and business sites. The first case is the Heineken Brewery (100 ha) in Zoeterwoude (NL) where more than 20 ha of intensively managed urban green was turned into a wildlife habitat resulting in a sudden abundance of rare orchids, hover flies and other pollinators. The second case is a nature-inclusive residential area in Eindhoven (NL). The chapter concludes with some lessons learned about how to turn residential and business sites into green, wildlife-friendly living and working places.

Chapter 5 addresses a shift in urban planning and urban water management. Cities around the world need to rethink the way they deal with water and redesign cities to become water smart cities using three strategies: (1) restore degraded ecosystems in natural surroundings of cities; (2) large-scale implementation of nature-based solutions; (3) close the urban water cycle and treat water as a resource. The water smart city approach utilises, where possible, nature-based solutions combined with smart technology, actions that are inspired by, supported by or copied from nature. It integrates urban planning and water management on a district, city and catchment scale. Furthermore, innovation and close collaboration between local stakeholders, businesses and research institutes is a key part of this approach, to continuously learn and improve by collecting and analysing data and creating an evidence base for nature-based adaptation in order to become less vulnerable to floods and droughts. The case studies presented are Oslo, New Delhi and Ho Chi Minh City.

Chapter 6 departs from the city as a complex system that can be made more resilient in the long term. Resilience is the capacity of a system to absorb disturbance and reorganise while undergoing change so as to retain essentially the same function, structure, identity, and feedbacks. It elaborates how climate change can be a driver for resilient urban development with the help of climate services. Climate services are tools, products and processes that can help to integrate climate information into urban policymaking and design for improving resilience. Because of the long-term nature of climate change, climate services force urban developers to adopt a long-term, transformative perspective in urban planning and design. Two illustrative cases are presented: Nijmegen (in the Netherlands) and Lisbon (in Portugal). The cases illustrate the modalities of climate services to trigger action: climate services have to be flexible and tailored to the specific local situation and they should be integrated with other urban challenges in order to develop innovative multifunctional urban solutions.

Chapter 7 argues that City Region Food Systems must depart from the natural and social characteristics of a particular city and its surrounding region. Following the Milan Urban Food Policy Pact, cities around the world are increasingly taking the lead in designing their own City Region Food Systems. A City Region Food System (CRFS) approach aims to foster the development of resilient and sustainable food systems within urban centres, peri-urban and rural areas surrounding cities by strengthening rural-urban linkages. In order to make the transformation more resilient, healthy and sustainable, it needs to be nature-based within the particular rural-urban region. A CRFS therefore needs three building blocks: (1) it includes a wide range of food system actors; (2) it fosters good governance arrangements; and (3) it focuses on decision-making, learning and leadership.

Chapter 8 focuses on the urgency that City Region Food Systems need to prevent natural resources from reaching their limits. Cities and their regions generate large 'sources' of residual material and waste flows with a high organic component. These flows offer prospects for 'circular valorisation', either recycling or replacing other (partly fossil) resources. Urban areas therefore represent attractive starting points in a transition towards circularity. In this chapter we look at the transition needed in the use of phosphorus (P), essential for life and an integral part of our food system (Cordell *et al.*, 2011), and yet it is a finite resource, obtained by mining. Reserves are limited and whilst estimates of global reserves vary between one hundred and several hundred years, it is believed that most current producers will have depleted their stocks within a century (Cooper *et al.*, 2011).This resource is used to analyse the challenges of circularity in urban areas. This chapter discusses two examples of P flows. The first explores the potential of a circular City Region Food System on the regional scale based on a scenario for the urban area of Almere, part of the Metropolitan Region of Amsterdam (MRA), selected because of its location adjacent to a large agricultural hinterland, the Flevo Polder. The second explores a case study that encompassed the entire MRA region. It explores the actual urban P-flow: what are the actual quantities of the P flows and to what degree is the P loop already closed? The chapter concludes that a circular food system is technically feasible and ends with recommendations on how to enhance circular urban food production that does not deplete but preserves its natural resources.

Chapter 9 focuses on the importance of restorative economics and food hubs as catalysts for a new urban economy. The common conception of the economy is a throughput model where inputs are converted into useful goods and services. Emissions and waste are the unfortunate by-product of this conversion process. Inputs and emissions and waste by-products seemingly come from nowhere and disappear into nowhere. In other words, the common conception is that the economy has no context. As a result, the unintended side effects of economic activity disappear from the purview of economics. Economists refer to these unaccounted side effects of economic activity as 'externalities'. The deficiencies of this context-less model have long been recognised by ecological and social economists. This chapter outlines a new economic model developed by the lead author called restorative economics. Like the doughnut model developed by Kate Raworth, it places the economy in its social and environmental context and makes visible long-neglected social and environmental impacts of economic activity. The restorative economics model, however, makes a further distinction regarding the context of economic activity that makes the model more operational. It distinguishes between sources and sinks. Sinks are the counterpoints to the resources that feed the front end of the economic process. Sinks deal with the tail end of the process. The restorative economics model therefore places value not only on the resources used to produce goods and services and on these outputs of goods and services themselves, but also on the many social and environmental sink factors that absorb emissions, process waste, reduce stress, and alleviate social tension. Restorative economics pays attention to both the resources and the sinks needed to sustain economic activity in the long run. We illustrate the value of these sinks by reviewing several local and regional initiatives from Washington DC, namely the Urban Food Hubs model, the Five Pillars of Economic Development, the Food Connects vision and the ambitious agenda of the Sustainable DC plan. We conclude the chapter by reviewing policies that support a restorative economy.

Chapter 10 provides a review on the relation between nature in cities and the health of citizens. It gives an overview of studies that have shown that the presence of greenery reduces illness. The first time the effect of nature on stress relief was properly investigated, the results were so remarkable that the top journal Science (Ulrich, 1984) made room for them. American patients who had a view of trees from their room recovered measurably faster from surgery than patients with a view of a blank wall. In the Netherlands, researchers showed that people are healthier if the living environment contains more greenery. First, it addresses which greenery is particularly important for the well-being of city residents. Is green in the living environment equally important for everyone, or do certain population groups benefit more from it? And how big are the benefits exactly? Second, it addresses the question of whether the biodiversity of that greenery plays a role in this.

Chapter 11 focuses on three pathways to enhance the social capital needed for greening cities; communal learning, new types of ownership for citizens' initiatives; making the impact of citizens' activities and practices more explicit. Three cases demonstrate each of these pathways. First, Operation Steenbreek, an initiative of universities, municipalities and urban ecologists, exemplifies communal learning through local actions (Stobbelaar *et al.*, 2021). Operation Steenbreek aims to engage citizens and local municipalities in greening public and private gardens by educating them about and demonstrating the added value of greening. In 2021 over 140 municipalities and local initiatives joined Operation Steenbreek. Second, communal ownership is exemplified by the Kaskantine in Amsterdam, where people in the neighbourhood organise their own urban farm café and use and design all kinds of technologies for off-grid living in the city. Third, communal impact is exemplified by the Doughnut Coalition in Amsterdam, that has developed their own bottom-up measurement technique to measure the social impact of their activities. The chapter will conclude with recommendations that local governments and urban planners need to take into account in order to involve communities in their greening strategies.

Chapter 12 argues that citizens need to be part of the design process to co-create their neighbourhoods, wards and districts. In the coming years, it is essential for landscape and urban designers, planners as well as architects to take up inclusive planning and design processes that acknowledge human needs and dignity, and foster participation in shared decision-making. Underpinning these processes are values of empathy, placation, accessibility, and identity essential to imagining and creating inclusive green cities. This chapter provides a deeper understanding of the different dimensions of an inclusive green city and how social equity is an integral part of any design effort. We begin by exploring our vision and the values necessary for green urban design. A framework for three aspects of equity – recognitional equity, procedural equity, and distributional equity – is presented with selected case studies that serve as evolving good practices for equitable green urban design. From this framework and resulting values, we built on a value-inclusive design canvas and present design principles that planners, designers and architects can adopt as their own.

Chapter 13 focuses on how to become a voice of nature in the transformation process needed to achieve the symbiotic city. This transformation requires a whole subset of transformative changes in cities. These range from re-embedding cities in the biotic and abiotic foundations of water, soil and organisms to building new institutional, technical, circular and socio-ecological networks in the city. Examples of these transformative practices have been described in the previous chapters with case studies in which people express different voices of nature. First, they point to the importance of the agency of urban actors, i.e. their ability to change existing urban practices in symbiotic ones. Second, they stress the importance of new networks, including governance arrangements that favour nature-based thinking. Third, economics needs to be re-embedded into society and ecology, resulting in new economic narratives that embrace nature. Finally, the inclusion of all people and their expressions of the values of nature as well as their heterogeneous knowledge in the transformation processes are central to enforcing the transformations. Grouped in six categories of citizens, businesses, NGOs, corporations, municipalities and knowledge organisations, actors can work together on the transformation to make the symbiotic city happen.

In this chapter we introduce these actors and we address the challenges that every type of actor can work on starting from tomorrow, from strategic, to tactical, to operational and reflexive activities that can be performed to become a voice of nature in urban transformations.

References

Aliste, E., Folchi, M. and Núñez, A., 2018. Discourses of nature in new perceptions of the natural landscape in Southern Chile. Frontiers in Psychology 9. https://doi.org/10.3389/fpsyg.2018.01177

Anderson, C.R., Bruil, J., Chappell, M.J., Kiss, C. and Pimbert, M.P., 2021. Agroecology now!: transformations towards more just and sustainable food systems. Springer Nature, Berlin, Germany.

Andersson, E., Barthel, S., Borgström, S., Colding, J., Elmqvist, T., Folke, C. and Gren, Å., 2014. Reconnecting cities to the biosphere: stewardship of green infrastructure and urban ecosystem services. Ambio 43: 445-453. https://doi.org/10.1007/s13280-014-0506-y

Arnaiz-Schmitz, C., Schmitz, M.F., Herrero-Jáuregui, C., Gutiérrez-Angonese, J., Pineda, F.D. and Montes, C., 2018. Identifying socio-ecological networks in rural-urban gradients: diagnosis of a changing cultural landscape. Science of the Total Environment 612: 625-635. https://doi.org/10.1016/j.scitotenv.2017.08.215

Artmann, M., Inostroza, L. and Fan, P., 2019. From urban sprawl to compact green cities – advancing multi-scale and multi-dimensional analysis. Ecological Indicators 96: 1-2. https://doi.org/10.1016/j.ecolind.2018.10.058

Arts, K., Bakker, L. and Buys, A., 2022. Rewilding in Nederland. KNNV Uitgeverij, Zeist, the Netherlands.

Bauman, Z., 1995. Searching for a centre that holds. In: Featherstone, M., Lash, S. and Robertson, R. (eds.) Global modernities. Sage, London, UK.

Bauman, Z., 1997. Postmodernity and its discontents. Polity Press, Oxford, UK.

Bauman, Z., 2009. Does ethics have a chance in a world of consumers? Harvard University Press, Cambridge, MA, USA. https://doi.org/10.4159/9780674033948

Bauman, Z. and Tester, K., 2001. Conversations with Zygmunt Bauman. Polity Press, Oxford, UK.

Beck, U., 1992. Risk society. Sage, London, UK.

Berkes, F. and Folke, C., 1998. Linking social and ecological systems: management practices and social mechanisms for building resilience. Cambridge University Press, Cambridge, UK.

Biermann, F., Abbott, K., Andresen, S., Bäckstrand, K., Bernstein, S., Betsill, M.M., Bulkeley, H., Cashore, B., Clapp, J., Folke, C., Gupta, A., Gupta, J., Haas, P.M., Jordan, A., Kanie, N., Kluvánková-Oravská, T., Lebel, L., Liverman, D., Meadowcroft, J., Mitchell, R.B., Newell, P., Oberthür, S., Olsson, L., Pattberg, P., Sánchez-Rodríguez, R., Schroeder, H., Underdal, A., Vieira, S.C., Vogel, C., Young, O.R., Brock, A. and Zondervan, R., 2012. Navigating the anthropocene: improving earth system governance. Science 335: 1306-1307. https://doi.org/10.1126/science.1217255

Bourdieu, P., 1979. Symbolic power. Critique of Anthropology 4: 77-85. https://doi.org/10.1177/0308275x7900401307

Breheny, M., 1997. Urban compaction: feasible and acceptable? Cities 14: 209-217. https://doi.org/10.1016/S0264-2751(97)00005-X

Brenner, N. and Schmid, C., 2015. Towards a new epistemology of the urban? City 19: 151-182. https://doi.org/10.1080/13604813.2015.1014712

Calthorpe, P., 1993. The next American Metropolis: ecology, community, and the American dream. Princeton Architectural Press, New York, NY, USA.

Carson, M., Sommerkorn, M., Behe, C., Cornell, S.E., Gamble, J., Mustonen, T., Peterson, G.D. and Vlasova, T., 2016. An arctic resilience assessment. In: Carson, M. and Peterson, G. (eds.) Arctic resilience report. Stockholm Environment Institute and Stockholm Resilience Centre, Stockholm, Sweden, pp. 2-26.

Clos, J., 2014. Compact cities to address climate change. Available at: https://tinyurl.com/4s2jwa8t.

Cooper, J., Lombardi, R., Boardman, D. and Carliell-Marquet, C., 2011. The future distribution and production of global phosphate rock reserves. Resources, Conservation and Recycling 57: 78-86. https://doi.org/10.1016/j.resconrec.2011.09.009

Corbusier, L., 1935. La ville radieuse, éléments d'une doctrine d'urbanisme pour l'équipement de la civilisation machiniste: Paris, Genève, Rio de Janeiro, Sao Paolo, Montevideo, Buenos-Aires, Alger, Moscou, Anvers, Barcelone, Stockholm, Nemours, Piacé. Éditions de l'architecture d'aujourd'hui.

Cordell, D., Rosemarin, A., Schröder, J.J. and Smit, A.L., 2011. Towards global phosphorus security: a systems framework for phosphorus recovery and reuse options. Chemosphere 84: 747-758. https://doi.org/10.1016/j.chemosphere.2011.02.032

Crutzen, P.J. and Stoermer, E.F., 2013. The 'Anthropocene (2000). In: Libby, R., Sverker, S. and Paul, W. (eds.) The future of nature: documents of global change. Yale University Press, pp. 479-490. https://doi.org/10.12987/9780300188479-041

Dietz, M., 1991. `Hannah Arendt and Feminist Politics'. In: C. Pateman and M.L. Shanley (eds.), Feminist interpretations and political theory. Polity Press, Oxford.

Duany, A., Plater-Zyberk, E. and Speck, J., 2000. Suburban Nation: The Rise of Sprawl and the Decline of the American Dream. North Point Press, New York.

Faber, A., 2018. De gemaakte planeet: Leven in het Antropoceen. Amsterdam University Press, New Haven, CT, USA.

FAO, 2021. Indigenous peoples' food systems: insights on sustainability and resilience from the front line of climate change. FAO, Rome, Italy. Available at: https://doi.org/10.4060/cb5131en.

Folke, C., Hahn, T., Olsson, P. and Norberg, J., 2005. Adaptive governance of social-ecological systems. Annual Review of Environment and Resources 30: 441-473. https://doi.org/10.1146/annurev.energy.30.050504.144511

Giddens, A., 1990. The consequences of modernity. Polity Press, Oxford, UK.

Giddens, A., 1991. Modernity and self-identity: self and society in the late modern age. Polity Press, Oxford, UK.

Gunderson, L.H. and Holling, C.S., 2002. Panarchy: understanding transformations in human and natural systems. Island Press, Washington DC, USA.

Haase, D., Frantzeskaki, N. and Elmqvist, T., 2014. Ecosystem services in urban landscapes: practical applications and governance implications. Ambio 43: 407-412. https://doi.org/10.1007/s13280-014-0503-1

Haase, D., Kabisch, S., Haase, A., Andersson, E., Banzhaf, E., Baró, F., Brenck, M., Fischer, L.K., Frantzeskaki, N. and Kabisch, N., 2017. Greening cities – to be socially inclusive? About the alleged paradox of society and ecology in cities. Habitat International 64: 41-48.

Hölscher, K., Wittmayer, J.M., Hirschnitz-Garbers, M., Olfert, A., Walther, J., Schiller, G. and Brunnow, B., 2021. Transforming science and society? Methodological lessons from and for transformation research. Research Evaluation 30: 73-89. https://doi.org/10.1093/reseval/rvaa034

Hölscher, K., Wittmayer, J.M. and Loorbach, D., 2018. Transition versus transformation: What's the difference? Environmental Innovation and Societal Transitions 27: 1-3. https://doi.org/10.1016/j.eist.2017.10.007

IPCC, 2022. Climate Change 2022: Impacts, adaptation, and vulnerability contribution of Working Group II to the Sixth Assessment Report of the Intergovernmental Panel on Climate Change. Cambridge University Press, Cambridge, UK.

Issar, S., 2018. Editorial: We have never been urban. City 22: 1-4. https://doi.org/10.1080/13604813.2018.1443594

Jacobs, J., 1961. The death and life of great American cities. Random House, New York, NY, USA.

Kemp, R., Rip, A. and Schot, J., 2001. Constructing transition paths through the management of niches, Path dependence and creation. Lawrence Erlbaum, Mahway, NJ, USA, pp. 269-299.

Kruize, H., van der Vliet, N., Staatsen, B., Bell, R., Chiabai, A., Muiños, G., Higgins, S., Quiroga, S., Martinez-Juarez, P., Aberg Yngwe, M., Tsichlas, F., Karnaki, P., Lima, M.L., García de Jalón, S., Khan, M., Morris, G. and Stegeman, I., 2019. Urban green space: creating a triple win for environmental sustainability, health, and health equity through behavior change. International Journal of Environmental Research and Public Health 16: 4403.

Latour, B., 1987. Science in action: how to follow scientists and engineers through society. Harvard University Press, Cambridge, MA, USA.

Latour, B., 1996. On actor-network theory: a few clarifications. Soziale Welt 47: 369-381.

Lehmann, S., 2010. The principles of green urbanism: transforming the city for sustainability. Earthscan London, UK.

Lenton, T., Rockström, J., Gaffney, O., Rahmstorf, S., Richardson, K., Steffen, W. and Schellnhuber, H., 2019. Climate tipping points – too risky to bet against. Nature 575: 592-595. https://doi.org/10.1038/d41586-019-03595-0

Loorbach, D., Wittmayer, J., Avelino, F., von Wirth, T. and Frantzeskaki, N., 2020. Transformative innovation and translocal diffusion. Environmental Innovation and Societal Transitions 35: 251-260. https://doi.org/10.1016/j.eist.2020.01.009

Mazzucato, M., 2018. The value of everything: making and taking in the global economy. Penguin Books Limited, London, UK.

Morrison, P.S. and Weckroth, M., 2018. Human values, subjective well-being and the metropolitan region. Regional Studies 52: 325-337. https://doi.org/10.1080/00343404.2017.1331036

Morton, T., 2010. The ecological thought. Harvard University Press, Cambridge, MA, USA.

Mouratidis, K., 2018. Is compact city livable? The impact of compact versus sprawled neighbourhoods on neighbourhood satisfaction. Urban Studies 55: 2408-2430. https://doi.org/10.1177/0042098017729109

Olsson, P., Folke, C. and Hahn, T., 2004. Social-ecological transformation for ecosystem management the development of adaptive co-management of a wetland landscape in Southern Sweden. Ecology and Society 9: 2. Available at: https://www.ecologyandsociety.org/vol9/iss4/art2/.

Pascual, U., Adams, W.M., Díaz, S., Lele, S., Mace, G.M. and Turnhout, E., 2021. Biodiversity and the challenge of pluralism. Nature Sustainability 4: 567-572. https://doi.org/10.1038/s41893-021-00694-7

Polanyi, K., 1944. The great transformation: the political and economic origins of our time. Farrar and Rinehart, New York, NY, USA.

Putnam, R.D., 2000. Bowling alone: the collapse and revival of American community. Proceedings of the 2000 ACM conference on Computer supported cooperative work. Association for Computing Machinery, Philadelphia, Pennsylvania, USA, pp. 357. https://doi.org/10.1145/358916.361990

Raworth, K., 2012. A safe and just space for humanity: can we live within the doughnut? Oxfam International. Available at: https://www.oxfam.org/en/research/safe-and-just-space-humanity.

Raworth, K., 2017. Doughnut economics: seven ways to think like a 21st-century economist. Random House, London, UK.

Rip, A., 2002. Science for the 21st century. The future of science and the humanities. Amsterdam University Press, Amsterdam, the Netherlands, pp. 99-148.

Roggema, R., Tillie, N. and Keeffe, G., 2021. Nature-based urbanization: scan opportunities, determine directions and create inspiring ecologies. Land 10: 651. https://doi.org/10.3390/land10060651

Routley, V. and Routley, R., 1980. Social theories, self management, and environmental problems. In: Mannison, D.S., McRobbie, M.A. and Sylvan, R. (eds.) Environmental philosophy. Dept. Of Philosophy, Research School of Social Sciences, Australian National University, pp. 217-332.

Schumacher, E.F., 1973. Small is beautiful: a study of economics as if people mattered. Blond and Briggs, London, UK.

Smith, S.B., 2016. Modernity and its discontents: making and unmaking the bourgeois from Machiavelli to Bellow. Yale University Press, New Haven, CT, USA.

Soper, K., 1998. Representing nature. Capitalism Nature Socialism 9: 61-65. https://doi.org/10.1080/10455759809358833

Spash, C., 1999. The development of environmental thinking in economics. Environmental Values 8: 413-435. https://doi.org/10.3197/096327199129341897

Stobbelaar, D.J., Van der Knaap, W. and Spijker, J., 2021. Greening the city: how to get rid of garden pavement! The 'Steenbreek' program as a Dutch example. Sustainability 13: 3117.

Stuiver, M., 2008. Regime change and storylines. A sociological analysis of manure practices in contemporary Dutch dairy farming. Wageningen University and Research, Wageningen, the Netherlands.

Tzoulas, K., Korpela, K., Venn, S., Yli-Pelkonen, V., Kaźmierczak, A., Niemela, J. and James, P., 2007. Promoting ecosystem and human health in urban areas using Green Infrastructure: A Literature review. Landscape and Urban Planning 81: 167-178. https://doi.org/10.1016/j.landurbplan.2007.02.001

Ulrich, R.S., 1984. View through a window may influence recovery from surgery. Science 224: 420-421. https://doi.org/10.1126/science.6143402

UN-Habitat III, 2017. The new urban agenda. United Nations, New York, NY, USA.

UN, 2015. Transforming our world: the 2030 Agenda for Sustainable Development. United Nations, New York, NY, USA.

UN DESA, 2019. World population prospects 2019: highlights. ST/ESA/SER.A/423., United Nations Department of Economic and Social Affairs, Population Division, New York, NY, USA.

Van Stiphout, M., 2019. First guide to nature inclusive design. Available at: https://nextcity.nl/first-guide-for-nature-inclusive-design/.

Walker, B., Holling, C.S., Carpenter, S. and Kinzig, A., 2004. Resilience, adaptability and transformability in social-ecological systems. Ecology and Society 9. https://doi.org/10.5751/ES-00650-090205

WCED, 1987. Our common future. Oxford University Press, Oxford, UK.

WEF, 2020. Global risk report. World Economic Forum, Geneve, Switzerland. Available at: https://www.weforum.org/reports/the-global-risks-report-2020/.

Westerink, J., Haase, D., Bauer, A., Ravetz, J., Jarrige, F. and Aalbers, C.B.E.M., 2013. Dealing with sustainability trade-offs of the compact city in peri-urban planning across European city regions. European Planning Studies 21: 473-497. https://doi.org/10.1080/09654313.2012.722927

Westerink, J., Kiers, M.A., Kok, K. and Aalbers, C.B.E.M., 2011. Scenarios and enhanced, strategies. Case study The Hague Region, the Netherlands.

Westley, F.R., Tjornbo, O., Schultz, L., Olsson, P., Folke, C., Crona, B. and Bodin, Ö., 2013. A theory of transformative agency in linked social-ecological systems. Ecology and Society 18. https://doi.org/10.5751/ES-05072-180327

White, L., 1967. The historical roots of our ecologic crisis. Science 155: 1203-1207. https://doi.org/10.1126/science.155.3767.1203

Wiskerke, J.S., 2015. Urban food systems. In: De Zeeuw, H. and Drechzel, P. (eds.) Cities and agriculture. Routledge, London, UK, pp. 19-43.

WUR, 2021. Agroecological agriculture requires tailored biodiversity. Available at: https://tinyurl.com/4rn2f4zj.

Yusoff, K., 2018. A billion black anthropocenes or none. University of Minnesota Press, Minneapolis, MN, USA.

Chapter 2

The symbiotic city as the sum of beneficial relationships between people and nature

Marian Stuiver

Abstract

This chapter guides the reader towards 'symbiotic thinking' in the future design of cities. Symbiotic cities are dynamic urban landscapes of beneficial relationships between people and nature. Urban actors can work on these beneficial relationships with nature in many different forms. Assumptions of the symbiotic city are therefore optimism, ecocentrism, inclusion, nature based thinking and the importance of embedded networks. The chapter sets out a three-pronged framework with which to understand the symbiotic city. It differentiates between the foundation, the networks and the symbiotic human practices that build the future symbiotic city. The foundation of the city, composed of biotic and abiotic entities such as soil, water and living organisms, needs to re-enter the networks and human practices of the city to meet the definition of the symbiotic city.

Keywords: symbiosis, dependencies, foundation, networks, human practices, nature-based thinking, embedded networks, ecocentrism, values of nature

2.1 Introduction

In modern, densely built-up cities, ecosystems have moved to the background of people's day-to-day lives and nature has taken a place outside the urban living environment (Lehmann, 2010). There is little space for greenery at ground level, soils are sealed and hardly available for any natural function or service as their main task is providing physical support and infrastructures such as cables, pipes, tunnels, etc. The urban networked infrastructure has negative effects on water regulation (Chung *et al.*, 2021), biodiversity and carbon sequestration (De Kimpe and Morel, 2000; Du *et al.*, 2015; EC, 2012). Even today, when new high-rise buildings, apartments or houses are built, the loss of soil, water and biodiversity is hardly compensated for by creating more greenspace elsewhere (Haaland and Van den Bosch, 2015; Tappert *et al.*, 2018). The open green spaces that are left in the modern urban areas mostly have a social, recreational and aesthetic function; parks, flowerbeds and hedges do not link up to ecological networks in the wider natural environment in or beyond the city and are not working towards a naturally functioning region. Consequently, city designs are bad for the health of ecosystems as well as the quality of life for citizens within the urban areas (Hansen *et al.*, 2019; Hedblom *et al.*, 2017).

10.3920/978-90-8686-935-0_2, © Wageningen Academic Publishers 2022

The good news is that it is possible to build cities in symbiosis with nature (Mata *et al.*, 2020) (The Nature Conservancy, 2016). In order to understand that this is possible, the first thing is to acknowledge that cities at present are the result of human practices of the past (Hardcastle *et al.*, 2005) that need to change and be transformed into practices that reflect a symbiotic relationship with nature. Humans possess agency, the capability to act and change. Unlike structuralist notions, which see human practice as expressions and humans themselves often as victims of the deeper structures of market economics, this book takes a more optimistic stance. In line with Anthony Giddens' theory of structuration (Turner, 1986), it views society as the result of interactions between human practices and existing structures. Humans and manmade structures such as the economy and technology, mutually change and influence each other (Hardcastle *et al.*, 2005; Turner, 1986). People can decide to replace old habits with new practices or alter their daily routines and practices. Actors can alter existing practices through strategic, tactical, operational (Loorbach *et al.*, 2015; Rotmans and Loorbach, 2009) and reflexive activities.

In today's cities, two types of human practices in relation to nature are dominant: impact reduction and risk management (Veeneklaas, 2012); and a third relationship is becoming more visible. This third relationship is called symbiosis and is the sum of beneficial dependencies between city dwellers and their natural environment. First, impact reduction concerns human activities to reduce the effect of their practices on nature. It is measured in environmental impact scores, biodiversity measurements and water quality scores. Second, risk management concerns human activities to improve people's dependence on nature, framed as external risks. For example, risk analyses looks at what external natural circumstances mean for heat waves and rain patterns due to climate change. Third, in symbiosis, nature becomes an intrinsic partner in human design efforts. In recent years there has been an increasing recognition in the literature, for example in theories of Complex Adaptive Systems, that the relationship between nature and humans is mutual (Berkes and Folke, 1998; Gunderson and Holling, 2002). The relationship is outward-looking: what can citizens do to protect the values of biodiversity, nature and the environment? The relationship is also inward-looking: what does the natural environment mean for the future of the city? The purpose of the symbiotic city is therefore mutual: to prevent the city from collapsing from external stress (Walker *et al.*, 2002) and to make use of what nature offers the city, without reducing these resources so that the natural system can repair itself.

In Section 2.2 assumptions of the symbiotic city are presented. Section 2.3 will set out a three-pronged framework for understanding the symbiotic city that differentiates between the foundation, the networks and the practices that build the symbiotic city. Section 2.4 will explain the foundation of the symbiotic city. Section 2.5 will elaborate on the networks of the symbiotic city. Section 2.6 presents eight sets of practices that can give expression to different values of nature. Section 2.7 presents the conclusions.

2.2 Assumptions of the symbiotic city

The symbiotic city is inspired by landscape-based approaches (Pedroli *et al.*, 2021; Van Rooij *et al.*, 2021), the biophilic movement (Schwab, 2019), the organic city, the doughnut economy (Raworth, 2017) the natural city (Reichow, 1948), rewilding cities (Arts *et al.*, 2022; Pereira and Navarro, 2015), neighbourhood thinking (Hajer *et al.*, 2021), and nature-based solutions for urban areas (Tzoulas *et al.*, 2021). Symbiosis is a term that stems from the biological sciences. Symbiosis can be defined as any kind of relationship or interaction between two dissimilar organisms, each of which may receive benefits from their partners that they did not have while living alone (Angelard and Bever, 2013).

The first assumption of the symbiotic city is optimism about today's cities, societies and its inhabitants: beyond the dark clouds of climate change and biodiversity loss, citizens, mayors, home owners, project developers and scientists are actively uniting themselves in new coalitions and organisations to build cities and their regions in a way that they become meaningful for nature, the environment as well as humans.

New symbiotic relationships between nature and the economy are being explored by real estate companies and commercial businesses. Greenery cools and dresses up buildings and increases the attractiveness of an area and as a result creates a better climate for both citizens and businesses. Greenery boosts the values of houses: housing prices are higher in a green neighbourhood than in a neighbourhood without green space when one controls for the size of the house and other qualitative features. When one greens the areas in business parks (see for an example Chapter 4) employees have healthier environments to live in and have a break during work. There are also many coalitions in circular or regenerative agriculture that seek a symbiotic relationship with their environment. The combination of food production, while using fewer inputs and practising soil management, serves as a way of increasing biodiversity as well as sustainable food production for people. Moreover, it has a large educational and communicative value for people because they get to know the natural environment and the origins of food. Vice versa, if citizens engage in urban food production in an ecologically sound way, it can contribute to biodiversity and ecosystem restoration of the city.

This book shows that the new dependencies between humans and nature are already being transformed by urban actors in so-called urban eco-innovation networks. Governments, corporations and science, businesses, NGOs and citizens' initiatives develop strategic visions for nature-based urban transformations. They also make tactical and operational decisions to develop socio-ecological networks, for instance in real estate development, water management, food production, but also in neighbourhoods, with new green community and business activities. Moreover, the actors develop reflexivity how to make social-ecological practices where nature and humans mutually benefit from each other.

One example is the Doughnut coalition in Amsterdam, where urbanites in line with the theory of the Doughnut Economy (Raworth, 2017) focus on economic and social activities to meet the needs of everyone within the carrying capacity of the Earth and live within the (planetary) boundaries of nature (Stefanovich, 2011). In 2021, eight community organisations, based in four neighbourhoods in Amsterdam Nieuw West, decided to create 'Neighbourhood Doughnuts'. The use of the Doughnut coalition is to mobilise/monitor civic engagement in taking care of social and natural resources in the neighbourhood. In many cases, natural resources have been in the hands of the government for such a long time that citizens need to relearn to recognise natural resources such as public land, roofs, food waste, parks and gardens. The neighbourhood doughnut is an investment in growing social capital, as it helps to bind people together in trust. With its focus on communal values, it enables mutually beneficial cooperation between groups that would normally not cooperate so easily. In this way it helps to overcome polarisation in the public debate. It also focuses on building new relationships between people so that citizens can start to cooperate and thus optimise the local resources in the neighbourhood for their own, as well as for the communities' sake (see also Chapter 11).

The second assumption is that we need to explore the diverse values of nature into the urban architectural or landscape design. Instead of the centuries-long practice of controlling and managing nature for human needs, nature should be allowed to reclaim urban areas (Pereira and Navarro, 2015). Therefore it is necessary to explore a more ecocentric or 'wild' worldview (Blok, 2022). According to this ecocentric worldview, nature has an intrinsic value, a value in itself. The city blends harmoniously into the natural world and is conceptualised as part of nature; it must therefore be organised and managed accordingly (Randrup *et al.*, 2020). Such cities that stress the interplay between the built and the natural environment are also known as 'biophilic cities'. Biophilic design is more than just adding natural elements to the urban environment. It is an ethos that views design in the built environment not merely as an aesthetic or functional discipline, but also recognises the positive impact of nature on improved mental and physical well-being (Schwab, 2019).

For example, in the Netherlands both the beaver and the wolf have re-entered the stage in the last few decades. Wolves, although often viewed with anxiety and fear, have a symbiotic relationship with their natural surroundings. They control other animal populations such as deer, that in return have an influence on the vegetation, which in return remains a viable ecosystem, allowing soils to absorb rain, which in return prevents droughts and fires (Jepson and Blythe, 2020). Beavers have been reintroduced, as their activities around the water in building dams and gnawing trees creates biodiversity and aids in water management. The gnawing of trees creates new space for herbs, butterflies and other insects. And the dams that the beavers build help the areas to retain water better after heavy rainfall.

The third assumption of the symbiotic city is that it can only be built with the inclusion of all people (Cook and Hegtvedt, 1983). Inclusion shares an intense relationship with its opposite term: exclusion (Levitas, 2005). In most cities exclusion is visible in specific districts or wards. The interplay between the physical environment, socio-economic and cultural conditions leads to a spatially concentrated exclusion (Gan *et al.*, 2017). Exclusion has many causes and many forms, including cultural, social, economic and environmental aspects, which are all interconnected (Bertrand *et al.*, 2015). It is essential to address and overcome exclusion, as everyone has a role to play in the design process to co-create symbiotic neighbourhoods, wards, and districts. The symbiotic city therefore develops biodiversity as well as social diversity (Bourdieu, 1972; Putnam, 2000). Inclusive urban design means the ongoing process of designing solutions to accommodate the viewpoints, experiences, and situations of people that were not accommodated before (Nussbaum, 2011).

An example of inclusive symbiotics is the development of greenery around school yards, churches, hospitals and sports facilities for the production of fruit and vegetables. These mini-orchards can be adopted and maintained by patients, inhabitants, teachers, pupils, parents, clubs, teams, corporate partners or other parties. The harvested fruit and vegetables would then serve as the backbone of the supply line of healthy food to the inhabitants, supplemented by new local-for-local fruit chains. Urban farming can establish significant vegetable and fruit production in the urban environment, creating mini biodiversity hotspots and promoting social cohesion in urban neighbourhoods (Canal Vieira *et al.*, 2021; Stuiver and O'Hara, 2021; Wiskerke, 2015). By making more room for (good quality) green in the city in the form of parks, playing fields and public gardens, city inhabitants can meet more easily in a low-threshold manner. These public spaces can prove an essential means for citizens to participate in the development of their own neighbourhoods (Chapter 11).

The fourth assumption closely follows this argument of inclusion and embraces nature-based thinking (Randrup *et al.*, 2020) which adds to the more utilitarian narrative of 'Nature-based Solutions' (NBS). Nature-based solutions have set the international urban agenda for further knowledge development and an evidence base (Nesshöver *et al.*, 2017) as '... solutions that are inspired and supported by nature, which are cost-effective, simultaneously provide environmental, social and economic benefits and help to build resilience' (EC, 2021). They are often expert-driven solutions, developed in scientific circles where nature has instrumental value for society (Tzoulas *et al.*, 2021). Examples of nature-based solutions are green roofs and roof gardens, vertical vegetation walls, rain gardens, green corridors, bioswales, retention ponds, urban wetlands and urban forestry projects, water squares and floating islands.

Timmermans and colleagues (2022) developed a future vision of Arnhem, a city in the Netherlands with nature-based solutions as their starting point. The solutions they propose are, for instance, giving way to the river as a form of climate adaptation, developing a new economy of sawmills to make materials to build houses and floating homes. Also, new lakes have been constructed for energy storage and specially designed breeze blows come from between the greenery-covered buildings to counter heat waves. Another example of NBS from the Netherlands is the 'Building with Nature' project that focused on developing nature-based solutions for larger infrastructural works with multifunctional designs that include environmental aspects, natural systems and technological capabilities to enable new approaches. Examples are: (1) the Sand Motor that combines the primary function of coastal protection with the creation of a new natural landscape that also provides new leisure opportunities; (2) the Sand Engine for a more sustainable, cheaper and ecologically robust coastal protection of the Friesian IJsselmeer coast; and (3) Tidal Parcs where a vegetated slope of the river bank breaks the waves when water levels are high, contributing to a future-proof Rotterdam region.

Nature-based solutions have a great advantage, as they bring more natural features and processes into cities through locally adapted, resource-efficient and systemic interventions (EEA, 2015). However, they remain expressions of and firmly embedded in a utilitarian framework of nature where nature has mostly instrumental value for society as a capital good. Nature-based thinking is therefore introduced to allow for the social diversity of ways in which people connect with biodiversity and can rediscover, understand, detect, organise and redistribute biodiversity, greenery and other urban natural resources available in their day-to-day environment (Pascual *et al.*, 2021). It transforms the abundant supply of nature and natural resources into all sorts of social and ecological assets different values beyond the capital goods value.

Economists distinguish between at least four concepts of value including market (exchange) value, use value, future (options) value, and intrinsic value (Kakovitch and O'Hara 2018). It also captures value concepts such as the relational value between people and nature connections for personal experience and health. Nature-based thinking also recognises that the value of a resource may be in its non-use or resource-saving rather than its use. Kakovitch and O'Hara (2021), for example, demonstrated this with respect to the non-use of water, which proved to be essential to the ability to maintain productive output, while higher water use and reduced water savings lowered output levels as measured in GDP. Kakovitch and O'Hara, however, also show in their analysis of the hydrological cycle and its impact on economic activity that the non-use of water has economic benefits and can therefore be viewed as utilitarian as well (Kakovitch and O'Hara, 2021). Beyond its expression in individual valuations, nature can also be captured in discursive processes (Hajer, 1995), especially when those whose voices often go unheard are at the table to represent the voice of nature (O'Hara, 1996). Yet even a nature-centred perspective must be careful to recognise that the non-human members of the nature-humans relationship depend on humans to articulate their value.

Urban green environments with a diverse vegetation will attract a variety of life, in particular soil organisms, insects, birds and mammals. These organisms are crucial for the services nature renders as well as having a value of their own. Developing urban greening has great potential to deliver on the climate mitigation goals while at the same time providing important benefits to soils, air, water and biodiversity. The win-win potential is tremendous. Green spaces in urban environments make people feel better. Children benefit from it when they can play in green (and safe) public areas. Being close to greenery lowers stress and supports recovery after illness (Maas *et al.*, 2006). City vegetation reduces fine dust and hence its very harmful effect on human health. Vegetation cools the city in summer: vegetated soils can reduce city heat due to vegetation evaporation, and trees provide shadow. In this way green soils can contribute to human health and well-being (De Vries *et al.*, 2003).

The fifth assumption is that we need to overcome dualistic notions, such as social versus ecological, human versus nature, economy versus ecology, technology versus society and the urban versus the rural and replace it with a networked approach to urban transformations. The notions are intrinsically linked and their mutual dependency becomes clear in the perspective of embedded (Polanyi, 1944) networks in the symbiotic city. The urban system of the symbiotic city potentially incorporates a wide range of networked connections including social-ecological networks, technical networks, institutional networks and circular networks (see Section 2.5. for an elaborate explanation).

It is also important to focus on the networked connections between landscape regions. In his book on organic urban planning, Reichow proposed the establishment of a new-style urban landscape (in German: Stadtlandschaft) built up from citizens' neighbourhoods that were kept closely attuned to their surrounding natural landscapes (Reichow, 1948). Circular networks evidently have to be integrated on a city-regional landscape level. Circular networks pay close attention to the regionally available sinks that buffer, absorb, and process the emissions and waste products that are the result of any production, resource use, and consumption process (see Chapter 7 for an example). Pedroli *et al.* (2021) propose to develop a circular landscape perspective where the urban core revitalises the rural region in a way to counteract climate change and biodiversity decrease. The networked approach of the symbiotic city perspective will become evident in Chapter 3, where the visions connect several seemingly disparate domains including nature, culture, economy, policy and technology in new embedded networks.

Urban soil multifunctionality (FAO and ITPS, 2022; O'Riordan *et al.*, 2021; Wang *et al.*, 2015) is an indispensable dimension of developing new circular networks on an urban landscape level. Soil enhancements can improve soil quality and carbon sequestration; green coverage of trees and shrubs can absorb heat, improve oxygen levels and reduce carbon dioxide. Nutrient recycling in production processes can reduce nutrient losses and reduce pollution levels. These practices can save natural resources and reduce pollution pressures on other resources (Anderson *et al.*, 2021; Pascucci, 2020; Stuiver and O'Hara, 2021). Developing open soils also means that rainfall water can have a natural pathway to the groundwater system. In this way it prevents flooding through the limited capacity of city drainage systems during large rainfall events. This is becoming increasingly important for citizens as current climate change is expected to increase the frequency of heavy rainfall events.

2.3 The three-pronged symbiotic city

The symbiotic city is a three-pronged approach (Figure 2.1) to understand and develop cities' and citizens' beneficial relationships with nature. It distinguishes between the foundation, the networks and symbiotic human practices.

This three-pronged approach of the symbiotic city is inspired by the work of Dutch landscape architects and adds a socio-ecological and agency perspective to their work. The landscape based approach introduces a tripartite structure that determines the designs of our environments: the layers of foundation, network and occupation (CRA, 2021). They focus on the technical and ecological aspects of the three layers of the landscape. Their approach is also a hierarchical structure; the first layer, the foundation of soil, water and animals, with its biotic and abiotic functions, determines the second layer of greenery, waterways, railroads, highways and ditches. These networks in return determine the third layer, the occupation layer of (high-rise) buildings and farms, sport facilities and wind turbines, factories and public parks (Hagens, 2006).

The symbiotic city approach adds a socio-ecological perspective to the landscape based approach. It add elements of socio-ecological systems theories (Gunderson and Holling, 2002; Walker *et al.*, 2004; Westley *et al.*, 2013) to the third layer. This occupation layer is built up from symbiotic

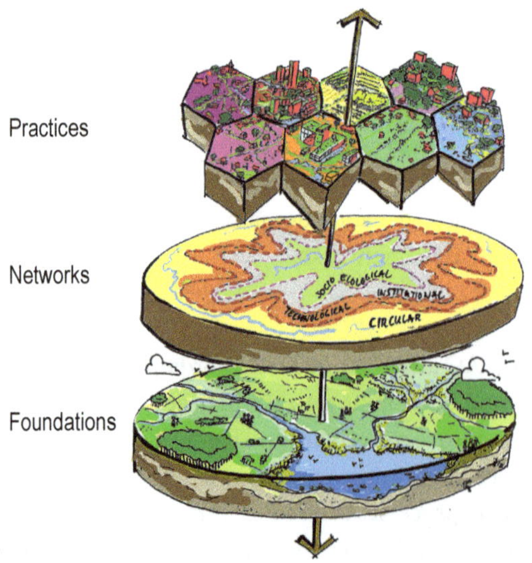

Figure 2.1. Three layers of the symbiotic city.

human practices. These symbiotic practices express different human values of nature (and are embedded) in the second layer of the socio-ecological, circular, institutional and technical networks. Human practices are structured by these networks but human agency also makes it possible that new networks are built. The symbiotic city approach therefore also introduces an agency perspective on how humans are able to transform their urban practices into symbiotic ones (Albrechts and Mandelbaum, 2005; Giddens, 1990; Hardcastle *et al.*, 2005; Turner, 1986). The human practices can restructure existing networks of urban areas in order to become more symbiotic with nature. The third layer of the symbiotic practices is inspired in iteration by the research in Chapter 3, where we have determined how different human values of nature are visible in hyperlocal initiatives among stakeholders in Washington DC and in the Netherlands.

2.4 The foundation: biotic and abiotic systems

The symbiotic city is embedded in the foundation (Figure 2.2) of the Earth that encompass biotic and abiotic factors. Biotic factors are living things such as plants, animals, and bacteria, while abiotic factors are non-living components, such as water, soil and atmosphere. For a long time, the abiotic soil and water systems have been looked at from their functions for physical and chemical processes in agriculture and physical infrastructure; the biotic factors and processes were not so relevant (Sonneveld and Bouma, 2003). However, the biotic processes are an important aspect of the foundation of the city (Guilland *et al.*, 2018; O'Riordan *et al.*, 2021). The Earth is no longer dead matter with resources or ecosystems to exploit, but a living system whose health depends on the health of its organs and tissues – its wetlands, forests, seagrass, mangroves, fish, corals, and more (Eisenstein, 2018).

The way in which biotic and abiotic factors can interact is critical in the first layer of the symbiotic city: it gives space to the different values of the biotic and abiotic foundation of the Earth (Tarr, 2002). Cities and their regions need to prevent these natural resources from reaching their limits (Chapter 9) and ensure that harmful substances or lack of groundwater do not damage the foundation (Chapter 5). When new houses are built, the loss of soil, water and biodiversity needs to be compensated for by creating more green space elsewhere. Urban systems need to be

Figure 2.2. First layer of the symbiotic city.

founded on principles of restoration and regeneration of resources, reducing emissions and waste, and respecting the material and energy cycles of nature, as these are crucial for creating healthy and natural urban livelihoods.

2.5 The networks

The symbiotic city perspective considers four types of networks (Figure 2.3) that are important for building the symbiotic city. Social-ecological networks, institutional networks, technical networks and circular networks are closely connected and embedded in an urban systems perspective. These four types of networks are explained in the following paragraphs.

2.5.1 Social-ecological networks

The social-ecological network perspective focuses on the actions needed to provide and restore the abundance of essential biotic and abiotic factors in the social-ecological networks. The term social-ecological networks stresses that the delineation between social and ecological networks is artificial and arbitrary. Social-ecological systems are neither humans embedded in an ecological system nor ecosystems embedded in human systems (Westley *et al.*, 2002). Although the social and ecological principles are identifiable, they cannot easily be parsed for either analytic or practical purposes. Urban areas are 'coupled social and ecological networks' whose performance is shaped by the interaction of ecological and social principles at multiple levels (Walker *et al.*, 2004). These theories are also very useful for the purpose of understanding the urban development discourse (Gibbs *et al.*, 2005; Lagendijk and Cornford, 2000; Morgan, 2004) as it appears that this is key to enhancing both ecological as well as social and cultural diversity (Bélair *et al.*, 2010).

First, ecological diversity plays a crucial role in increasing resilient urban networks. For example, a large diversity of vegetation in urban forests and grassland on soils can store essential resources such as water and nitrogen and carbon, which are essential for urban climate adaptation and carbon sequestration. Second, when the management of social-ecological networks such as gardens, parks, waterways and other ecological landscapes are shared by a diverse group of

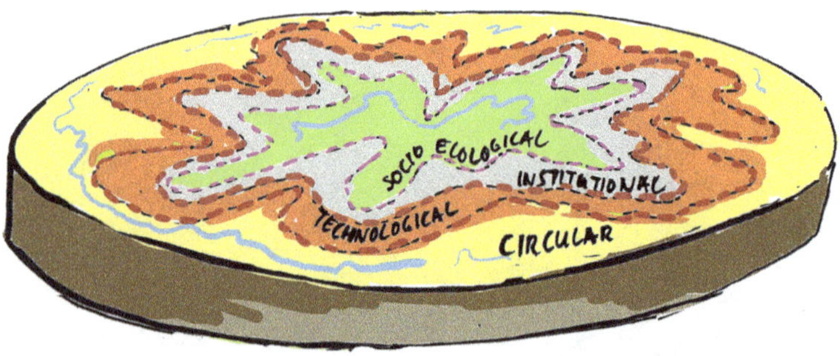

Figure 2.3. Second layer of the symbiotic city.

stakeholders (e.g. local resource users, community members and government workers), decision-making is better informed and more options exist for testing urban greening policies (Berkes and Folke, 1998; Olsson *et al.*, 2004).

The management of these social-ecological networks is complex and therefore necessarily adaptive (Buizer *et al.*, 2011). The social-ecological network model is based on a three-step approach (Arnaiz-Schmitz *et al.*, 2018). The first step is to define the social-ecological interdependencies for a particular urban area. This could, for example, be direct resource extraction, for instance for urban agriculture or water management. The second step is to define the social actors and ecological resources that form the basis for the social-ecological system (e.g. farmers and water organisations in the urban core, and their produce). The third step is to define social-to-social and ecological-to-ecological links. This could, for example, be exchange of tools and contracts between municipalities, the waterboards and citizens, but also the interactions between the soil, water management practices and the flora and fauna networks in the city and the urban core (Ulanowicz, 2021).

Social diversity is equally important if these networks are to flourish. A wide range of stakeholders must be engaged in envisioning, designing, and implementing the new social-ecological networks. For instance, a neighbourhood network can exchange future scenarios for greening the neighbourhoods with (social) entrepreneurs, professionals, government representatives and citizens. Such networks of stakeholders can incorporate the goals of their municipalities, as well as those of citizens groups and private sector businesses. They can inspire and connect, strengthen and accelerate new nature-based goals and solutions.

2.5.2 Institutional networks

Institutional networks form the new coalitions, organisations and arrangements that are needed to build the symbiotic city. Actors' collective agency to develop new institutional networks determines whether they can succeed in their efforts (Chung *et al.*, 2021). In order to be successful, these new institutional networks need to facilitate cooperation between companies, governments, knowledge institutions (of education and research) as well as the involvement of consumers, citizens, social and food organisations (Canal Vieira *et al.*, 2021).

There are novel coalitions emerging in the Netherlands that aim to include nature in their new urban plans. In 2020 a network called Climate Adaptive Building with Nature (KAN) started that consists of more than 50 constructors and municipalities working together on understanding climate adaptive building with nature. During a webinar with 110 participants, KAN was launched by the trade associations Bouwend Nederland, NVB-bouw and NEPROM. The KAN platform focuses on the one hand on climate-related problems, such as summer heat stress and flooding from extreme downpours. On the other hand, working with nature is primarily an opportunity for new ways of construction real estate.

Developers, builders and municipalities work together to develop and share knowledge and experience. If participants in the KAN platform want to further explore a specific bottleneck, they form a new network, consisting of experts and practitioners.

Building institutional networks needs to be done in an inclusive manner. Institutional networks can enable distributive, participative and recognitional equity for all urban actors (Bryant and Callewaert, 2008; Bullard, 2001). The inclusion of stakeholders may result in the integration of different types of knowledge and values of nature and can give space for the more informal and local perspectives of non-credentialed experts in the institutional networks.

Bloei & Groei in Amsterdam South East, the Netherlands is an organisation that aims to improve access to green space for all residents in the neighbourhoods. Bloei & Groei started as a community of people in Amsterdam South East who felt the need to create green safe spaces for those that identify as women. They set up a community garden to give the locals of the neighbourhood a place to meet and relax and invest in personal skills and development. After that they started to involve other target groups: children, youth and neighbourhood organisations. They aim to develop green spaces that are inclusive for many citizens, not just the privileged few. Through knowledge exchange sessions, the participants teach each other new skills. Their dream is to be a centre for the neighbourhood where everyone can meet to exchange their knowledge and ideas. The gardens also aim to contribute to solving local food shortages and to the development and the well-being of the community as a whole. They try to have a spin-off in other urban areas to develop high-quality greenery, regardless of the origin and income level of the residents. Bloei & Groei is working towards an agreement with the municipality that the urban gardens are taken care of by the residents in the neighbourhood. They aim to have a constant place in the community to do their work, in a good collaboration with the municipality. Therefore they also promote their work in the firm belief that neighbourhood organisations such as themselves have a large role to play for the well-being of the community. They aim to become more embedded in the institutional networks of the municipality.

Neighbourhoods are a central point of departure for developing symbiotic cities. Neighbourhood ecologies should be sustainable, sociable, inclusive and adaptable (Hajer *et al.*, 2021) These spatial and social units are closely connected places and people (Reichow, 1948) where the micro-unit is the neighbourhood, a settlement that is born, grows and matures just like organisms. Networks (Scrivens and Smith, 2013) such as neighbourhood coalitions or food policy councils are composed of the activities and practices with which people engage in community life. These practices could include voluntary work, learning practices, or neighbourhood membership in the surroundings where they live. Neighbourhoods can build new institutional networks as an important response to the decline of personal ties and the alienation and isolation of people in urbanised cities. It is a form of resistance against cities that 'have become anonymous places where people sleep and work without much sense of community' (Putnam, 2000). These networks refer to building new relationships between people and the necessary social behaviours that contribute to establishing and maintaining these relationships. All types of resources – emotional, material, practical, financial, intellectual or professional – are made available to each individual through the networks. These networks can motivate and mobilise people to get involved in initiatives or community efforts.

New institutional networks are the result of new forms of cooperation between actors in urban, peri-urban and rural areas to integrate nature in urban transformations. There are four levels of cooperation: strategic, tactical, operational (Loorbach *et al.*, 2015), and reflexive. Municipalities for instance can start new urban coalitions on a strategic level and invite private and non-profit sector organisations to participate in a strategic envisioning process for integrating nature in urban policies (see for examples Chapter 9 and 11). On the tactical level it is important to integrate nature into other governance domains themes such as health, urban planning, economic development, utilities (especially energy and water) and education. This coordination across municipal and regional governance structures requires active officials and distributive leaders that are willing to forge collaborations in an inclusive way (Chapter 9 and 12). On an operational level, local and regional initiatives such as organising festivals, neighbourhood garden, regional markets, and pop-up restaurants, can forge meaningful collaborations. Municipalities can facilitate these kinds of collaborations in various ways, ranging from granting subsidies, to making room in zoning regulations, to playing the role of broker and facilitator. The reflexive level is an important element of building urban coalitions since it makes room for stakeholders to meet and participate in learning networks outside of their own municipality or organisation. Municipalities are well positioned to take the lead in these reflexive learning processes (Fontein and Stuiver, 2011).

2.5.3 Technical networks

Cities are built on and connected by technical networks of infrastructure, such as houses, streets, roads, highways, water systems and digital connections. Animals, such as insects, bird and bat species, depend on these technical networks in the urban environment. By also building technical networks that make space for organisms, such as insects, birds and flowers, biodiversity can be increased. By building small rewilding areas (Arts *et al.*, 2022) in for instance construction works, road construction, railway construction and real estate development, nature can re-enter the city (Cities Biodiversity Center, 2017; Lanchester, 2018; Van Stiphout, 2019; Voskamp *et al.*, 2021). New technical networks can combine nature-inclusivity, water storage, combating heat stress, and in this way improve human health in urban areas (Hartig *et al.*, 2014; Kruize *et al.*, 2019). New technical networks need to close the urban water cycle and treat water as a resource (Brown and Clarke, 2007; Ligtvoet *et al.*, 2018; WWAP/UN-Water, 2018). For instance, through the water smart city approach as presented in Chapter 5 of this book (Van Hattum *et al.*, 2016).

In 2021 many Dutch organisations wrote a manifesto to strive for nature-inclusive construction, resulting in green cities and villages (NL Greenlabel, 2021). Their argument is that everywhere where technical construction works take place, nature has to have a presence: in neighbourhoods, gardens, and in or on the buildings. They propose measures that are beneficial for biodiversity, water storage, combating heat stress, health and therefore belong in every construction project. In addition, neighbourhoods must be made climate-adaptive and extra space must be created for nature. The authors of the manifesto want to realize nature-inclusive new buildings and in this way new technical networks in the city that enable nature to assume a place in the streets, homes and neighbourhoods. Examples of new technical networks are: (1) living areas or nesting stones in the facades and roofs for building-dependent species such as the house sparrow, swift and various bats; (2) green roofs and/or green facades (with ecological value); (3) ecological greenery in the garden; (4) green public space in the vicinity of every home; and (5) connecting routes for animals. See also the list of measures presented in Chapter 13 of this book.

2.5.4 Circular networks

To sustain the processing capacities of the biotic and abiotic foundations we must change from linear organised to circular networks that base themselves on nature and how nature works. Circular networks are founded on regenerative principles like minimising the loss of resources, reducing emissions and waste, reducing pressure on water, soil, and biodiversity, and maximising reuse and recycling. Circularity must also pay attention to the sinks that buffer, absorb, and process the emissions and waste products that are the result of any production, resource use, and consumption process. See Chapter 10 and 11 of this book for an elaborate discussion on the importance of circularity. The integrated nature of circular networks strengthens linked social-ecological networks (Agudelo-Vera *et al.*, 2011), and introduces smart technological networks that mimic nature.

Cities and their regions generate large 'sources' of residual material and waste flows. These flows offer prospects for 'circular valorisation', either recycling or replacing other (partly fossil) resources. Urban areas therefore represent attractive starting points in a transition towards circularity. Urban waste can be reused as a raw material as much as possible and as directly as possible, so that the accumulation of waste is prevented and raw material stocks are used as little as possible (see also Chapter 11).

2.6 Symbiotic practices

The third layer of the symbiotic city consists of symbiotic urban practices (Figure 2.4) where urban residents give expression to their values of nature in different practices. The symbiotic practices are visible in urban planning, in tactical and operational decisions made by actors in real estate development, water management, food production and in communal and business activities. These highly heterogeneous practices can be made visible through storylines narrated by different coalitions of agents (Gieryn, 1983; Hajer, 1995; Wynne, 1996) to attach significance to the new

Figure 2.4. Third layer of the symbiotic city.

practices. The storylines that will be presented in Chapter 3, are visible in existing practices of urban actors, from the double triple helix of governments, large cooperates and science, small business, NGOs and neighbourhood initiatives (Figure 1.5).

The eight storylines express shared values on the relation between humans and nature. These storylines can create the glue that binds people together while providing them with a sense of history, common ground, and future, thus enabling them to persist even in the context of resistance. It is through these symbiotic stories that people realise personal relationships with land, animals, rivers, air, and even bacteria as well as new technologies that impact the environment (Lejano and Fernandez de Castro, 2014). These storylines become visible in oral and written presentations of a sequence of activities that are combined in a plot (Bruner, 1986). The humans and non-human artefacts and their significant roles in the symbiotic practice are also combined within this plot. When the plot recurs in many different local stories it becomes a storyline and will govern and attach significance to the succession of the events (Czarniawska, 1998; Polkinghorne, 1988; 1995). The common storyline adds meaning to the new symbiotic practices and reduces the complexity in practices (Deuten and Rip, 2000; Mildorf, 2002). The patchwork of the symbiotic city consists of eight storylines (Figure 2.5) based on the values of: (1) adaptation; (2) community; (3) green neighbourhoods; (4) urban forest; (5) water smart city; (6) nature-based solutions; (7) personal lifestyle and health; and (8) restorative urban foodshed. These storylines will be narrated in detail in Chapter 3.

2.7 Conclusions

This chapter introduced symbiotic cities as living and dynamic landscapes of mutual dependencies build between people and nature. The first assumption is optimistic about today's cities, societies and its inhabitants: citizens, mayors, home owners, project developers and scientists are actively uniting themselves in new coalitions and organisations to build cities and their regions in a way that they become meaningful for nature, the environment as well as humans. The second

Figure 2.5. Eight storylines of symbiotic human practices in the symbiotic city.

assumption is that nature can be reintroduced into the urban environment through a diversity of symbiotic practices that represent different values of nature and natural resources in city design. The third assumption of the symbiotic city is that it can only be built with the inclusion of all people (Cook and Hegtvedt, 1983). The fourth assumption follows this argument of inclusion closely and embraces nature-based thinking (Randrup *et al.*, 2020), which adds to the more utilitarian narrative of nature-based solutions (NBS) and includes a variety of values of nature. The fifth assumption is that we need to overcome dualistic notions, such as human versus nature, economy versus ecology, technology versus society and the urban versus the rural. These categories are intrinsically linked and mutually dependent through the perspective of embedded (Polanyi, 1944) networks in the Symbiotic City.

The chapter sets out a three-pronged framework (Figure 2.1) that differentiates between the foundation, the networks and the human practices that build the future symbiotic city. The symbiotic city is embedded in the foundation (Figure 2.2) of the Earth that encompass biotic and abiotic factors. Biotic factors are living things such as plants, animals, and bacteria, while abiotic are non-living components, such as water, soil and atmosphere. In the symbiotic city urban actors restore the networks of the city (Figure 2.3) First, the social-ecological networks of greenery, forests, waterways, lakes, ditches, grasslands and meadows (Gunderson *et al.*, 1995; Olsson *et al.*, 2004). The parks, flowerbeds and hedges in the urban core link up to the ecological networks in the wider natural environment in or beyond the city and are working towards a naturally functioning region. The second types of networks that needs to be made more robust are institutional networks of coalitions, arrangements and organisations bound by rules and regulations that have to support and sustain the symbiotic city. The third types of networks that need to be adapted are technical networks that need to be designed in such a way that they give way to nature-based infrastructures such as buildings, streets, highways and railroads that are closely incorporated into the social-ecological networks. The fourth kinds of networks of crucial importance are the circular networks of resources in the area, such as soil, water and resources and sinks that are restoratively used and maintained by humans for building the technical networks and social-ecological networks. The third layer of the symbiotic city consists of symbiotic urban practices where urban residents give expression to their values of nature in different practices. These storylines are not invented in a blueprint scenario, but are visible in so-called re-rooting practices in cities already being undertaken by present urban actors, from the double triple helix of governments, large cooperates and science, businesses, NGOs and citizens and communal initiatives.

References

Agudelo-Vera, C.M., Mels, A.R., Keesman, K.J. and Rijnaarts, H.H., 2011. Resource management as a key factor for sustainable urban planning. Journal of Environmental Management 92: 2295-2303.

Albrechts, L. and Mandelbaum, S., 2005. The network society: a new context for planning. Routledge, London, UK.

Anderson, C.R., Bruil, J., Chappell, M.J., Kiss, C. and Pimbert, M.P., 2021. Agroecology now!: transformations towards more just and sustainable food systems. Springer Nature, Berlin, Germany.

Angelard, C. and Bever, J.D., 2013. Genetics of symbionts. In: Maloy, S. and Hughes, K. (eds.) Brenner's encyclopedia of genetics (2nd ed.). Academic Press, San Diego, CA, USA, pp. 595-597. https://doi.org/10.1016/B978-0-12-374984-0.01496-0

Arnaiz-Schmitz, C., Schmitz, M.F., Herrero-Jáuregui, C., Gutiérrez-Angonese, J., Pineda, F.D. and Montes, C., 2018. Identifying socio-ecological networks in rural-urban gradients: diagnosis of a changing cultural landscape. Science of the Total Environment 612: 625-635. https://doi.org/10.1016/j.scitotenv.2017.08.215

Arts, K., Bakker, L. and Buys, A., 2022. Rewilding in Nederland. KNNV Uitgeverij, Zeist, the Netherlands.

Bélair, C., Ichikawa, K., Wong, B. and Mulongoy, K., 2010. Sustainable use of biological diversity in socio-ecological production landscapes. Background to the 'Satoyama Initiative for the benefit of biodiversity and human well-being. Technical Series 52: 184.

Berkes, F. and Folke, C., 1998. Linking social and ecological systems: management practices and social mechanisms for building resilience. Cambridge University Press, Cambridge, UK.

Bertrand, N., De Haas, W., Jensen, A., Krellenberg, K., Peltonen, L. and Seifert-Dähnn, I., 2015. Urban Environmental equality. Position Paper, Partnership for European Environmental Research, Brussels, Belgiumn. https://doi.org/10.15200/winn.142960.01917

Blok, V., 2022. Van wereld naar aarde: filosofische ecologie van een bedreigde planeet. Boom, Amsterdam, the Netherlands.

Bourdieu, P., 1972. Esquisse d'une théorie de la pratique. Esquisse d'une théorie de la pratique. Librairie Droz, Genève, Switzerland, pp. 157-243.

Brown, R.R. and Clarke, J.M., 2007. Transition to water sensitive urban design: the story of Melbourne, Australia. FAWB research reports, 07/1. Facility for Advancing Water Biofiltration, Melbourne, Australia.

Bruner, E.M., 1986. The anthropology of experience. University of Illinois Press, Champaign, IL, USA.

Bryant, B. and Callewaert, J., 2008. Why is understanding urban ecosystems important to people concerned about environmental justice? Urban ecology. Springer, Heidelberg, Germany, pp. 597-605.

Buizer, M., Arts, B. and Kok, K., 2011. Governance, scale and the environment: the importance of recognizing knowledge claims in transdisciplinary arenas. Ecology and Society 16. https://doi.org/10.5751/ES-03908-160121

Bullard, R.D., 2001. Environmental justice in the 21st century: race still matters. Phylon 49: 151-171.

Canal Vieira, L., Serrao-Neumann, S. and Howes, M., 2021. Daring to build fair and sustainable urban food systems: a case study of alternative food networks in Australia. Agroecology and Sustainable Food Systems 45: 344-365.

Chung, M.G., Frank, K.A., Pokhrel, Y., Dietz, T. and Liu, J., 2021. Natural infrastructure in sustaining global urban freshwater ecosystem services. Nature Sustainability 4: 1068-1075. https://doi.org/10.1038/s41893-021-00786-4

Cities Biodiversity Center, 2017. How cities are using nature-based solutions for sustainable urban development. Available at: https://cbc.iclei.org/cities-using-nature-based-solutions-sustainable-urban-development/.

Cook, K.S. and Hegtvedt, K.A., 1983. Distributive justice, equity, and equality. Annual Review of Sociology 9: 217-241.

CRA, 2021. De 22e eeuw begint nu: Agenda College van Rijksadviseurs, College van Rijksadviseurs, Den Haag, the Netherlands.

Czarniawska, B., 1998. A narrative approach to organization studies. SAGE, Thousand Oaks, CA, USA. https://doi.org/10.4135/9781412983235

De Kimpe, C.R. and Morel, J.-L., 2000. Urban soil management: a growing concern. Soil Science 165: 31-40. https://doi.org/10.1097/00010694-200001000-00005

De Vries, S., Verheij, R.A., Groenewegen, P.P. and Spreeuwenberg, P., 2003. Natural environments – healthy environments? An exploratory analysis of the relationship between greenspace and health. Environment and Planning A: Economy and Space 35: 1717-1731. https://doi.org/10.1068/a35111

Deuten, J.J. and Rip, A., 2000. Narrative infrastructure in product creation processes. Organization 7: 69-93. https://doi.org/10.1177/135050840071005

Du, S., Shi, P., Van Rompaey, A. and Wen, J., 2015. Quantifying the impact of impervious surface location on flood peak discharge in urban areas. Natural Hazards 76: 1457-1471. https://doi.org/10.1007/s11069-014-1463-2

EC, 2012. Guidelines on best practice to limit, mitigate or compensate soil sealing, European Commission, Publications Office of the European Union, Luxembourg.

EC, 2021. Nature based solutions: Horizon 2020 NBS research projects tackle the climate and biodiversity crisis. European Commission: Executive Agency for Small Medium-sized Enterprises, Publications Office of the European Union. https://doi.org/10.2826/59817

EEA, 2015. The European environment — state and outlook 2015: synthesis report, European Environment Agency, Copenhagen, Denmark.

Eisenstein, C., 2018. Climate: a new story. North Atlantic Books, Berkeley, CA, USA.

FAO & ITPS, 2022. Urbanisation and soil sealing. Food and Agriculture Organization of the United Nations and Intergovernmental Technical Panel on Soils, Rome, Italy.

Fontein, R.J. and Stuiver, M., 2011. Naar duurzame regionale voedselsystemen. Alterra, Wageningen Universtity & Research, Wageningen, the Netherlands.

Gan, Q., Jongen, L., Riharsya, A., Van Noorden, T. and Vaspintra, P., 2017. A study about the distribution of urban green space vs people's perception and exploitation in two neighborhoods of Amsterdam: Bijlmermeer and Amsterdam Nieuw-West, Wageningen University & Research, Wageningen, the Netherlands.

Gibbs, D., Deutz, P. and Proctor, A., 2005. Industrial ecology and eco-industrial development: A potential paradigm for local and regional development? Regional Studies 39: 171-183. https://doi.org/10.1080/0034340052000599959

Giddens, A., 1990. The consequences of modernity. Polity Press, Oxford, UK.

Gieryn, T.F., 1983. Boundary-work and the demarcation of science from non-science: strains and interests in professional ideologies of scientists. American Sociological Review 48: 781-795. https://doi.org/10.2307/2095325

Guilland, C., Maron, P.A., Damas, O. and Ranjard, L., 2018. Biodiversity of urban soils for sustainable cities. Environmental Chemistry Letters 16: 1267-1282. https://doi.org/10.1007/s10311-018-0751-6

Gunderson, L.H. and Holling, C.S., 2002. Panarchy: understanding transformations in human and natural systems. Island Press, Washington DC, USA.

Gunderson, L.H., Holling, C.S. and Light, S.S., 1995. Barriers and bridges to the renewal of ecosystems and institutions. Barriers & bridges. Columbia University Press, New York, NY, USA.

Haaland, C. and Van den Bosch, C.K., 2015. Challenges and strategies for urban green-space planning in cities undergoing densification: A review. Urban forestry & urban greening 14: 760-771. https://doi.org/10.1016/j.ufug.2015.07.009

Hagens, J.E., 2006. De lagenbenadering in de ruimtelijke planning: over de waarde van de Nederlandse club sandwich. Topos : periodiek over landschapsarchitectuur, ruimtelijke planning en sociaal-ruimtelijke analyse 16: 24-27.

Hajer, M.A., 1995. The politics of environmental discourse: ecological modernization and the policy process. Clarendon Press, Oxford, UK.

Hajer, M.A., Pelzer, P., Van den Hurk, M., Dam, C. and Buitelaar, E., 2021. Neighbourhoods for the future: a plea for a social and ecological urbanism. trancity*valiz, Amsterdam, the Netherlands.

Hansen, R., Olafsson, A.S., Van der Jagt, A.P.N., Rall, E. and Pauleit, S., 2019. Planning multifunctional green infrastructure for compact cities: what is the state of practice? Ecological Indicators 96: 99-110. https://doi.org/10.1016/j.ecolind.2017.09.042

Hardcastle, M.A., Usher, K.J. and Holmes, C.A., 2005. An overview of structuration theory and its usefulness for nursing research. Nursing Philosophy 6: 223-234. https://doi.org/10.1111/j.1466-769X.2005.00230.x

Hartig, T., Mitchell, R., De Vries, S. and Frumkin, H., 2014. Nature and health. Annual Review of Public Health 35: 207-228.

Hedblom, M., Andersson, E. and Borgström, S., 2017. Flexible land-use and undefined governance: from threats to potentials in peri-urban landscape planning. Land Use Policy 63: 523-527.

Jepson, P. and Blythe, C., 2020. Rewilding: the radical new science of ecological recovery. Icon Books, Limited, London, UK.

Kakovitch, T.S. and O'Hara, S., 2021. Water and the circular economy: learning from nature. Sustainability 13: 12597. https://doi.org/10.3390/su132212597

Kruize, H., Van der Vliet, N., Staatsen, B., Bell, R., Chiabai, A., Muiños, G., Higgins, S., Quiroga, S., Martinez-Juarez, P. and Aberg Yngwe, M., 2019. Urban green space: creating a triple win for environmental sustainability, health, and health equity through behavior change. International Journal of Environmental Research and Public Health 16: 4403.

Lagendijk, A. and Cornford, J., 2000. Regional institutions and knowledge – tracking new forms of regional development policy. Geoforum 31: 209-218. https://doi.org/10.1016/S0016-7185(99)00031-7

Lanchester, A., 2018. Cities: nature's new wild, Outcasts. BBC.

Lehmann, S., 2010. The principles of green urbanism: transforming the city for sustainability. Earthscan, London, UK.

Lejano, R.P. and Fernandez de Castro, F., 2014. Norm, network, and commons: the invisible hand of community. Environmental Science & Policy 36: 73-85. https://doi.org/10.1016/j.envsci.2013.07.012

Levitas, R., 2005. The inclusive society?: social exclusion and new labour. Palgrave Macmillan, Basingstoke, UK.

Ligtvoet, W., Bouwman, A., Knoop, J., De Bruin, S., Nabielek, K., Huitzing, H., Janse, J., Van Minne, J., Gernaat, D., Van Puijenbroek, P., De Ruiter, J. and Visser, H., 2018. The geography of future water challenges. PBL Netherlands Environmental Assessment Agency, The Hague, the Netherlands.

Loorbach, D., Frantzeskaki, N. and Huffenreuter, R.L., 2015. Transition management: taking stock from governance experimentation. Journal of Corporate Citizenship: 48-66. https://doi.org/10.9774/GLEAF.4700.2015.ju.00008

Maas, J., Verheij, R.A., Groenewegen, P.P., De Vries, S. and Spreeuwenberg, P., 2006. Green space, urbanity, and health: how strong is the relation? Journal of Epidemiology & Community Health 60: 587-592. https://doi.org/10.1136/jech.2005.043125

Mata, L., Ramalho, C.E., Kennedy, J., Parris, K.M., Valentine, L., Miller, M., Bekessy, S., Hurley, S. and Cumpston, Z., 2020. Bringing nature back into cities. People and Nature 2: 350-368. https://doi.org/10.1002/pan3.10088

Mildorf, J., 2002. 'Opening up a can of worms': physicians' narrative construction of knowledge about domestic violence. Narrative Inquiry 12: 233-260. https://doi.org/10.1075/ni.12.2.02mil

Morgan, K., 2004. Sustainable regions: governance, innovation and scale. European Planning Studies 12: 871-889. https://doi.org/10.1080/0965431042000251909

Nesshöver, C., Assmuth, T., Irvine, K.N., Rusch, G.M., Waylen, K.A., Delbaere, B., Haase, D., Jones-Walters, L., Keune, H., Kovacs, E., Krauze, K., Külvik, M., Rey, F., Van Dijk, J., Vistad, O.I., Wilkinson, M.E. and Wittmer, H., 2017. The science, policy and practice of nature-based solutions: an interdisciplinary perspective. Science of the Total Environment 579: 1215-1227. https://doi.org/10.1016/j.scitotenv.2016.11.106

NL Greenlabel, 2021. Manifest Bouwen voor natuur, drie bouwstenen voor natuurlijke dorpen en steden. Available at: https://tinyurl.com/3pct9pfk.

Nussbaum, M.C., 2011. Creating capabilities: the human development approach. Harvard University Press, Cambridge, MA, USA. https://doi.org/10.4159/harvard.9780674061200

O'Hara, S.U., 1996. Discursive ethics in ecosystems valuation and environmental policy. Ecological Economics 16: 95-107. https://doi.org/10.1016/0921-8009(95)00085-2

O'Riordan, R., Davies, J., Stevens, C., Quinton, J.N. and Boyko, C., 2021. The ecosystem services of urban soils: a review. Geoderma 395: 115076. https://doi.org/10.1016/j.geoderma.2021.115076

Olsson, P., Folke, C. and Hahn, T., 2004. Social-ecological transformation for ecosystem management the development of adaptive co-management of a wetland landscape in southern Sweden. Ecology and Society 9.

Pascual, U., Adams, W.M., Díaz, S., Lele, S., Mace, G.M. and Turnhout, E., 2021. Biodiversity and the challenge of pluralism. Nature Sustainability 4: 567-572. https://doi.org/10.1038/s41893-021-00694-7

Pascucci, S., 2020. Circular food economies. Routledge handbook of sustainable and regenerative food systems. Routledge, London, UK, pp. 318-335.

Pedroli, G.B.M., Van Rooij, S.A.M., Roosenschoon, O.R., Sterk, M. and Timmermans, W., 2021. Landscape in the round – towards a circular rural-urban agenda. European Public Mosaic: 62-75.

Pereira, H.M. and Navarro, L.M., 2015. Rewilding european landscapes. Springer Nature, Berlin, Germany.

Polanyi, K., 1944. The great transformation: the political and economic origins of our time. Farrar and Rinehart, New York, NY, USA.

Polkinghorne, D.E., 1988. Narrative knowing and the human sciences. Suny Press, Albany, NY, USA.

Polkinghorne, D.E., 1995. Narrative configuration in qualitative analysis. International Journal of Qualitative Studies in Education 8: 5-23. https://doi.org/10.1080/0951839950080103

Putnam, R.D., 2000. Bowling alone: the collapse and revival of American community. Proceedings of the 2000 ACM conference on Computer supported cooperative work. Association for Computing Machinery, Philadelphia, PA, USA, pp. 357. https://doi.org/10.1145/358916.361990

Randrup, T.B., Buijs, A., Konijnendijk, C.C. and Wild, T., 2020. Moving beyond the nature-based solutions discourse: introducing nature-based thinking. Urban Ecosystems 23: 919-926. https://doi.org/10.1007/s11252-020-00964-w

Raworth, K., 2017. Doughnut economics: seven ways to think like a 21st-century economist. Random House, London, UK.

Reichow, H.B., 1948. Organische Stadtbaukunst von der Grossstadt zur Stadtlandschaft. Georg Westermann Verlag, Braunschweig, Germany.

Rotmans, J. and Loorbach, D., 2009. Complexity and transition management. Journal of Industrial Ecology 13: 184-196. https://doi.org/10.1111/j.1530-9290.2009.00116.x

Schwab, K., 2019. What is biophilic design, and can it really make you happier and healthier? Available at: https://tinyurl.com/3tk25k82.

Scrivens, K. and Smith, C., 2013. Four interpretations of social capital. OECD, Paris, France. https://doi.org/10.1787/5jzbcx010wmt-en

Sonneveld, M.P.W. and Bouma, J., 2003. Effects of combinations of land use history and nitrogen application on nitrate concentration in the groundwater. NJAS – Wageningen Journal of Life Sciences 51: 135-146. https://doi.org/10.1016/S1573-5214(03)80030-1

Stuiver, M. and O'Hara, S., 2021. Food connects Washington DC in 2050 – a vision for urban food systems as the centerpieces of a circular economy. Sustainability 13: 7821. https://doi.org/10.3390/su13147821

Tappert, S., Klöti, T. and Drilling, M., 2018. Contested urban green spaces in the compact city: the (re-) negotiation of urban gardening in Swiss cities. Landscape and Urban Planning 170: 69-78. https://doi.org/10.1016/j.landurbplan.2017.08.016

Tarr, J.A., 2002. The metabolism of the industrial city: the case of Pittsburgh. Journal of Urban History 28: 511-545. https://doi.org/10.1177/0096144202028005001

The Nature Conservancy, 2016. The power of nature in cities. Available at: https://tinyurl.com/8fdpjdds.

Timmermans, W., Lenzholzer, S., Voskamp, I., Struckman, L., Maagdenberg, G., Weppelman, I., Mashhoodi, B., Dill, S., Cortesão, J., De Haas, W., Van Hattum, T., Luo, S., Van Rooij, S., Roosenschoon, O., Sterk, M., Stremke, S., Wamelink, W., Verzandvoort, S. and Harkema, T., 2022. De stad van 2120: natuurlijk! Project number: BO-60-003-004, Wageningen University & Research, Wageningen, the Netherlands. https://doi.org/10.18174/565411

Turner, J.H., 1986. The theory of structuration. American Journal of Sociology 91: 969-977.

Tzoulas, K., Galan, J., Venn, S., Dennis, M., Pedroli, B., Mishra, H., Haase, D., Pauleit, S., Niemelä, J. and James, P., 2021. A conceptual model of the social-ecological system of nature-based solutions in urban environments. Ambio 50: 335-345. https://doi.org/10.1007/s13280-020-01380-2

Ulanowicz, R.E., 2021. Socio-ecological networks: a lens that focuses beyond physics. Frontiers in Ecology and Evolution 9. https://doi.org/10.3389/fevo.2021.643122

Van Hattum, T., Blauw, M., Jensen, M. and Bruin, K., 2016. Towards water smart cities: climate adaptation is a huge opportunity to improve the quality of life in cities, Wageningen University & Research, Wageningen, the Netherlands.

Van Rooij, S., Timmermans, W., Roosenschoon, O., Keesstra, S., Sterk, M. and Pedroli, B., 2021. Landscape-based visions as powerful boundary objects in spatial planning: lessons from three Dutch projects. Land 10: 16.

Van Stiphout, M., 2019. First guide to nature inclusive design. Available at: https://nextcity.nl/first-guide-for-nature-inclusive-design/.

Veeneklaas, F.R., 2012. Het natuurdebat ontward: natuur als erfgoed, als kapitaalgoed en als consumentengoed, InnovatieNetwerk, Utrecht, the Netherlands.

Voskamp, I.M., De Luca, C., Polo-Ballinas, M.B., Hulsman, H. and Brolsma, R., 2021. Nature-based solutions tools for planning urban climate adaptation: state of the art. Sustainability 13: 6381.

Walker, B., Carpenter, S., Anderies, J., Abel, N., Cumming, G., Janssen, M., Lebel, L., Norberg, J., Peterson, G.D. and Pritchard, R., 2002. Resilience management in social-ecological systems a working hypothesis for a participatory approach. Ecology and Society 6: 14. Available at: https://www.ecologyandsociety.org/vol6/iss1/art14/.

Walker, B., Holling, C.S., Carpenter, S. and Kinzig, A., 2004. Resilience, adaptability and transformability in social-ecological systems. Ecology and Society 9. https://doi.org/10.5751/ES-00650-090205

Wang, M., Faber, J.H., Chen, W., Li, X. and Markert, B., 2015. Effects of land use intensity on the natural attenuation capacity of urban soils in Beijing, China. Ecotoxicology and Environmental Safety 117: 89-95.

Westley, F., Carpenter, S., Brock, W.A., Holling, C.s. and Gunderson, L.H., 2002. Why systems of people and nature are not just social and ecological systems. In: Gunderson, L.H. and Holling, C.S. (eds.) Panarchy: understanding transformations in human and natural systems. Island Press, Washington DC, USA, pp. 103-119.

Westley, F.R., Tjornbo, O., Schultz, L., Olsson, P., Folke, C., Crona, B. and Bodin, Ö., 2013. A Theory of Transformative agency in linked social-ecological systems. Ecology and Society 18. https://doi.org/10.5751/ES-05072-180327

Wiskerke, J.S., 2015. Urban food systems. In: De Zeeuw, H. and Drechzel, P. (eds.) Cities and agriculture. Routledge, London, UK, pp. 19-43.

WWAP/UN-Water, 2018. The United Nations World Water Development Report 2018: Nature-Based Solutions for Water, UNESCO, Paris, France.

Wynne, B., 1996. May the sheep safely graze? A reflexive view of the expert-lay knowledge divide. In: Lash, S., Szerszynski, B. and Wynne, B. (eds.) Risk, environment and modernity: towards a new ecology. Sage, London, UK, pp. 44-83.

Chapter 3

Envisioning the symbiotic city in 2050: two visions of Washington DC and the Netherlands

Marian Stuiver and Sabine O'Hara

Abstract

This chapter presents two visions of two different urban areas as symbiotic cities in 2050. Both visions seek to capture the full potential of a symbiotic city that effectively integrates the value of nature into the urban scape and its social, built, and geographic characteristics. We based our visions on a methodology committed to community engagement, thoughtfulness about the process, and the inclusion of multiple small initiatives that constitute the structuring elements of the visions. The first is a vision for a City Region Food System in Washington DC in 2050. The vision focuses on food as the connecting link for a nature-based and socially inclusive urban economy embedded in the larger Washington DC metropolitan region. The eight wards of Washington DC in 2050 value natural resources, social and cultural diversity, and a shared commitment to nature-based innovation. We argue that small, hyperlocal, neighbourhood-based initiatives can become crucial game changers and catalysts of change for entire neighbourhoods, cities, and regions. The second is a vision for a region in the middle of the Netherlands in 2050. The area consists of eight neighbourhoods that all constitute different storylines of nature-based thinking and give space to different values that people attach to natural resources, including water, soils and biodiversity. Both the visions of Washington DC and the Netherlands paint a picture of what is possible in 2050. Further specification is needed to advance the visions toward their implementation. In addition, the visions raise the question of how different actors in the urban innovation ecosystem are crucial for the development of the symbiotic city. This question will be addressed in further detail in the following chapters of this book.

Keywords: visions of hope, human practices, future vision, Washington DC, the Netherlands, food economies, systems perspective, values of nature

3.1 Building visions of hope

What exactly can a symbiotic city look like in 2050 and how does it fit within its larger regional and physical/environmental context? This chapter presents two visions that illustrate the process of envisioning urban areas as a symbiotic city. The range of social, cultural, environmental, and geographic characteristics of the city are captured in eight values of nature: adaptation, communal values, personal natural lifestyle, green neighbourhoods, rewilding, water smart cities, nature-based solutions and restorative urban foodshed.

10.3920/978-90-8686-935-0_3, © Wageningen Academic Publishers 2022

Our approach shows how more function-oriented notions of nature in a city, must be supplemented by participatory bottom-up approaches in order to effectively integrate the heterogeneity of social and cultural values of nature in the make-up of future cities. We outline a three-pronged methodology of: (1) scaling up existing small initiatives; (2) stakeholder consultation and engagement; (3) process-focused transformation to include all relevant voices of nature, in the design of the city of the future.

In 2019 we created a 2050 Vision for a Washington DC City Region Food System (CRFS) based on two local initiatives: (1) the Urban Food Hubs Model of the College of Agriculture, Urban Sustainability and Environmental Sciences (CAUSES) of the University of the District of Columbia (O'Hara, 2015; 2017); and (2) the Five Pillars of Economic Development, a community-based development visioning process that engaged residents from two of the most underserved neighbourhoods in Washington DC in writing a story about a sustainable future for their neighbourhoods (O'Hara, 2018a). The Urban Food Hubs demonstrate ways to create a circular urban food economy that can integrate seamlessly with green infrastructure and an alternative energy economy. The visions developed by local residents in the Five Pillars process include such an urban food economy focused on health and wellness, off-the-grid energy generation and water treatment, and a commitment to local innovation building on a proud history of African American entrepreneurship. Building on these grass-roots visions, we developed a more detailed vision rooted in the eight Wards of Washington DC and embedded it into the larger Washington DC metropolitan region. The 2050 Washington DC vision deliberately engages the local expertise of residents as well as the credentialed expertise of engineers, urban planners, and private sector partners who share their perspectives on a symbiotic city of the future.

Using the experience from Washington DC, we developed a second vision of the 2050 symbiotic city for the Netherlands. Like the Washington DC vision, the vision for the Netherlands was also built on existing practices that we captured by analysing various case studies, and by engaging a range of stakeholders in our vision process. The result was a cohesive vision that seeks to capture a longer-term urban development vision for the Netherlands. In developing the two symbiotic city visions we intentionally move beyond existing doomsdays scenarios of flooded coastal areas, energy shortages, and excessive heat, to offer solutions that can inspire citizens, businesses and governments to a renewed commitment to a sustainable urban future. This is in line with recent climate reports that stress the importance of finding urban solutions in a continuously urbanising world (IPCC, 2021, 2022).

The current reality of Washington DC is different from our 2050 vision. The downtown area of the nation's capital is a major tourist destination and culinary mecca. Washington DC is organised into eight Wards of approximately 70,000 residents each. Parts of Ward 2 and Ward 3 in the northwest quadrant of the city are home to some of the best-educated workforce and highest household incomes in the United States. These same Wards have an array of amenities and services. In contrast, Wards 7 and 8, located east of the Anacostia River, have limited amenities, high unemployment rates, and low household incomes (see Table 9.1 and 9.2 in Chapter 9 of this book). Sadly, the stark disparities in socio-economic characteristics translate into dramatically different health outcomes and life expectancies that follow clear racial divides. Wards with high household incomes, high life expectancy, and low unemployment rates have predominantly white populations, while those with low household incomes, low life expectancy, and high unemployment, have predominantly non-Hispanic black populations. Washington DC is located in the Mid-Atlantic region of the eastern United States and is the capital of a vast and geographically diverse country that spans the American continent from the Atlantic to the Pacific Ocean.

In contrast, the Netherlands is a small country that can be compared to a metropolitan area. The DC metro area is 14,000 km^2 and has 5.3 million people, compared to the 41,500 km^2 and 17.4 million residents of the Netherlands. Given the small size of the country, commitments to resilience, climate adaptation and ecosystems restoration are essential for the well-being of its population. The decline in biodiversity and the deterioration of ecosystems in the Netherlands is occurring at an astonishing rate (CBS *et al.*, 2021; Hallmann *et al.*, 2017; Hüttl and Schneider, 1998; Kemmers, 2011). The main causes of these losses include intensifying land use, loss of permeable surfaces and green cover, water-related issues (Sterk and Wamelink, 2019), tourism, nutrient (esp. nitrogen) and pesticide run-off, expanding invasive species, and climate change (CBD, 2014; Kramer *et al.*, 2019). In the Netherlands habitat loss has often occurred because nature has become fragmented due to land-use pressures and high contamination rates (PBL, 2021). Additional land-use pressures, as well as habitat and biodiversity losses are expected as an additional 1 million homes will be built in the Netherlands by 2030. A lot of challenges are anticipated as these new neighbourhoods are being built. In addition to providing much-needed additional living space for humans, storm water run-off must be managed to prevent flooding, planting is needed to mitigate heat island effects, and habitats are needed to preserve plant and animal species. The more than three thousand business parks in the Netherlands also form an important part of the built environment (approx. 30%). Most of the urbanised land is covered by impervious material, such as buildings, streets, parking lots, etc. Urban surfaces are therefore becoming increasingly 'sealed'. Soil sealing is defined as the permanent loss of permeable surfaces as more and more areas are paved over and covered with impervious materials such as concrete, tar, buildings and infrastructure that cannot easily be removed (FAO & ITPS, 2022). The urgent need for climate adaptation, biodiversity recovery, energy transition, and habitat restoration requires a new way of thinking about urban areas in the Netherlands.

The two 2050 visions of Washington DC (Section 3.2) and the Netherlands (Section 3.3) paint a picture of what is possible in each of the two locations through eight different storylines. While both areas are remarkably similar in terms of their population density, the DC metropolitan area has minimal food production capacity while the Netherlands are known for their highly productive agricultural sector and are one of the world's largest exporters of vegetables and horticulture products (Jukema *et al.*, 2022). Section 3.4 presents some conclusions and pathways for achieving symbiotic cities in the future.

3.2 Envisioning a 2050 City Region Food System (CRFS) for the Washington DC metropolitan region

In 2020 we developed a vision for a regional environmentally and socially sustainable food system for the Washington DC Metropolitan Region (Figure 3.1). Our vision for the 2050 City Regional Food System (CRFS) of the Washington DC metropolitan region is the result of a three-pronged approach: (1) community engagement; (2) thoughtfulness about the process; and (3) implementing multiple small initiatives that move toward the overall vision of a future symbiotic city (Stuiver and O'Hara, 2021).True to these three principles, our CRFS vision for Washington DC starts with two prior local initiatives, namely the award-winning Urban Food Hubs model of UDC CAUSES, and the community-based vision process entitled 'The Five Pillars of Economic Development' (O'Hara, 2017, 2018a).

The CAUSES Urban Food Hubs models has four integrated components: Food Production, Food Preparation, Food Distribution and closing the loop through Waste and Water Management. These components are shaped by the environmental/physical and social/cultural context of the Washington DC neighbourhoods where the food hubs are located: (1) the food production component takes the form of bio-intensive soil based, or hydroponic and aquaponic systems, located on the ground and on roofs; (2) the food preparation component utilises kitchens and

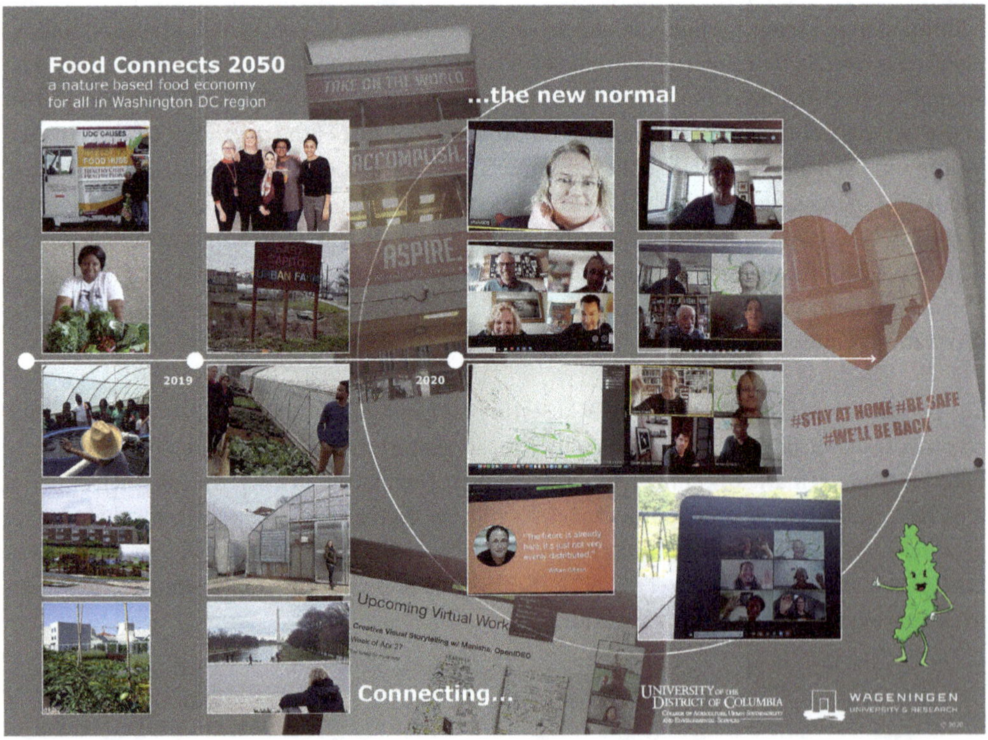

Figure 3.1. Local initiatives in Washington DC.

food trucks to add value and offer nutrition education; (3) the food distribution component uses traditional venues like farmers markets and CSAs and innovative ones like direct distribution to ethnic food markets and restaurants, and 'veggie prescriptions' where health providers prescribe coupons to farmers markets and CSAs rather than medications to improve health outcomes; (4) and the closing-the-loop component utilises a range of soil, water and energy focused activities including composting, rain water harvesting, rain gardens to reduce stormwater run-off and bio-digesters to generate energy from food waste.

Each hub also comprises a cluster of business opportunities for entrepreneurs who ideally come from the neighbourhood where a Food Hub is located. These entrepreneurial activities go well beyond food production and value addition through food processing like growing micro-greens and herbs for high-end restaurants, and making salsa for local and regional markets. They may also include activities like nutrition counselling and lifestyle coaching; growing on green roofs to mitigate water run-off and to offer a green event space for everything from business lunches to yoga classes; raising native plant seedlings for urban parks and rain gardens; composting plant waste to distribute compost to local school gardens and public parks; and turning food waste into bio-energy to heat greenhouses and cool storage facilities. The Urban Food Hubs seek to build a decentralised network of food-related activities that improve food security, nutritional health, employment, green infrastructure, and resilience, especially in neighbourhoods that have deficits on all counts.

The second local initiative that forms the basis for our Washington DC CRFS is the so-called Five Pillars project. Based on her localised and highly interdisciplinary economic development framework that focuses on the social determinants of successful sustainable development outcomes (O'Hara, 2018a) two focus groups were convened in Ward 7 (Deanwood) and Ward 8 (Congress Heights) of Washington DC. The focus groups of local residents engaged in writing a story about a sustainable future for their neighbourhoods around five pillars of sustainable community-based development: education, health, social and cultural amenities, environmental quality and recreation, and access to information and transportation. The collective story starts with the statement that, by 2030, the two DC neighbourhoods will be '... thriving communities that are popular among residents and visitors alike' (O'Hara, 2018b). The story points to the mix of young and old, singles and families, diverse ethnicities, and cultures that gives the two DC Wards their vibrancy. It also stresses the African American heritage of the so-called 'chocolate city' by envisioning a museum featuring African American inventors, scientists, and entrepreneurs. The story is articulate about developing a vision that connects the rich entrepreneurial history of DC's most food-insecure neighbourhoods to cutting-edge energy generation facilities, water filtration systems, and soil-less food production facilities. It also stresses the importance of social connections in the culturally diverse tapestry of Washington DC. The story lines which form an indispensable basis for our CRFS vision were therefore developed to a large extent by the participants in the two story-writing focus groups in Wards 7 and 8, which are the two Wards with the highest percentage of African American residents, the lowest household incomes, and the lowest life expectancy.

Both the Urban Food Hubs and the Five Pillars initiatives satisfy our principles of a successful CRFS, namely: (1) community engagement; (2) thoughtfulness about the process; and (3) utilising multiple small initiatives to move toward the implementation of the overall vision. Progress has been slow in the wake of the two initiatives. While local food options in Washington DC are expanding the food divide between neighbourhoods with easy access to a wide range of food options and those with very limited options persists.

Given the administrative structure of Washington DC, which consists of eight Wards of roughly equal population size, we paint a future in which all of the city's eight Wards are equally strong and rooted in a nature-based food system that uses the strongest assets of each Ward as its basis. Our Washington DC vision therefore is structured around eight distinct dreams for a food system linked to the nature-based characteristics of each of the eight Wards of Washington DC.

A brief description of each of our eight hyper-local dreams (Figure 3.2) follows.

3.2.1 Ward 1: Washington DC automated fresh food mobility centre

Food connects the wards: it's 2050 and hundreds of drones have alleviated the traffic jams in downtown DC. They zip about making just-in-time deliveries of strawberries fresh from the urban farms in Ward 4 for breakfast. Seedlings for plant-based water filtration systems are distributed

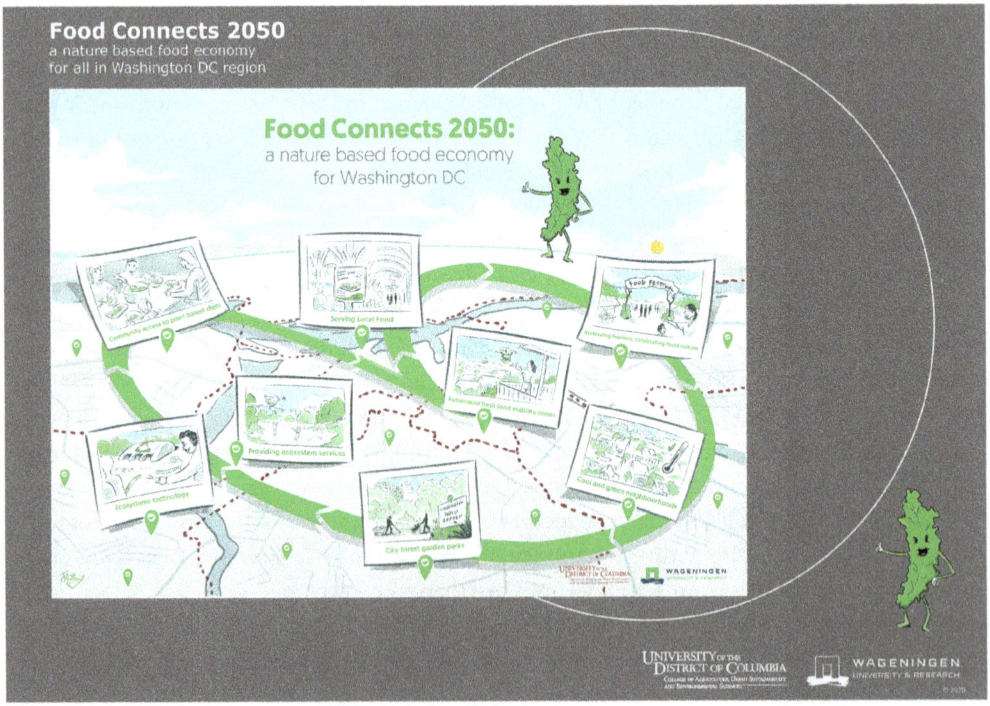

Figure 3.2. The eight Wards of the CRFS of Washington DC.

from Ward 7 to downtown offices. Personal 'silent natural wings' drones surprise tourists on the Mall and entertain those waiting for entrance to one of the Smithsonian museums. The lower traffic congestion has opened up former traffic lanes and pavements for added green space that makes a meaningful contribution to reducing the summer heat in downtown DC.

3.2.2 Ward 2: Hotels and restaurants serve local and regional food

Food connects economy and the regional: visitors to the city in 2050 are proudly presented with a guide to plant-based regional cuisine in the area upon their arrival at area airports and train stations. Passionate youth movements started the revolution that brought about the change needed to implement a local and regional food system. There is a network of hotels, government buildings, hospitals, community centres and schools that produce their vegetables and fruits in energy-neutral climate rooms and urban greenhouses. A network of outdoor food systems in raised gardens encourages wellness and community. A new generation of students and employees are working in the circular food economy that features a diverse locally grown and culturally sensitive cuisine. The area around Washington DC produces crops for the urban core, thus creating a truly regional food shed.

3.2.3 Ward 3: Removing barriers – celebrating food cultures

Food connects people and the world: people living in the affluent neighbourhoods of Ward 3 embrace and encourage the cultural diversity expressed in the food traditions of other DC Wards. By 2050 food has become a vehicle for learning languages, traditions, poetry and music, and food festivals highlight food niches that give new residents and immigrants the opportunity to share their roots and traditions. By 2050, DC's education, health and non-profit sectors have established strong partnerships with the food sector. Schools, churches and community centres give access to their kitchen facilities to provide healthy meals for children, the elderly and those with medical conditions, while also serving as teaching and training facilities for scores of children, young adults and older residents. All nutrition and food safety education programmes are culturally sensitive and aware of the social pressures associated with food preparation traditions and eating habits. Cultural sensitivity to food origins is especially important as changing eating habits are promoted that move residents and visitors toward diets that are less meat-intensive and more plant based.

3.2.4 Ward 4: Cool and green neighbourhoods

Food connects people and nature: Ward 4 has long been a diverse neighbourhood sometimes characterised by contrasts and contradictions. Its close connection to Rock Creek Park, the largest park in Washington DC, make it the perfect location for a wide variety of gardens and landscapes. Grape arbours, medicinal herbs, berries and ethnic crops serve as a rich source of vegetables, herbs and fruits for communities throughout the DC metro areas. Various food production techniques are shared and management practices constantly improve to increase soil quality, reduce nutrient loss, enhance freshwater reservoirs and improve air quality. The diverse landscapes of the rural Mid Atlantic Coastal plain, the foothills of the Appalachian Piedmont and the Folded Appalachians/Blue Ridge to the Appalachian Plateau bring richness and diversity to the innovative gardens rooted in the diverse neighbourhoods of Ward 4. Open spaces with native plants, grasses and shrubs also enhance the landscape and contribute to air quality enhancement, heat island mitigation, biodiversity, and pollinator protection.

3.2.5 Ward 5: City forest garden parks

Food connects economy and nature: Ward 5 has become the lung of Washington DC in 2050. Its neighbourhoods are comprised of urban forests featuring eatable fruits, nuts and berries. The highly productive eatable landscapes emit oxygen and enhance the air quality of the city while also contributing to a much-needed cooling effect during the summer months. Some sections of the urban forests deliver wood for wood products and a thriving artisan community. In addition to the eatable forests, greenhouses and vertical farms are interspersed throughout the diverse neighbourhoods of Ward 5. New attractive urban and peri-urban pathways with multifunctional green-blue veining provide a sustainable basis that links the city to the wider regional foodshed, based on a well-coordinated system of innovative sustainable urban agriculture methods informed by the culturally diverse dietary needs of the DC metro population. Outdoor growing spaces are interspersed with cutting-edge soil-less systems growing food at maximum intensity in vertical hydroponic and aquaponic systems that blend seamlessly into the formerly industrial landscape of Ward 5.

3.2.6 Ward 6: Providing ecosystems services

Food connects natural resources and the world: given its close proximity to the rivers that cut through the city, Ward 6 has become the natural resource hub of Washington DC. In 2050 the Anacostia River is swimmable and fishable providing many wellness and food system-related resources to the city. Moreover, the river has become the centrepiece for state-of-the-art plant-based water filtration systems that have taken considerable pressure off the city's water treatment facility in blue plains. Water is recirculated through a series of sediment and plant-filled growing beds that are arranged in sequence and installed in greenhouses to allow for year-round water filtration. Any excess water from the intensive food production systems in Wards 4 and 5 is also channelled to Ward 6. Here the water run-off is tested for nutrient content and any nutrient rich run-off is diverted for fertigation purposes (irrigating with nutrient-rich water) to ensure all nutrients are used to their maximum.

Some of the areas along the river serve as aquaculture centres for the city, raising fish in pristine conditions to serve as high-end protein source for local restaurants and food festivals. The annual fish fry is an especially popular event. The fish refuse generated in the aquaculture centres is used as nutrient rich organic fertiliser for agriculture and gardening projects throughout the region. The green-blue veining that characterises Ward 6 combines multiple benefits of climate services, energy production, health and food, and is highly valued by DC and metro region residents as well as visitors to the region. Nature trails serve as study areas and outdoor classrooms throughout the Ward. This fantastic landscape has become a benchmark for other cities, illustrating how they can positively contribute to larger meta-region and global environmental quality improvements, with a special focus on climate change mitigation.

3.2.7 Ward 7: Nature-based technology centres

Food connects technology and economy: a nature-based lifestyle meets high-tech in Ward 7. By 2050, technological advances have merged with sustainable solutions to form the backbone of a thriving network of technology businesses. The Ward also prides itself on the rich history of African American entrepreneurship. Nature-based technology and biomimicry techniques established in the Ward have made it possible to move the entire energy generation for 75,000 residents off the grid, making households independent of the local utility. Ward 7 also builds the sediments and plant-based living machines for the water filtration systems in neighbouring Ward 6. STEM-focused technology entrepreneurship classes are a staple of the businesses in Ward 7 and many have become a mecca for small-scale green technology entrepreneurs from around the world. General practitioners provide citizens with personalised health and dietary advice combining data from exercise, eating behaviour, food choices and shopping. Sensors in homes and commercial buildings link everything from smart toilets, to wastewater, and air quality monitoring to a user's medical condition and metabolism. The results of these cutting-edge technology applications have been improved health outcomes and better environmental quality.

3.2.8 Ward 8: Community access to plant-based diets

Food connects people with people: the much-needed shift to personalised diets that embrace less meat and more plant and plant-protein-based food while maintaining culturally embedded tastes and textures has been fully realised in Ward 8. Local community farms, restaurants, bars, and music venues form a thriving network of culturally diverse go-to places in 2050. True to its rich African American history, hip-hop and bee-bop are the beating heart of the food and people scene of Ward 8 which has become a popular destination.

Wellness too is an important feature of this Ward, with walking and biking trails, and dancing and performance venues all over the city's far south-east neighbourhoods. Residents of all ages and backgrounds share their knowledge of healthy lifestyles and food cultures. Community-based greenhouses, incubator kitchens and canning sheds dot the neighbourhoods. The public space surrounding these food and wellness hubs has contributed to shared expertise, reduced crime, and increased trust in this culturally rich community where many native Washingtonians live alongside neighbours from every continent.

The eight hyper-local dreams of the eight Wards of Washington DC illustrate the connective potential of food. People have access to locally and sustainably grown food throughout the city. They are connected to their neighbours and neighbourhoods, work in circular business parks, preserve natural resources and enjoy an urban environment that is nature-centred with restored ecosystems and abundant green infrastructure. Food is the connecting link throughout the greater

Washington region. Diets are largely plant-based and offer opportunities to share cultural roots and innovation; technology strengthens ecosystems rather than depleting them; economic activity is circular and recharges rather than exploiting; policies emphasise cross-regional collaboration rather than siloed operations; and urban, peri-urban and rural strategies form an interconnected system. For example, highly perishable, nutrient-dense crops are grown in the urban core; less perishable, high-demand high-revenue vegetables are produced in the peri-urban ring; and the less densely populated areas where land values are lower are better suited to grow grains and root vegetables. These distinct place-based complementarities are linked to up- and downstream sectors of suppliers and buyers creating additional benefits for the entire region. Figure 3.3 illustrates our regional vision and Figure 3.4 shows the DC Metro CRFS in the larger bio-region.

While our vision builds on existing initiatives that engage a broad range of local and regional actors, its implementation is a work in progress. Much coordination and continued collaborations with a range of food and sustainability-based initiatives are needed to advance the vision toward implementation. This may include new governance structures that emphasise collaboration and coordination and recognise that restorative activities require compensation just as much as resourcing and production activities (Chapter 9).

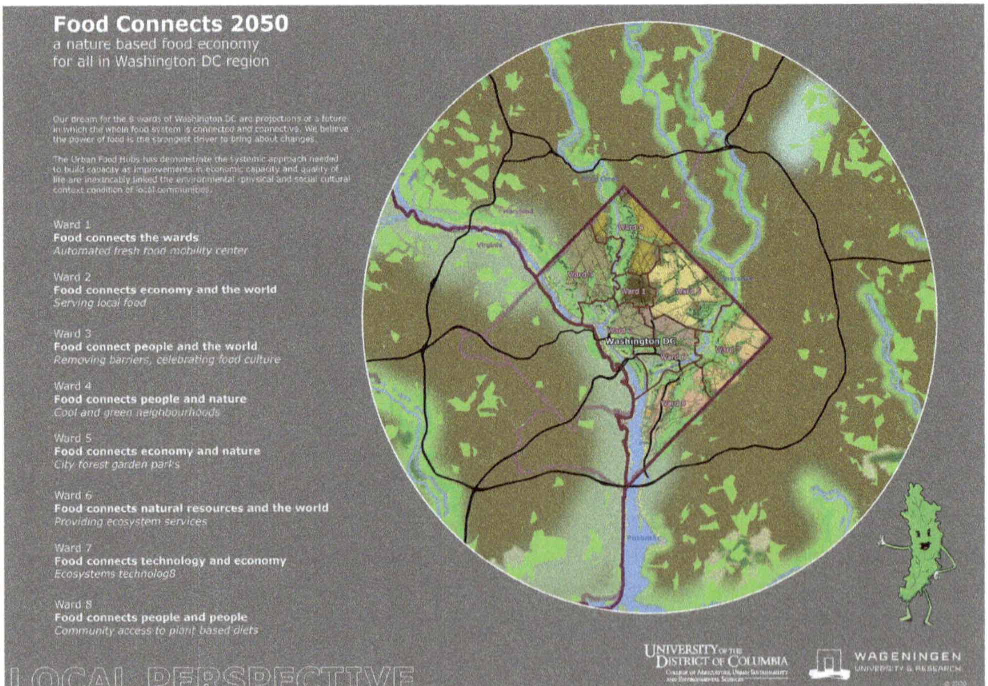

Figure 3.3. Food connects the eight DC Wards to the DC Metro Region.

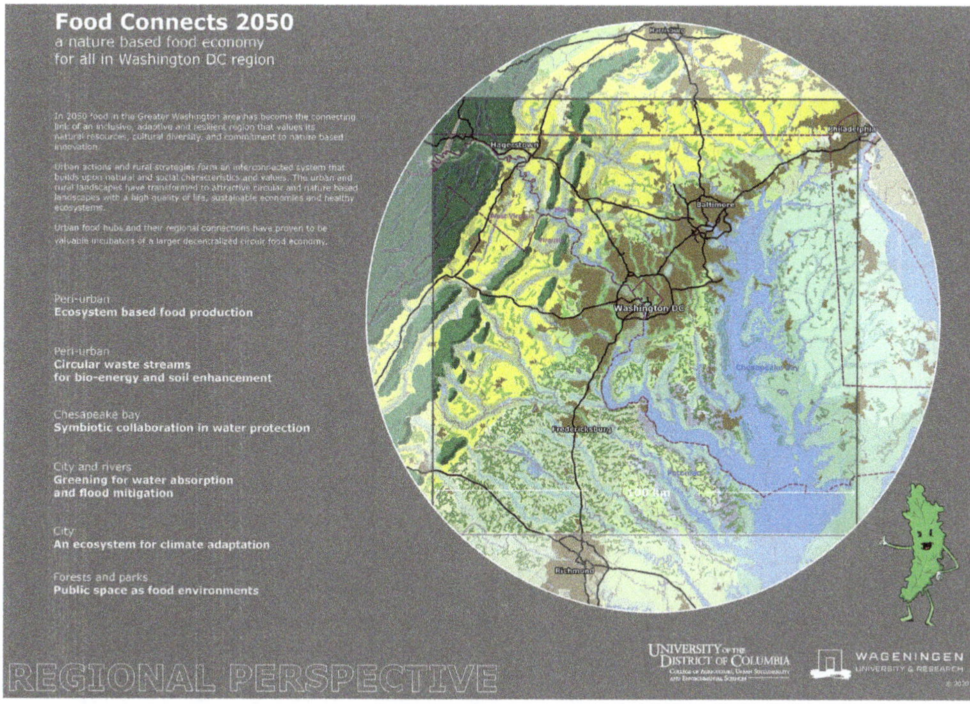

Figure 3.4. The DC metro CRFS in the wider bio-region.

3.3 The 2050 Symbiotic City in the Netherlands

The second vision is a 2050 Symbiotic City for the centre of the Netherlands with Zwolle, Almere and Apeldoorn as connecting cities and the provinces of Flevoland, Gelderland and Overijssel as their metro region context.

Table 3.1 provides some background information about the region and its demographics. The area not only differs in terms of population density but also in backgrounds, age and household composition. People from all walks of life and cultural and (non) religious backgrounds live in this area, and in this way it is representative of the total population of the Netherlands. Figure 3.5 shows the main land use at present, with three major cities with a high population density, agricultural land in the Flevopolder and natural areas in Gelderland, among them the Oostvaardersplassen and the Hoge Veluwe. Together IJsselmeer, Markermeer, IJmeer and Gouwzee cover around 4% of the total surface of the Netherlands. IJsselmeer is a landscape used for freshwater management, drainage of surplus water, protection against floods, for agriculture on the manmade drained polders, and as a habitat for migrating birds. The lake is the outcome of the Zuiderzee project, a huge pool that resulted from damming of the Zuiderzee by building the Afsluitdijk (Beenakker, 2022).

Table 3.1. General information about the region and it's demographics (data retrieved from CBS, 2022).

Neighbourhood	Total population	Female	Male	Age 0-25	Age 65+	Western immigrant	Non-Western Immigrant	Population density	One person households	Municipalities
		%	%	%	%	%	%	km^{-2}	%	
Hanza	129,840	50.8	49.2	30.2	16.2	6.8	10.8	1,087.8	41.4	*Zwolle*
Nieboerhof	47,189	49.7	50.3	30.5	19.6	3.5	3.0	286.4	26.3	*Elburg, Oldebroek*
Woenige	123,002	50.3	49.7	29.5	20.9	5.0	5.1	326.8	29.6	*Ermelo, Nijkerk, Nunspeet, Putten*
Boschlust	216,755	50.2	49.8	26.7	22.2	7.6	8.7	374.4	34.8	*Apeldoorn, Epe, Heerde*
Blauwe Land	146,107	49.5	50.5	33.6	16.5	5.2	5.6	153.0	29.2	*Kampen, Noordoostpolder, Urk, Zwartewaterland*
Nova Tiempo	263,441	50.4	49.6	30.7	13.5	10.3	29.7	886.9	32.2	*Almere, Harderwijk*
Heemwijk	54,239	50.0	50.0	29.2	19.2	8.0	7.8	121.1	30.3	*Dronten, Hattem*
Bruchemerpolder	102,690	49.6	50.4	29.8	17.0	11.6	18.7	99.3	33.7	*Lelystad, Zeewolde*
The Netherlands	17,475,415	50.3	49.7	27.8	19.8	10.6	14	519	38.5	

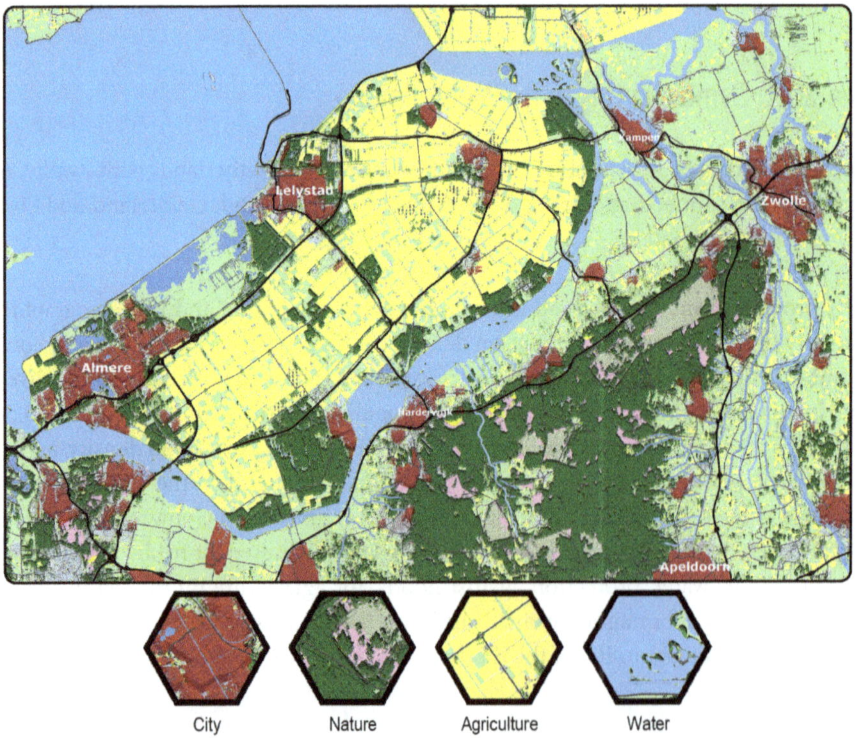

Figure 3.5. The present situation of the Netherlands region.

Our approach for developing the 2050 vision for the Netherlands is the result of a three-pronged methodology: (1) case study analysis of present symbiotic practices; (2) stakeholder engagement and consultation including community leaders, municipal agencies, entrepreneurs, students, and researchers; and (3) nature-based thinking that incorporates the multiple values of nature expressed through the practices of diverse actors.

3.3.1 Case study analysis

A background study among different stakeholders in the centre of the Netherlands provided the empirical basis for analysing pre-existing storylines. A centre piece of these storylines is the 2022 Floriade in 2022 in Almere with its compelling theme: 'Growing Green Cities' that has given an impulse in this specific area (Floriade Expo, 2022).

In creating our overall vision of the 2050 symbiotic city in the Netherlands we discovered eight storylines where nature is valued in different ways: (1) adaptation; (2) community; (3) green neighbourhoods; (4) urban forest; (5) water smart city; (6) nature-based solutions; (7) personal lifestyle and health; and (8) restorative urban foodshed. Adaptation is for instance exemplified by an investigation of Zwolle in the region (Dona Stedenbouw & Studio Bereikbaar, 2020). One other important focal point is the local projects for climate adaptation (Gemeente Zwolle, 2022). Communal values are visible in the practices of residents in the Almere area as discussed by Oosterwold (Oosterwold, 2022), the Floriade initiatives (Van der Plas *et al.*, 2022) and initiatives in Lelystad (Gemeente Lelystad, 2022) and Harderwijk (Steenbreek, 2021) (Floriade Expo, 2022; Van der Plas *et al.*, 2022). Additional green neighbourhood initiatives have been documented in Harderwijk and Steenbreek (Steenbreek, 2021) and at the Veluwe where the focus has been on green schoolyards (Maas *et al.*, 2021). An urban forest initiative is inspired by rewilding scenarios (Kopnina *et al.*, 2019) that have also been applied in the Oostvaardersplassen (Smit *et al.*, 2010), on the Veluwe at the Hierdense Beek region (Verdonschot *et al.*, 2021) and at the Veluwe (Stichting Wisent op de Veluwe, 2022). Water management is inspired by Zwartewaterland (Gemeente Zwartewaterland, 2019), the Lowlands festival (Elshof, 2018) and Zwolle initiatives on climate and water measures (Dona Stedenbouw & Studio Bereikbaar, 2020; Gemeente Zwolle, 2022). The nature-based solutions storyline has at its core the city of Almere that enforces technological innovations in the circular economy sector with several examples from the province of Flevoland (GCA, 2022; Jutte *et al.*, 2018; LIDL, 2021; PRICE-CE, 2022). Personal lifestyle and natural health stories are inspired by Kampen's initiative on Nature for Each Other (Gemeente Kampen, 2021) and the Lowlands Festival taking place every year in the Flevopolder. The restorative food shed storyline is based on initiatives in controlled environment farming (Butturini and Marcelis, 2019; Roobeek, 2018), as well as research by Kranendonk *et al.* (2018) who examined the potential for an urban food shed in the Almere area (Sukkel *et al.*, 2010). The study on Oosterwold (Oosterwold, 2022) and Lelystad highlighted the numerous community groups engaged in local farming who articulated the desire for urban and peri-urban populations to reconnect with their food and engage at least to some degree in growing their own food (Forssell and Lankoski, 2015).

3.3.2 Stakeholder engagement

The overall visioning process for the eight different storylines for the Netherlands was based on intensive stakeholder consultation dating primarily to 2021. First, the lead author consulted engaged citizens, civic organisations, government officials, businesses and real estate agencies through in-depth interviews about their social-ecological practices. Secondly, a series of meetings with stakeholders who are at the forefront of developing inclusive green cities in the Netherlands, identified relevant characteristics of a nature-focused development agenda. Being a member of the core team of the Green Circles project (Lemstra *et al.*, 2020) initiated in 2021 in the area of Zuid Holland, inspired the lead author to dream of a radically green city. Meetings were attended by the lead author in Rotterdam and Leiden, where a multitude of actors expressed their versions of a radical green city. The Wageningen Biodiversity Initiative which started in 2021, included preparations for the Floriade (Floriade Expo, 2022) as well as the participation in working groups of the Biodiversity Platform of the Netherlands (Bouma, 2017). Thirdly, the lead author set up a project (Anonymous, 2021) with colleagues from Amsterdam, Washington DC, Utrecht and Eindhoven (EWUU, 2022) to make cities more inclusive in their greening strategies (see also Chapter 11 and 12 of this book).

3.3.3 Nature-based thinking

Both the empirical analyses of multiple existing initiatives and the stakeholder consultations revealed a multitude of values of nature. Stakeholders expressed that they often wrestle with how to adapt more technology-focused solutions to their specific contexts and expressed some unease with the fact that the nature-based solutions leave little room for expressions of the intrinsic value of nature and the relationship value of nature. Nature-based thinking was expressed through (1) a more nature-centred perspective that views nature as having intrinsic rights of its own, while human stakeholders can act as advocates for nature (Burgers and Outer, 2021); (2) nature contributing to a healthy relationship between people and nature and between people and people (Manon, 2020); and (3) the value of nature in its non-use or resource saving rather than use. A useful example of this non-use priority is urban waste. In mainstream economics waste has negative value, because its processing as waste product is costly while it has no immediate market exchange value or even use value. Yet if a waste product turns out to be a resource, its value must be reconsidered as the non-use value of waste can turn into a future resource that can be brought back in the local value chain.

As a result, the concept of nature-based thinking was advanced as an alternative to nature-based solutions to leave room for non-utilitarian and non-use valuation concepts. Nature-based thinking acknowledges diverse human-nature relations that encompass symbiotic commitments to nature and human communities as well as human-nature-technology relationships. A commitment to nature-based thinking is neither anti-social, nor anti-technology. Instead, it offers opportunities to engage by creating human-human and human-nature connections, and engaging in meaningful and supportive collaborations. Everyone has a role to play in creating the symbiotic city of the future.

Our vision of the symbiotic city of the Netherlands thus presents eight diverse yet interconnected storylines (Figure 3.6) that capture different value concepts of nature which surfaced in the case study analysis and stakeholder processes. What follows is a brief description of the eight distinct stories that make up the overall vision of the symbiotic city in the Netherlands (Figure 3.7).

3.3.4 District 1: Adapt in Hanza – resilience

When one enters Hanza by electric train from Amsterdam, the first thing that comes into view while crossing the river IJssel is the beautiful city park called 'The English Works' (In Dutch: het Engelse Werk) followed by a network of bike lanes leading to the open green and blue spaces in the buzzing city centre of Hanza. The Hanza area has converted roads and sidewalks into added green space significantly increasing permeable surfaces in the area. This has made a meaningful contribution to cooling in the summer and regulates stormwater in the fall and winter.

By 2050 Hanza has evolved into a distribution hub where regional produce from the North of Europe arrives by CO_2 neutral transport. Thriving businesses sell sustainable building components needed to build and renovate a climate neutral and circular construction sector. Tourists that visit the city are impressed by the retail venues in the city centre, where all sorts of innovative local produce and nature-based products are available along with an array of outdoor amenities. Students and workers in the outdoor venues give the district a real buzz. The outdoor sports museum close to the river is well known for its interactive experiences that offer contact with nature, including water, mud, and tree canopy emersion tours of the area that represents the history of the water and sandy soils of the region and its adaptive capacities.

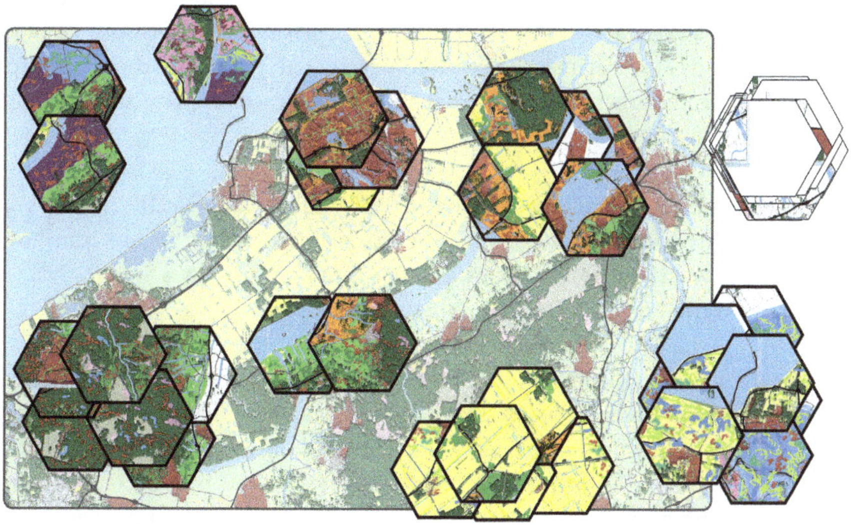

Figure 3.6. Expressions of eight symbiotic practices in the Netherlands.

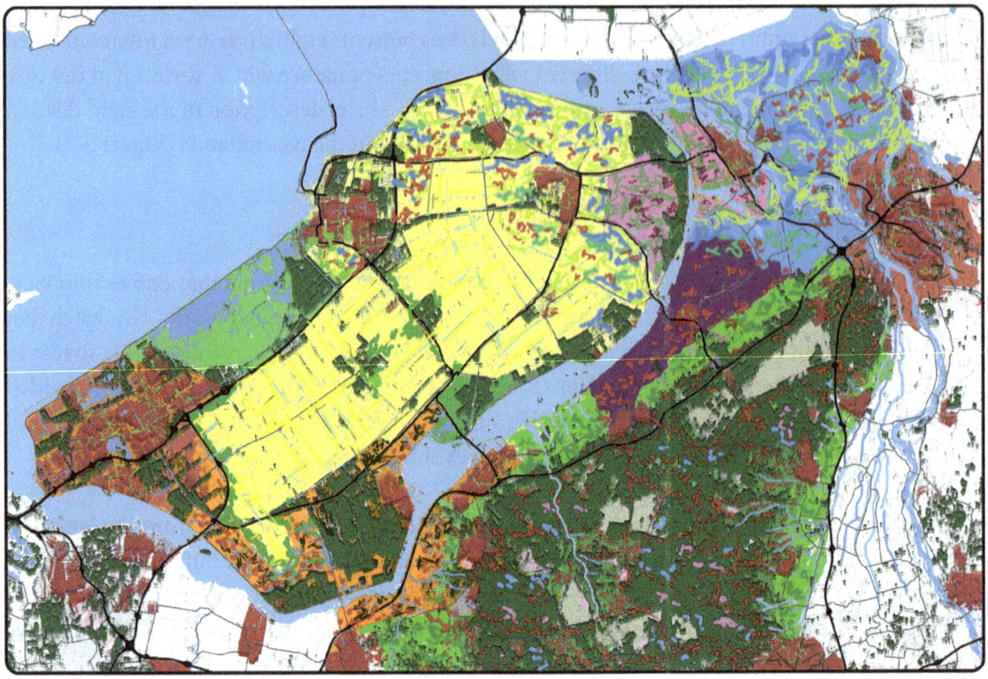

Figure 3.7. The symbiotic city 2050 in the Netherlands.

3.3.5 District 2: Meet in Nieboerhof – communal values of nature

District 2 embraces communal values of nature where old and new residents and immigrants mix easily to share their roots, traditions, and new-found expressions of biodiversity. The neighbourhoods have transformed previously restricted spaces like schoolyards, sports facilities, church yards, and market squares, into open access green public spaces and meeting venues where people from all walks of life can meet.

Outdoor camping is a popular activity in District 2 where Saxenheim, the country's oldest campsite is famous for its openness and hospitality. Every year two million visitors come to the District to stay in the natural cottages and enjoy outdoor activities, such as a food festival with game and deer, lupine varieties, and other nature-based attractions.

District 2 also has a range of housing concepts that express innovative ways of urban living, not designed by urban planners but by citizens themselves who want to create their own public spaces and streetscapes. Inspired by citizen groups like Steenbreek (Stobbelaar *et al.*, 2021), and the Doughnut coalition (Raworth, 2017), municipal governments have become highly decentralised to allow local groups to develop their own integrated nature based and circular economy solutions. Residents organise their activities through local neighbourhood rights agendas that design, implement, and maintain their own urban neighbourhood practices and amenities. Sharing and donating goods and services is seen as a form of constructive action outside of the money economy. Local solutions include citizens' initiatives and businesses that develop highly efficient recycling efforts of natural resources and pursue multiple value creation goals rather than focusing on profit maximisation alone. These businesses develop a social innovation and entrepreneurship climate that drives ecosystems restoration and resource management.

3.3.6 District 3: Create greenery in Woenige – biodiverse neighbourhoods

In District 3 the neighbourhoods are fully developed into green landscapes and gardens by the master gardeners of the district. Nature makes the area attractive and resilient and the landscape consists of networks of green sponge gardens and blue-green veins including wadis, rain gardens, and climate squares. Paving is minimised and permeable surfaces are maximised. Wherever hard surfaces are necessary, water permeable pavement options are used and cutting-edge methods counteract soil compaction. Citizens have recognised the importance of soil, and compost their own organic waste in neighbourhood composting sites to add organic matter to fields and gardens. By increasing organic material, soils have developed higher fertility, higher capacity to retain water, and higher rates of carbon sequestration to contribute to climate neutrality.

District 3 embraces the nature-inclusive construction of homes made from wood and recycles building materials. In addition to its ecological value, the nature-inclusive construction also adds social value to the district. Each house has an outdoor space such as a green roof, garden, park or a balcony. This adds outdoor living space to private indoor living spaces. Green spaces and walking and cycling routes form a network of safe and attractive public spaces that bring recreation and exercise into daily routines and a National Park merges with the city itself. Shade, and cooling through evaporation, has reduced mortality rates among the elderly. The inclusion of useable green spaces into daily lives and routines has reduced stress-related disorders and inspires healthy lifestyle.

3.3.7 District 4: Reclaim space for nature in Boschlust – rewilding cities

District 4 is the lung of the city where nature has room and value of its own. Parks and avenues on the Hoge Veluwe are open to everyone with Palace het Loo as the centre piece in the middle of the district. A 'green umbrella' of thousands of old and new trees in residential and commercial areas alike provides shade in the summer and attractive living and working environments year around. The green umbrella provides a home to birds, pollinators, and insects. The more mature sections of the urban forests deliver wood for construction and furnishings but are also the home of mammals, such as wolves and foxes. Urban greening encompasses a range of initiatives including tree canopies, parks, and gardens (Givoni, 1991). The district's houses and neighbourhoods are built in balance with nature, by paying attention to plant and animal species constructing habitats that complement human living spaces. All green spaces feature indigenous plants and plants exchanges have become a major social activity in the area. Many residents are master naturalists and master gardeners and are keenly aware of the value nature adds to their living environment.

The district has achieved an ecological balance in which different species and their habitats balance each other. Due to a diversity in species, age and genetic material, trees, plants and animals are more resistant against shocks and diseases. The city has become a member of a worldwide initiative called 'National Park Cities' that shares a deep commitment to a cityscape that is rich in greenspace and wildlife. District 4 has won several prizes among its fellow national park cities including London, Milan, Washington and Singapore.

3.3.8 District 5: Float in the Blue Land – water smart cities

Given its close proximity to the rivers that cut through the city, District 5 has become the water smart city hub. By 2050 the river is the water reservoir of the city providing freshwater and many wellness and water-related resources to the city. District 5 has put a lot of effort into restoring its previously degraded water ways into natural ecosystems and securing the water resources for the city. River systems have been transformed into new transportation routes; aquaculture and aquaponic facilities dot the landscape; floating homes, offices, and gardens are connected by electric ferries; and fish hatcheries integrated into the water habitats serve as local protein source. Harvesting and recycling water, water retention hollows to recharge groundwater and watershed protection practices are commonplace in District 5. The IJsselmeer forms the blue heart of the north of the Netherlands. This district reflects the diverse ways in which citizens in the Netherlands live with water in river systems, deltas, and below sea level. District 5 has taken this rich symbiotic history of living with water and has re-envisioned it for the future. Thousands of visitors come to the Bluelands festival to celebrate the joy and beauty of living with the water. Bluelands is a floating performance venue for all kind of social groups who meet for music, theatre, dialogue, learning, shared stewardship, and celebration. Bluelands is a welcoming place for different cultures and art forms. It has provided special inspiration for culinary and artistic practices that focus on reconnecting humans and nature. Bluelands has also sparked many new water-based practices including a ferry connection from Amsterdam Central Station to the festival terrain.

3.3.9 District 6: Restore in Nova Tiempo – nature-based solutions

In District 6 nature-based thinking meets high tech. Businesses use smart, cutting-edge nature-based technologies that range from circular waste management to nature-inclusive building techniques and waste-free food systems. The innovations that are developed form an experimental base for the entire Netherlands and beyond. Green design firms constantly improve the green roofs, green walls, rain gardens and green buildings that dot the neighbourhoods. Green high-tech businesses stream the data that optimises the energy generation and plant-based water treatment facilities located throughout the city. Private and public-sector agencies have long ago moved off the grid with support from the neighbourhood-scale green infrastructure ventures that have stabilised the city's utilities. Most of the buildings in this District are off the grid, making households and commercial enterprises independent of the local utility. A whole new line of research has emerged within the academia of Almere in connection with Wageningen University & Research with projects that connect technology, geodata, AI, circularity and nature. Advanced technologies are tested here and result in nature-based innovations both in terms of products and methods. Techniques and management practices are constantly shared and improved to increase soil quality, reduce nutrient loss, enhance freshwater reservoirs and improve air quality.

3.3.10 District 7: Celebrate life in Heemwijk – natural lifestyles

District 7 encourages healthy lifestyles and the diverse expressions of citizens of a personal relationship with nature. The district enables outdoor exercise, communal and cultural diverse activities and encourages the much-needed shift to personalised diets that embrace less meat and more plant and plant-protein-based food while maintaining culturally embedded tastes and textures. Residents maintain a significant part of the public green spaces themselves, which means they are actively engaged in opportunities for social connections and healthy exercise. With the help of tools, such as virtual applications made by start-up companies in District 7, citizens can check the health of the flora and fauna that surrounds their living and working environment. Residents also benefit from the (global) visitors who frequent hotel lobbies that have been transformed into green (outdoor) meeting spaces for visitors and residents alike. The health and environment focus of District 7 also provides meaning and satisfaction for the elderly and those who do not have a working environment that would normally provide social connections. Children learn about nature through their education and by playing in their green school yards.

3.3.11 District 8: Harvest in Bruchemerpolder – restorative urban food shed

There are many storylines that merge to form the vision of District 8. It is a land once claimed from the sea and in 2050 it forms the restorative food shed for the larger Metropolitan Region of Amsterdam. As in the vision of the Washington DC area CRFS, food in District 8 is produced in a series of concentric circles around the urban core. In the city centre, urban agriculture focused on highly perishable, highly nutrient-dense crops. The peri-urban ring focuses on less perishable and bulkier crops that will benefit from a shorter supply chain enabled by keeping production close to market. Production will also focus on direct marketing to consumers, restaurants and food processors. Throughout the region, commercial food production will be supplemented by community and home gardens that will make a meaningful contribution to healthy plant-based diets of residents. Farms grow nutrient-rich healthy and fresh foods, as well as plant-based raw materials for processed new foods. Farm businesses diversify their production by producing new crops such as seaweed and uncommon crops desired by customers while also providing a regional stock of healthy foods in case of food shortages. Farmers have a strong working relationship with government agencies, knowledge institutions, food processing companies and suppliers to form a learning ecosystem and ensure an optimal exchange of knowledge from nature-based farming and circular farming to expertise in agri-food, governance, and regulations.

3.4 Conclusions and pathways for developing the symbiotic city

The two visions of Washington DC and the Netherlands are composed of imaginary storylines that can outline the agenda for new land-use alternatives. The visions illustrate how nature-based thinking concepts can be incorporated in urban practices to restore the foundation of soil, water and sustain ecosystems health in urban networks. Urban, peri-urban and rural practices form interconnected regional networks of technologies and institutions, that build upon the characteristics of nature and the societies and cultures embedded within them. The two visions engage a wide range of actors and embraces diverse values of nature. The diverse storylines that together embrace the total vision of the Symbiotic City express heterogeneous values of community, culture and inclusion. These values find expression in the eight presented storylines of symbiotic practices that people can develop in their own local neighbourhoods and regions.

Our visions for both Washington DC and the Netherlands therefore contain eight possible pathways for urban regimes that are poised to become more symbiotic in engaging their social contexts and natural surroundings. These include: (1) adaptation; (2) community; (3) green neighbourhoods; (4) urban forests; (5) the water smart city; (6) nature based solutions: (7) personal lifestyle and health; and (8) restorative urban food sheds.

The first pathway focuses on the value of nature for adaptation. The symbiotic city offers connections between soil, water, biodiversity and technical networks that improve the adaptivity of cities to future shock events, such as pandemics and climate challenges (Voskamp *et al.*, 2021). This includes the necessary social-ecological and institutional networks between rural, peri-urban and urban regions.

The second pathway focuses on the communal values of nature where inclusion of all people and social diversity is needed to stabilise communities in the same way that biodiversity stabilises ecosystems (O'Hara 1995). It involves the participation of all citizens in institutional networks of government and science. It allows space for transparent and informed discourse on 'nature-inclusiveness' as a standard for urban development and policy instruments to move toward its implementation. It incorporates different modes of knowledge production, for instance through citizens' science, in collecting information on the progress toward the nature-inclusive symbiotic city starting with observations of best practices defined by citizens themselves.

The third pathway focuses on green neighbourhoods and districts that adopt nature-based thinking in urban practices to replenish resources and restore the social-ecological networks of soil, water, flora and fauna. It embraces green spaces and forests and the nature-inclusive construction of the buildings made from wood and recycles building materials that add ecological value in addition to social value.

The fourth pathway makes room for the intrinsic value of nature through urban forestry, restoration and regeneration practices, which creates the conditions for creating biodiversity hotspots in cities. These practices focus on rewilding in urban areas, including reintroducing native plant species, building parks on empty lots and allowing nature to reclaim space. Residents are trained master naturalists and are keenly aware of the intrinsic value that nature adds to their living environment. City neighbourhoods are dotted with urban forests where a range of nature-focused activities takes place and a variety of species find their home.

The fifth pathway gives space to water smart cities, where citizens engage in smart water management to re-root the foundations of the city. Adopting smart water management is necessary to create diversity in land use and biodiversity in urban areas. Part of this pathway is to develop powerful icons for citizens to experience nature and water celebrations and develop their personal story lines of living with nature, with water, below sea level, along rivers or in a delta.

The sixth pathway focuses on nature-based technological networks and circular networks in the urban areas that enhance natural resources and strengthen sink capacities rather than depleting them (see also Chapter 9). This gives space to the soil of the city, avoids soil sealing and compaction and increases the organic matter content of soils so that the city itself can restore its capacity to retain water and restore carbon. This pathway aims at a resilient urban economy that values its natural resources, sink capacities and commitments to nature-based innovation. This circular economy is based on principles of restoration and regeneration rather than exploitation. It includes a commitment to the non-use of natural resources rather than their use or even reduced use.

The seventh pathway focuses on individual values of nature, personal lifestyles and health. It develops practices that support lifestyles that integrate nature in the daily living experience of urban citizens. This includes diets that are largely plant-based and have positive health implications by addressing diet-related health impacts and food insecurity associated respectively with too much and not enough food. The goal of these pathways is to find a balance between the individual needs of citizens to experience nature and making a natural surroundings that is good for all kinds of human and non-human visitors to the area.

The eighth pathway focuses on urban food production and consumption and creates food production networks in the urban scape that are nature based, sustainable, and able to link urban, peri-urban, and rural communities. The food economies are embedded in local biodiverse landscapes and green technology networks.

The two visions we present in this chapter take the hyperlocal initiatives of Washington DC and the Netherlands as their starting points to narrate future pathways that look past short-term interests and focus on what is possible in the long term. When urbanites join in these hyperlocal innovations and start developing future visions and pathways of a symbiotic city, the niches have the potential to transform entire regions, continents, and our world into connected communities that will attain a sustainable and just nature-based future.

References

Anonymous, 2021. White Paper Circular and Inclusive Cities. Alliance EWUU, Utrecht, the Netherlands.

Beenakker, C., 2022. The Zuiderzee project. Available at: https://lorentz.leidenuniv.nl/history/zuiderzee/zuiderzee.html.

Bouma, J., 2017. Boeren, natuurorganisaties en wetenschappers in overleg om de biodiversiteit te redden. Trouw. DPG Media BV, Amsterdam, the Netherlands. Available at: https://tinyurl.com/4r3cpfpx.

Burgers, L. and Outer, J., 2021. Rights of nature – case-studies from six continents: Compendium, Ambassade van de Noordzee.

Butturini, M. and Marcelis, L.F.M., 2019. Vertical farming in Europe: present status and outlook. In: Kozai, T., Niu, G. and Takagaki, M. (eds.) Plant factory. Elsevier, New York, NY, USA.

CBD, 2014. Global biodiversity outlook. Secretariat of the Convention on Biological Diversity, Montreal, Canada.

CBS, 2022. Statline. Available at: https://opendata.cbs.nl/statline/#/CBS/nl/.

CBS, PBL, RIVM and WUR, 2021. Fauna van het land, 1990-2019 [17] (indicator 1579, versie 05, 1 april 2021), Centraal Bureau voor de Statistiek (CBS), Den Haag; PBL Planbureau voor de Leefomgeving, Den Haag; RIVM Rijksinstituut voor Volksgezondheid en Milieu, Bilthoven; Wageningen University and Research, Wageningen, the Netherlands.

Dona Stedenbouw & Studio Bereikbaar, 2020. Eindrapport analysefase, Provincie Overijssel, Gemeente Zwolle, Regio Zwolle, Ministeries van BZK en IenW, RWS – Oost-Nederland, Waterschap WDOD.

Elshof, L., 2018. Duurzaam plassen bij 100 beats per minuut, H2O. KNW, Den Haag, the Netherlands, pp. 20-23.

EWUU, A., 2022. Circular society. Available at: https://ewuu.nl/nl/onderzoek/circular-society/.

FAO & ITPS, 2022. Urbanisation and soil sealing. Food and Agriculture Organization of the United Nations and Intergovernmental Technical Panel on Soils, Rome, Italy.

Floriade Expo, 2022. Amsterdam-Almere. Available at: https://floriade.com/nl/op-de-expo/nationale-paviljoens/amsterdam-almere/.

Forssell, S. and Lankoski, L., 2015. The sustainability promise of alternative food networks: an examination through 'alternative' characteristics. Agriculture and Human Values 32: 63-75.

GCA, 2022. Grondstoffen Collectief Almere. Available at: https://gca-almere.nl/.

Gemeente Kampen, 2021. Natuur voor elkaar. Available at: https://tinyurl.com/ms9yhc87.

Gemeente Lelystad, 2022. Operatie Steenbreek. Available at: https://tinyurl.com/2s4zz5bx.

Gemeente Zwartewaterland, 2019. Notitie bouwsteen klimaatadaptatie. Gemeente Zwartewaterland, the Netherlands.

Gemeente Zwolle, 2022. Hoe Zwolle zich in 2021 aanpaste aan het veranderende klimaat, Gemeente Zwolle, the Netherlands.

Givoni, B., 1991. Impact of planted areas on urban environmental quality: a review. Atmospheric Environment. Part B. Urban Atmosphere 25: 289-299. https://doi.org/10.1016/0957-1272(91)90001-U

Hallmann, C.A., Sorg, M., Jongejans, E., Siepel, H., Hofland, N., Schwan, H., Stenmans, W., Müller, A., Sumser, H. and Hörren, T., 2017. More than 75 percent decline over 27 years in total flying insect biomass in protected areas. PloS one 12: e0185809. https://doi.org/10.1371/journal.pone.0185809

Hüttl, R.F. and Schneider, B.U., 1998. Forest ecosystem degradation and rehabilitation1Presented at the 20th International Union of Forestry Research Organizations (IUFRO) World Congress, Tampere, Finland, August 5-12, 1995.1. Ecological Engineering 10: 19-31. https://doi.org/10.1016/S0925-8574(97)10023-4

IPCC, 2021. Climate Change 2021: The Physical Science Basis. Contribution of Working Group I to the Sixth Assessment Report of the Intergovernmental Panel on Climate Change. Cambridge University Press, Cambridge, UK.

IPCC, 2022. Climate Change 2022: Impacts, adaptation, and vulnerability contribution of Working Group II to the Sixth Assessment Report of the Intergovernmental Panel on Climate Change. Cambridge University Press, Cambridge, UK.

Jukema, G., Ramaekers, P. and Berkhout, P., 2022. De Nederlandse agrarische sector in internationaal verband – editie 2022. Wageningen Economic Research, Wageningen, the Netherlands.

Jutte, J.B., Mul, P., Huisman, C. and Stemann, B., 2018. Circulaire atlas provincie Flevoland. Projectnummer BF7588-101-100, Royal HaskoningDHV. Available at: https://tinyurl.com/b32zxec3.

Kemmers, R.H., 2011. Effecten van verzuring op bodemleven en stikstofstromen in bossen: verkenning van mogelijkheden voor herstelmaatregelen. 1566-7197, Alterra, Wageningen, the Netherlands.

Kopnina, H., Leadbeater, S. and Cryer, P., 2019. The golden rules of rewilding – examining the case of Oostvaardersplassen. ECOS 40: 1-16.

Kramer, N., Mens, M., Beersma, J. and Kielen, N., 2019. Hoe extreem was de droogte van 2018?, H_2O-Online, pp. 1-7.

Kranendonk, R., Wascher, D., Agricola, H., Jeurissen, L., van Bossum, J., Rademakers, K.W., ten Brug, L. and Wertheim-Heck, S., 2018. Evidence-based food system design (EFSD). Amsterdam Institute for Advanced Metropolitan Solutions (AMS), Amsterdam, the Netherlands.

Lemstra, A.-B., Noordermeer, R., Slob, T.J., Gremmen, M., Bosch, C.v.d., Rees, A.v., Biesmeijer, K. and Westerink, J., 2020. Groene cirkels kaas en bodemdaling. Available at: https://www.online-broadcast.com/kaasenbodemdaling.

LIDL, 2021. Lidl opent duurzaamste supermarkt van Nederland. Available at: https://tinyurl.com/yc2h5a7w.

Maas, J., Van den Bogerd, N., Van Dam, S., Bouman, M. and Van der Veen, V., 2021. De waarde van een natuurrijke gezonde buitenruimte rond scholen, Jantje Beton, VU Amsterdam, IVN Natuureducatie, Amsterdam, the Netherlands.

Manon, 2020. Recap SYMBIOSIS Festival 2020 – how can we live in harmony with nature again? Available at: https://www.bluecity.nl/blog/recap-symbiosis-festival-2020/.

O'Hara, S., 2015. Food security: the urban food hub solution. Solutions 6: 42-52.

O'Hara, S., 2017. The urban food hubs solution: building capacity in urban communities. Metropolitan Universities 28: 69-93.

O'Hara, S., 2018a. The Five pillars of economic development: a story of a sustainable future for ward 7 and 8 in Washington DC. College of Agriculture, Urban Sustainability and Environmental Sciences, University of the District of Columbia, Washington DC, USA.

O'Hara, S., 2018b. Production in context: the concept of sustaining production. In: Farley, J. and Malghan, D. (eds.) Beyond uneconomic growth. Edward Elgar, Cheltenham, UK, pp. 75-106.

Oosterwold, 2022. Maak Oosterwold. Available at: https://maakoosterwold.nl/.

PBL, 2021. Inleiding natuur en biodiversiteit. Available at: https://www.pbl.nl/Inleiding_natuur_en_biodiversiteit.

PRICE-CE, 2022. Praktijk en innovatie centrum circulaire economie. Available at: https://price-ce.nl/.

Raworth, K., 2017. Doughnut economics: seven ways to think like a 21st-century economist. Random House, London, UK.

Roobeek, A., 2018. Collaboration in ecoystems to unleash radical innovation in vertical farming. Richland Sources. Available at: https://tinyurl.com/4hy3pc4x.

Smit, R., Dillen, J.D. and Vera, F., 2010. De oostvaardersplassen. Staatsbosbeheer, Driebergen, the Netherlands.

Steenbreek, 2021. Twee winnaars van de Steenbreektroffee. Available at: https://steenbreek.nl/twee-winnaars-van-de-steenbreektrofee/.

Sterk, M. and Wamelink, G.W.W., 2019. Case study: de droogte van 2018: op zoek naar een succesvolle systeemaanpak om te reageren op droogte. Water Governance: 57-60. Available at: https://library.wur.nl/WebQuery/wurpubs/fulltext/498512.

Stichting Wisent op de Veluwe, 2022. Wisent op de Veluwe. Available at: https://wisentopdeveluwe.nl/.

Stobbelaar, D.J., Van der Knaap, W. and Spijker, J., 2021. Greening the city: how to get rid of garden pavement! The 'Steenbreek' program as a Dutch example. Sustainability 13: 3117.

Stuiver, M. and O'Hara, S., 2021. Food connects Washington DC in 2050 – a vision for urban food systems as the centerpieces of a circular economy. Sustainability 13: 7821. https://doi.org/10.3390/su13147821

Sukkel, W., Stilma, E.S.C. and Jansma, J.E., 2010. Verkenning van de milieueffecten van lokale productie en distributie van voedsel in Almere: energieverbruik, emissie van broeikasgassen en voedselkilometers, Praktijkonderzoek Plant & Omgeving B.V., Lelystad, the Netherlands.

Van der Plas, M., Van Thiel, L. and Pieterse, O., 2022. Growing Green Almere 3. Gemeente Almere: Team Growing Green Cities, Almere, the Netherlands.

Verdonschot, R., Penning, E., Berends, K., Schoelynck, J., Reitsema, R. and Verdonschot, P., 2021. Aangepast beheer en onderhoud en kleinschalige maatregelen in beken. VBNE, Vereniging van Bos- en Natuurterreineigenaren, Driebergen, the Netherlands.

Voskamp, I.M., de Luca, C., Polo-Ballinas, M.B., Hulsman, H. and Brolsma, R., 2021. Nature-Based Solutions Tools for Planning Urban Climate Adaptation: State of the Art. Sustainability 13: 6381.

PART 2
NATURE IN THE SYMBIOTIC CITY

Chapter 4
Biodiversity in cities – creating spaces for the coexistence of humans and wildlife

Robbert Snep and Wolfgang Weisser

Abstract

This chapter departs from the view that cities are urban landscapes accommodating a variety of different ecosystems to harbour wildlife. Creating wildlife habitats in an urban context provides opportunities to conserve biodiversity and reconnect humans with nature. Metropolitan regions are, however, not yet designed and managed for the needs of wildlife. Addressing wildlife habitat development in an urban context first means raising awareness among the involved city stakeholders, getting them enthusiastic about the benefits of a greener and more natural city environment, and providing clues on how to implement habitat development in the concrete case of a built-up location. Each part of the city thereby has its own group of actors and potential for habitat development. Residential districts and business sites are unlikely places for wildlife conservation. However, in times of more attention to sustainable living, social entrepreneurship, business climate and health, both citizens and companies are becoming more interested in how to develop (natural) green in their living and working environment. In this chapter we discuss the overall phenomenon of green, biodiverse residential and business sites. We describe which characteristics of such locations provide opportunities for which type of wildlife, identify relevant stakeholder groups and accompanying arguments, and demonstrate the added value of biodiverse residential and business sites. One of our cases is the Heineken Brewery (100 ha) in Zoeterwoude (NL) where more than 20 ha of intensively managed urban green was turned into a wildlife habitat resulting in a sudden abundance of rare orchids, hover flies and other pollinators. Another case is a nature-inclusive residential area in Eindhoven (NL). We conclude the chapter with some lessons learned from studying the collaboration between site owners, local governments, NGOs and other actors in turning residential and business sites into green, wildlife-friendly living and working places.

Keywords: urban landscapes, wildlife, greenery, habitat development, businesses, real estate development

10.3920/978-90-8686-935-0_4,

4.1 Introduction

It is only recently that urban areas have been considered landscapes which accommodate a variety of different ecosystems that harbour wildlife (Barthel *et al.*, 20005). Urban areas are man-made landscapes, designed for human needs. However, because of their geographical location, often in biodiversity hotspots, their size, their general lay-out (with green infrastructure) and their land-use heterogeneity, they act as ecosystems, which include ecological niches, complex ecological interactions and a high diversity of plants and animals (Snep and Clergeau, 2020).

Residential districts and business sites are unlikely places to think of as sites for wildlife conservation. However, in times of more attention for issues such as sustainable living, social entrepreneurship, business, climate adaptation and health, governments, citizens and companies are becoming more interested in how they can develop (natural) greenery in their cities, living and working environments. With the increased awareness of and attention to this positive aspect of urban greenery by residents, its potential contribution to wildlife conservation can be considerably enhanced (Apfelbeck *et al.*, 2020).

As worldwide urbanisation is increasing at an astonishing rate, it raises questions within governments about the role cities can play in terms of biodiversity conservation. In 2012, the Convention on Biological Diversity presented their 'Cities & Biodiversity Outlook' (Campbell *et al.*, 2012). This white paper emphasises the relevance of cities in conserving wildlife and linking human and natural life. The UN Sustainable Development Goals (SDGs) also address urban biodiversity, as explained in their 'Getting Started with the SDGs in Cities' report (Kanuri *et al.*, 2016). On the EU level, the added value of cities in conserving wildlife and bridging the people's and natural world was discussed at a conference called 'Biodiver-City: Enhancing urban biodiversity and ecosystem services to make cities more resilient' (2018).

The impact of urban land use on the possibilities for biodiversity is different from the impact of land use in non-urban areas on biodiversity (Biggs *et al.*, 2015). Cities differ radically from non-urban areas due to their built-up and paved structures, population density and human activities, the levels of noise, temperature, moisture, light and many other environmental factors such as the design and management of green, etc. Thus, rewilding in these spaces will look different from rewilding non-urban areas. Rewilding in the city could thrive due to the abundance of specific food and waste resources, the abiotic conditions (often resembling their natural habitat) and the ability of plants and animals to cope with urban stress factors (like human disturbance or light pollution) (Snep and Clergeau, 2020).

Some species really want to be in the city. Think of the Feral pigeon (descending from the Rock pigeon) which has already coexisted with humans in cities for a very long time. Recently, a large increase in the number of species colonising cities has been observed. Among them are pioneering species like the Natterjack toad, which are attracted to the early-successional biotopes at vacant lots in business and industrial areas at the city's edge (Snep and Ottburg, 2008). Why would such a species want to live in a city, anyway? Natterjack toads are opportunistic animals and cities provide a good pioneering habitat. However, in other cases, completely new species have evolved through adaptation to the urban context, like the metro mosquito.

Cities provide an increasing potential to support wildlife, and their contribution to wildlife conservation is increasingly being recognised. Creating wildlife habitats in an urban context provides opportunities to conserve biodiversity and reconnect humans with nature. Furthermore, creation of these habitats can be combined with solutions addressing other pressing urban needs, such as climate adaptation and smart water management (Butt *et al.*, 2018).

For this chapter it is important to mention that, although there is growing awareness of their potential to conserve biodiversity, cities are not yet designed and managed for the needs of wildlife (Weisser and Hauck, 2017). Addressing wildlife habitat development in an urban context means first and foremost that urban stakeholders need to get involved and enthusiastic about the benefits of a greener and more natural urban environment. Their involvement can provide insights into how to implement habitat development in specific and concrete cases, which could be a park, a pond, a street or a neighbourhood.

However, with ongoing urbanisation, cities are both growing and becoming more dense, thus putting pressure on the availability of nature in and around cities. This means there is a need to incorporate biodiversity in urbanisation plans, to safeguard human and non-human species, as well as to avoid a lack of access to nature for residents (called 'nature-deficit disorder', Louv, 2005). In addition, urban green spaces provide excellent opportunities for and contributions to so-called Nature-based Solutions (NBS) for other metropolitan challenges like climate change adaptation and health issues. In some cases, NBS may also have an added value for urban biodiversity, e.g. smart green roofs and wadis can be good places for plants and insects and can also be good for climate adaptation (Snep *et al.*, 2020). Aside from their benefits for humans, these places can also support wildlife, as cool spots or nectar resources (as flowers and vegetation on such green roofs will not dry up) (Snep *et al.*, 2016).

Urban planning, design, and development are processes that determine the layout of cities, at the scale of the entire city, down to that of a single building. It is in these processes that biodiversity should get serious attention. Until now, urban planning/design/development professionals and site owners have only occasionally shown interest in the biodiversity topic.

Two cases demonstrate how to bring together a multitude of stakeholders at green, biodiverse residential and business sites, something which the lead author of this chapter has played a crucial role in developing. In the two cases assessed, we examine: (1) the relevant stakeholder groups and accompanying arguments; (2) which characteristics of residential and business locations provide opportunities for which type of wildlife; (3) how stakeholder processes result in real habitat development in city environments; and (4) the added value of biodiverse residential and business sites.

The first case study is the Heineken Brewery (100 ha) in Zoeterwoude (NL) where more than 20 ha of intensively managed urban green was turned into wildlife habitat resulting in a sudden abundance of rare orchids, hover flies and other pollinators. The second case study is a nature-inclusive residential area in Eindhoven (NL). The chapter ends with conclusions and lessons learned from studying the cases and the collaboration between site owners, local governments, NGOs and other actors in turning residential and business sites into green, wildlife-friendly, living and working places.

4.2 Case 1: pollinator habitat development at the Heineken Brewery, Zoeterwoude (NL)

4.2.1 Stakeholders & arguments

Heineken is a large player in the Dutch and international food industry. It is the second largest brewer in the world and has its origins in the Netherlands. The Heineken Brewery in Zoeterwoude is mainly used for exporting beer to the USA and other regions around the world. The Zoeterwoude site is approximately 100 ha and includes a brewery, the headquarters of Heineken NL, and accommodation (De Hooiberg) to host events and welcome guests. Heineken has a global sustainability strategy ('brewing a better future') that aims to make Heineken the number 1 global brand in sustainable beer.

To achieve its biodiversity conservation goals, Heineken asked Wageningen University and Research (WUR) how the Zoeterwoude brewery could be managed in a more ecological way. The request from Heineken NL to WUR to explore options for habitat development included additional requirements, specific to the brewery situation.

First of all, Heineken wanted to have habitat development that would support the 'brewing a better future' message and make people enthusiastic about the total sustainability strategy of the company. Second, they wanted to have habitat development that would lead to increasing biodiversity levels in the short term, in order to make quick gains and achieve visible results. Third, they had practical restrictions on the type of habitat development because it must not cause a nuisance due to wildlife in company buildings. The production site especially must not be invaded by bugs and other pests. No habitat development must affect the location's appearance in a negative way, especially the parts visited by staff and guests. Fourth, they did not want habitat development that would mean substantial extra costs in urban green maintenance. Finally, they did not want habitat development that would significantly increase the colonisation of the site by protected wildlife (Snep, 2014).

4.2.2 Stakeholder process

On the Zoeterwoude location, Heineken has cooperated since 2015 with WUR, the province of South Holland and other partners in elaborating and implementing sustainable practices. This collaboration, called 'Green Circles', aims to develop a regional approach in which Heineken is part of a circular economic system (closing cycles of water, materials, energy, etc.) that also enhances landscape and biodiversity levels (www.groenecirkels.nl).

The essence of the Green Circles form of cooperation is that the partners work together for system change through knowledge development & knowledge sharing. They acknowledge that developing a dream for a green future will require cooperation and collective work to make progress towards such a dream. This cooperation in itself can be considered a living laboratory where unusual parties experiment with solutions and innovation in a dialogue of trust and collective ownership for the solutions (Van der Steen *et al.*, 2019).

With Green Circle Heineken, nature is taken as the starting point from which to develop a climate neutral Heineken brewery, as part of a climate neutral supply chain, to enhance a sustainable economy and to develop a pleasant living environment where knowledge is developed for a circular society. Green Circles aim to take nature as a starting point and to realise programmes in the areas of energy, water, raw materials, mobility, and the living environment (Van der Steen *et al.*, 2019).

4.2.3 Characteristics of the site for wildlife

Altogether the urban green in between and surrounding the brewery is about 35 ha (Figure 4.1). The site was developed in the 1970s and has a 40-year history of traditional urban green management, with intensively mowed lawns, ornamental green and few native flowering plants. As a result, the site was only inhabited by the species commonly found in urban park settings. These consisted of garden and park birds, including exotic geese species, that were attracted by the combination of open water and surrounding lawns. Butterflies, wild bees, and other species of more native vegetation were only present in low numbers.

Figure 4.1. Business site green at the Heineken brewery: situation before transformation (left) and after (right).

As is the case in many countries, the Netherlands has experienced a serious decline in pollinator abundance over recent decades. Approximately 40% of the wild bee species is under threat, and honeybee, hover fly, and butterfly populations are also declining. The impact of a drop in insect abundance has huge consequences on the ecosystem as a whole, but also on food production. 70% of food crops need pollination. In the Netherlands, authorities, companies and conservation NGOs have addressed this challenge (Delta Biodiversity Plan) and started to make changes in land use and lowering pesticide use to improve habitat conditions for insects.

In 2014, Wageningen University and Research explored four habitat development options for the brewery location: derelict industrial estates (like the Ruhr area), peat meadow landscape (like the surrounding landscape), young forests, and small-scale agricultural landscape. The research recommended choosing the small-scale agricultural landscape (including flower-rich grassland) combined with natural ditches and other water bodies as known from the peat meadow district.

The advantage of this combination is that it could be developed in the short term, presented minimal risks to the brewery business, and would help endangered pollinator species like wild (bumble) bees. Supporting pollinators would be a recognisable aim for a large food and beverage company like Heineken and fit perfectly in their sustainability strategy. The Heineken NL executive board agreed because their conditions for the development of the habitat were met in this proposal.

In a next step a revised urban green management plan was outlined (Snep, 2014). A key action in the plan was the transformation of the lawns into flower-rich grasslands. As a visual start, 0.5 ha was sown with annual and biennial plants, resulting in a sea of flowers near the entrance and the events location. For the remaining lawn area, a different mowing scheme was introduced, which will ensure a more gradual transformation. The shrubs, trees, and ditch verges were also integrated into a more ecological management scheme.

4.2.4 Added value for wildlife

The results of the transformation into an ecologically-managed business site included a large increase in the number and abundance of pollinator species, including special hover flies and bees, according to the pollinator monitoring programme. There was also a change in perception of the brewery environment: employees liked the flower-rich grassland, and especially the high number of orchids that emerged after the green management regime change. The success came from a collaboration between the Heineken company and the involved landscaping firm and from the communication within the Heineken organisation and to all external stakeholders from surrounding business sites, agricultural land, municipalities, infrastructure companies and citizens. The pollinator landscape at Heineken was presented and considered as best practice for an adjacent SME business site (Grote Polder) and the regional pollinator habitat development programme. The pollinator habitat development also acts as a showcase for Heineken's Green Circles collaboration and a catalyst for subsequent actions like the wastewater purification swamp at the Heineken brewery Figure 4.2).

Figure 4.2. The Heineken Brewery in Zoeterwoude, the Netherlands (courtesy of Heineken Brewery, Zoeterwoude).

Spin-offs from this project include, amongst others (see Brand, 2021 for an overview), the Green Circles' 'Green Healthy City' that was officially launched in 2021. The ambition of this project is to sufficiently prepare cities in South Holland for extreme weather conditions by 2050. The parties have jointly created a connecting vision, a dream.

The dreamed-of green city has hundreds of thousands of new trees in the living and working areas of the cities of South Holland. Trees that have a cooling effect during the hot weather and a home to birds, butterflies and insects. Green gardens and enough green infrastructure make the cities resilient to heavy rainfall and give space to soil and water to find their way into the city. There is a better connection between the nature in and around the city, so that wildlife can migrate more easily. In the Green Circles' Green Healthy City, the coalition works on overarching issues, such as knowledge development, legislation, regulations, and financing. The parties also initiate new activities. They jointly ensure that the greening of cities is placed on the agenda of the provincial and national government. To tackle this enormous challenge, a coalition is needed in which the partners work together in an open and equal manner (Van der Steen, 2019).

4.3 Case 2: the nature-inclusive PURE residential development in Eindhoven (NL)

The previous example showed that in the Netherlands there is a growing awareness that nature-inclusive urban development has numerous benefits for people and wildlife. Nature-inclusive urban design and development in a strict sense means that habitat requirements of wildlife are incorporated in the design, construction, and maintenance phases of urban development (Hauck and Weisser, 2019). The built-up area thereby becomes a habitat for designated species. In a broader sense, creating a green and natural urban environment means that citizens may benefit from green views, recreation opportunities, human-wildlife interactions (nature experience), and supporting health and wellbeing. The green design may also decrease the impacts of climate change (pluvial flooding, heat stress), accommodate circular ways of processing rain and wastewater, and increase real estate value.

4.3.1 Stakeholders & arguments

The city region of Eindhoven is called the Brainport and was named Smartest Region in the World in 2011. A whole network of high-tech companies are situated in the Eindhoven urban area, including well-known international players, such as Philips, ASML, FEY, NXP and Brainport Industries. This specific industry is in demand among high-skilled knowledge workers and thus attracts people from all over the world to the Brainport region. For these expats, aside from an interesting job, an excellent living environment is key. Expats generally hope to start a family, enjoy culture and city life, and explore their working and living surroundings. A sustainable way of living is becoming increasingly popular and nature and healthy lifestyles are more and more connected (see Chapter 9 of this book). Cities that provide more greenery can contribute to all these aspects of the living environment. Eindhoven already has several business parks (High Tech Campus, Brainport Industries Campus) that can be considered working landscapes with biodiversity values. At present, the city council is introducing 'nature-inclusive' as a basis for new residential development sites. We now present the case of Bosrijk, a new residential development plan in the west of Eindhoven (Meerhoven) that won the bid in the tendering procedure.

4.3.2 Stakeholder process

A consortium of partners was set up to design the development plan in 2018. It was led by the building company BPD, who invited designers and engineering companies such as Sprangers, Marc Koehler Architects, Boom Landscape, Tom van Tuijn Urban Design, Superuse Studios, ARUP, YuconVR to join. Scientists contributed with measures for biodiversity, as well as to make the building plot climate-proof, and to reuse rain and wastewater, using a combination of physical and natural purification (helophyte filter), and storage in a water tower and a central pond. Together they developed the concept 'PURE' ('Zuiver') for the Bosrijk location and have submitted it to the municipality to enter the tendering process.

4.3.3 Characteristics of the site for wildlife

PURE is a radical green alternative to the typical Dutch residential district. The houses are clustered in three compact units, creating a green heart for social and natural outdoor life. The architecture and landscape design are developed in such a way that they include habitats for building-dwelling species like house sparrow, swifts, and dwarf bats. Nest boxes, as well as foraging and resting places for these species, will be provided. The grassland and swamp area in the green heart of the building plot will be managed in such a way that butterflies, bees, and amphibians find habitats. By doing so, the building location acts as a stepping stone and habitat spot in the larger network of natural areas including the nearby Ecopark Acht. Ecopark Acht, ecological city park, was created in 1998 in the old village centre of Acht, now a district of Eindhoven, based on a design by the Butterfly Foundation. This park has been carefully managed for more than 20 years and monitored for the development of natural values. On the whole, this nature is developing well in terms of plants, birds, amphibians and insects in comparison to an average park with lawn and trees (Aalbers, 2000).

4.3.4 Added value

The PURE concept is an integral living concept, designed as a circular ecosystem. New residents have a say in the configuration of their houses and surroundings, including the green heart where children may explore nature. Together with solar energy, heat from bio-energy, and upcycled and natural materials, the residents are stimulated to coexist in a sustainable way with nature. Elevation gradients shape the various layers of public and private, and biodiversity and ecology. The sightlines, the layered structure of the houses, and indirect approaches of the ensembles give the environment a human scale with which the buildings define themselves as sculptures in a new natural landscape.

4.3.5 Epilogue

The PURE concept was a winning design in a sustainable urban area development tender by the municipality of Eindhoven. Only after the contest was won, did it become clear that the basic conditions for the innovative design – as requested – were not in place. The idea of circular water management and local processing of household water using a helophyte filter was not approved by the water board. They were not involved in defining the tender ambitions, but had a final say in the water treatment plans. Also, the open greenspace in between the building was designed as an ecological flower-rich grassland, but the municipality – responsible for the maintenance – only wanted a regular, more traditional type of green management. Finally, the whole concept of biodiversity conservation and climate adaptation required close collaboration between the building development and the open space ... but in reality the developments in the private (building) and public (greenspace) areas were not closely interlinked (see for an impression of the end result Figure 4.3 and 4.4). All in all, one could say that the mismatch between local policy (sustainability ambitions) and local practice (regulation building processes and maintaining public space) is obvious, and resulted in a less sustainable practice than was requested.

Figure 4.3. First impression of Bosrijk, Eindhoven (courtesy of Nanda Sluijsmans).

Figure 4.4. Second impression of Bosrijk, Eindhoven (courtesy of Nanda Sluijsmans).

4.4 Conclusions and recommendations

This chapter argues for a new way of looking at cities, which would consider them potential hotspots for wildlife. Rewilding the city is a relatively new trend in the field of urbanisation and the debate about its pros and cons is only just beginning. However, with biodiversity declining at a rapid pace, there are groups of stakeholders from business, such as real estate and larger corporations, that feel the urgency to rebuild biodiversity in their real estate and the private and public spaces that are part of it.

In this chapter, two case studies exemplified the processes of stakeholders from business, government, and science, aiming to develop more biodiversity in their residential and business sites in the Netherlands. Both cases differ in various ways. One important aspect of the Heineken case is that there is one landowner, that is also the user and the beneficiary of the actions being taken. As a company, Heineken has a business approach: setting a target, taking into account the impacts on their core business and – if convinced that it's the right idea – then switching directly to action. How different the PURE case is! There are multiple stakeholders directly involved in the area development process, the future users are not yet known and between ambitions and actions there is a real obstacle in terms of fine-tuning regulations, maintenance issues, etc. The two cases illustrate the variety not only in spatial but also much more in organisational settings that can be seen when creating urban spaces for the coexistence of humans and wildlife.

These cases also demonstrate: (1) which characteristics of residential and business locations provide opportunities for which type of wildlife; (2) how the processes resulted in real habitat development in city environments; (3) how to assemble relevant stakeholder groups and how accompanying arguments were formed; and (4) the added value of biodiverse residential and business sites.

Green habitat development at business sites and residential districts is quite a new phenomenon, stimulated by several trends coming together: sustainable lifestyle and corporate social responsibility (CSR), the need for climate adaptation of buildings and properties, higher working place standards, more attention to health prevention via smart urban design, the alarming decline in insect life, the urgency to act on these issues, and more.

Natural green can be considered an effective urban element to contribute to all of this. However, the process of creating awareness and understanding among the relevant actors is one important aspect, while integrating natural green as a structural part in urban planning, design and development is another.

Although the momentum is there, professionals, procedures, legislation and materials are not necessarily already equipped for this job. The cases described above are best practices in which extra attention and ambition from all the involved stakeholders made the difference. To upscale the success of these type of best practices, an overall transition in cities is required. Such a transition is best driven by the urgency shared by broader society, with priorities recognised by all and supported by key stakeholders in urban development, as shown in the second case study, PURE (and in Chapter 2 and 13 of this book).

One could argue that urban biodiversity in itself is probably not enough of an emergency to justify calling for such a transition. However, recent developments show that more and more societal actors are expressing the need for urban biodiversity and nature-inclusive building. Climate resilience (Chapter 6), better health-supporting urban environments (Chapter 10) and the socio-economic status of cities (Chapter 2) are strong additional arguments that could have a larger impact on mainstreaming a nature-based way of urban development. And with the right attention, all measures taken for these issues can potentially have a spin-off value for biodiversity.

Urban planners, project developers and builders that would like to develop nature-inclusive cities should bear in mind the following lessons:

- If you have knowledge about the conditions of water, soil and ecology at an early stage of the building or planning process, you can solve all kinds of problems, such as water storage and how to guarantee a variety of species. Make these conditions of water, soil and ecology visible in the plans that are made.
- Think through all the steps in planning, design, development, and maintenance of location and the consequences for the nature-inclusive developments of this urban location: who should do what, and when and how to guarantee quality throughout the process? Address potential opportunities and bottlenecks in advance and involve actors in the use and maintenance phase early in the project (see Chapter 12 and 13 of this book).
- Take into consideration the involved stakeholders' perception of the place where you start building. Decide on the selection of target species and biotopes with all actors involved, such as architects and urban planners that develop the site as well as the residents and companies that will live and work there. Communicate to all future users of the site what 'nature-inclusive' means and organise support from the future residents and owners of the site.
- If you want to develop wildlife in cities it is important to link habitat development not only to biodiversity goals but also to broader interests, such as human needs (climate proof, health, etc.). Choose to add natural elements to the site that would fit both the natural space as well as the human use of the urban area.
- Climate-adaptive and nature-inclusive building are intertwined. When you start working on climate adaptation, it is a small step to also include biodiversity and nature inclusivity.

Finally, as we urbanise our planet, we should not forget to create places where people – especially the future generations – can be inspired and informed by nature in their direct living and working environments. Taking into consideration the lessons learned above, nature-inclusive urban development can become a reality far beyond the current best practices. It could ensure that humankind will continue to be a part of nature, if in a different context from before. Perhaps one day we could consider nature-inclusive cities as species-rich and amazing, in the way we now think of tropical forests and coral reefs. Cities as true biodiversity hotspots!

References

Albers, K., 2000. Acht Ontwerp, aanleg en monitoring van een ecologisch stadspark. Journal Groen: Vakblad voor Groen in Stad en Landschap 56(1): 46-51.

Apfelbeck, B., Snep, R.P., Hauck, T.E., Ferguson, J., Holy, M., Jakoby, C., MacIvor, J.S., Schär, L., Taylor, M. and Weisser, W.W., 2020. Designing wildlife-inclusive cities that support human-animal co-existence. Landscape and Urban Planning 200: 103817.

Barthel, S., Colding, J., Erixon, H., Grahn, S., Kärsten, C., Marcus, L. and Torsvall, J., 2013. Principles of social-ecological urbanism – case study: Albano Campus, Stockholm. Trita-ARK Forskningspublikationer 2013:3.

Biggs, R., Schlüter, M. and Schoon, M.L. (eds.), 2015. Principles for building resilience: sustaining ecosystem services in social-ecological systems. Cambridge University Press, Cambridge, UK.

Brand, J., 2021. Acht jaar Groene Cirkels: samenwerken met de natuur als partner. Nature Today. Available at: https://www.naturetoday.com/nl/nl/nature-reports/message/?msg=28263.

Butt, N., Shanahan, D.F., Shumway, N., Bekessy, S.A., Fuller, R.A., Watson, J.E.M., Maggini, R. and Hole, D.G., 2018. Opportunities for biodiversity conservation as cities adapt to climate change. Geo: Geography and Environment 5(1): e00052.

Campbell, K., Chan, L., Custot, J., Elmqvist, T., Galt, R., Katti, M., Mader, A., Persic, A., Puppim de Oliveira, J. and Rudd, A., 2012. Cities and biodiversity outlook: action and policy. Available at: 10.6084/M9.FIGSHARE.99889.

Hauck, T.E. and Weisser, W.W. (eds.), 2019. Animal-aided design in the living environment. Integrating the needs of animal species into the planning and design of urban open spaces. 62 pp. Available at: https://tinyurl.com/4pznkup9.

Kanuri, C., Revi, A., Espey, J. and Kuhle, H., 2016. Getting started with the SDGs in cities. Sustainable Development Solutions Network, New York, NY, USA.

Louv, R., 2005. Last child in the woods: saving our children from nature-deficit disorder. Algonquin books, Chapel Hill, NC, USA.

Snep, R., 2014. Biodiversiteit op de brouwerij – inrichting en beheer van de Heineken locatie Zoeterwoude voor biodiversiteitbehoud en -beleving. Alterra-rapport 2567. Alterra, Wageningen, the Netherlands. Available at: https://tinyurl.com/3dcr2yfa.

Snep, R.P.H. and Clergeau, P., 2020. Biodiversity in cities, reconnecting humans with nature. In: Loftness, V. and Haase, D. (eds.) Sustainable built environments. Springer, New York, NY, USA, pp. 251-274. https://doi.org/10.1007/978-1-4614-5828-9

Snep, R.P. and Ottburg, F.G., 2008. The 'habitat backbone'as strategy to conserve pioneer species in dynamic port habitats: lessons from the natterjack toad (*Bufo calamita*) in the Port of Antwerp (Belgium). Landscape Ecology 23(10): 1277-1289.

Snep, R.P., Kooijmans, J.L., Kwak, R.G., Foppen, R.P., Parsons, H., Awasthy, M., Sierdsma, H.L.K., Marzluff, J.M., Fernandez-Juricic, E., De Laet, J. and Van Heezik, Y.M., 2016. Urban bird conservation: presenting stakeholder-specific arguments for the development of bird-friendly cities. Urban Ecosystems 19(4): 1535-1550.

Snep, R.P., Voeten, J.G., Mol, G. and Van Hattum, T., 2020. Nature based solutions for urban resilience: a distinction between no-tech, low-tech and high-tech solutions. Frontiers in Environmental Science 8: 259.

Van der Steen, M., Van Delden, M., Scherpenisse, J. and Schram, J., 2019. Dromen voor doeners, een reflectie op de werkwijze van Groene Cirkels. Nederlandse school voor openbaar bestuur, the Hague, the Netherlands, 74 pp. Available at: https://tinyurl.com/2p86n457.

Weisser, W.W. and Hauck, T.E., 2017. Animal-aided design – using a species' life-cycle to improve open space planning and conservation in cities and elsewhere. biorxiv 150359. https://doi.org/10.1101/150359

Chapter 5
A nature-based approach to building Water Smart Cities

Tim van Hattum, Karianne de Bruin, Katarzyna Kujawa-Roeleveld and Ivo Demmers

Abstract

Cities around the world need to rethink the way they deal with water and develop into water smart cities. This chapter will address a shift in urban water management towards water smart cities using three strategies: (1) restore degraded ecosystems in natural surroundings of cities; (2) implement Nature-Based Solutions (NBS) on a large scale; (3) close the urban water cycle and treat water as a resource. NBS are solutions inspired by, supported by or copied from nature. The Water Smart City approach integrates urban water management on a district, city and catchment scale. Furthermore, innovation and close collaboration between local stakeholders, businesses and research institutes are a key part of this approach, to continuously learn and improve the solutions. By collecting and analysing data, cities can create an evidence base for nature-based adaptation. Case studies presented are Oslo, New Delhi and Ho Chi Minh City. The chapter concludes with a roadmap to become a Water Smart City.

Keywords: water management, water cycle, ecosystem, nature based solution, climate adaptation, urban infrastructure

5.1 Introduction – the challenge

Water is essential for all life in cities. It is a precious resource upon which all human activities and ecosystems depend. We need water for urban water supply, food production, energy supply and ecosystem preservation. And vice versa, we need biodiversity for water storage and conservation.

Water is an ecosystem service itself in the form of rivers, lakes, streams, wetlands, aquifers, estuaries and coastal waters but also depends on other ecosystems, such as forests, grasslands, soils, which provide services that influence the availability of water and its quality, flow regulation, filtering and flood and drought protection.

Cities around the world are currently facing major water-related challenges, such as water scarcity, water pollution and floods. Recent studies show that half a billion people already face severe water scarcity all year round (Mekonnen and Hoekstra, 2016). Water is becoming an increasingly scarce resource as a result of urbanisation, poor water management and increased competition

10.3920/978-90-8686-935-0_5, © Wageningen Academic Publishers 2022

between various economic sectors. Water scarcity will place cities in competition for water with a wide variety of other sectors, including agriculture, energy production and tourism (Nature Conservancy, 2014).

Rapid urban growth will put huge pressure on the demand for water. By 2050, the global population is expected to reach nine billion and the global water demand is expected to increase by 55% (UNESCO, 2015). Fast-growing cities of Asia, Africa and Latin America, but also cities in the developed world face these challenges.

Urbanised deltas belong to the most challenging regions of the world, considering their large concentrations of population, their role in the world's ecosystems and their significance to the world's economy (Meyer, 2016). These regions are dealing with extreme vulnerability and face multiple threats. Land subsidence as a result of intense drainage and over-extraction of groundwater is a major problem. As a result cities in many deltas are sinking below sea level and becoming more vulnerable to flooding. Increasing salinisation caused by the combination of intense dredging, falling ground water levels, land subsidence and sea level rises is causing a shortage in fresh water supply and leading to problems for drinking water supply and irrigation.

On top of these challenges, climate change will have a major impact on cities. Climate change will cause more weather extremes, such as droughts, excess precipitation and heat waves. As cities grow, demand and competition for limited clean water resources will therefore increase, and climate-linked impacts are very likely to make pressures worse in urban areas across the globe (UCCRN, 2018). The IPCC Special Report on the impacts of global warming concluded that there is a very high risk that under current emissions trajectories and current national pledges global warming will exceed 1.5 °C above pre-industrial levels. In a 1.5 °C warmer world, climate change and climate change responses will already affect people in countries at all levels of development.

Risks are not evenly distributed across the globe. The coastal cities and urbanised deltas will be hardest hit; they will face rising seas, salinisation of groundwater, extreme rainfall as well as droughts. The report notes that the Mediterranean and Southern Africa, where drought risks are substantial, are hot spots for increasing dryness but the prognoses for other regions are unclear (IPCC, 2018). Climate adaptation is key. The cost of adapting to climate change could hit $500 billion per year by 2050 (UNEP, 2016).

Cities around the world need to rethink the way they deal with water and transform themselves into Water Smart Cities in order to deal with these challenges. We need to find ways in which to manage water in a sustainable way that allows future generations of urban dwellers to have sufficient access to clean, safe freshwater, and where the built-up area of the city with all its physical assets can function despite a more extreme climate. This requires a holistic and integrated approach to the urban water management of cities and their hinterlands.

Future urbanisation programmes offer opportunities to create a resilient and water smart infrastructure. In fact, creating a water smart infrastructure is a prerequisite for the resilience of the cities of the future. Over US$90 trillion needs to be/will be invested in urban infrastructure globally over the next 15 years. It is estimated that half of the urban infrastructure in African cities by 2035 has yet to be built (WWAP, 2017).

In this chapter we introduce the Water Smart City, an approach that aims to increase natural urban resilience combined with nature-based technologies in order to manage climate extremes and maximise the water efficiency of the city and their hinterlands. The Water Smart City approach integrates urban planning and the urban water cycle for water supply, wastewater, and storm water.

We explain the Water Smart City approach and the three strategies for getting there. Furthermore, we highlight some examples from all over the world and we conclude our chapter with a roadmap to (re)design cities from drained (linear) cities to water smart (circular) cities.

5.2 The Water Smart City approach

5.2.1 Introduction

To create Water Smart Cities we present three interconnected strategies in the following paragraphs:

- Restore degraded ecosystems in natural surroundings of cities (landscape-based adaptation). Restoring degraded ecosystems is essential for securing water resources for cities and urbanised deltas and requires an approach that acknowledges the diversity and quality of aquatic ecosystems.
- Implement nature-based solutions (NBS) on a large scale. NBS play an important role in improving the supply and quality of water in water scarce and stressed cities.
- Close the urban water cycle and treat water as a resource.

5.2.2 Restore degraded ecosystems in natural surroundings of cities

There is a clear relationship between cities and their ecosystems. The latter are influenced by developments in river basins and deltas: urbanisation, agricultural production, water abstraction for irrigation, industries and drinking water, dams constructed for hydropower, and water pollution caused by emissions from agriculture, households and industries. This will affect water resources and ecosystem services, such as clean drinking water, irrigation water, carbon storage, and natural flood protection (Ligtvoet *et al.*, 2018). Ecosystem services in many river basins and deltas are under threat due to deforestation and wetland degradation. Since the year 1900, an estimated 64-71% of the natural wetland area worldwide has been lost due to human activity (WWAP, 2018).

Restoring degraded ecosystems is essential for securing water resources for cities and urbanised deltas and requires an approach that acknowledges the diversity and quality of aquatic ecosystems. Nature conservation plays an important role in keeping water sources reliable and clean.

To find sustainable solutions we need to consider cities in the perspective of their surrounding landscape. Below an overview of measures for the restoration of degraded ecosystems in the natural surroundings of cities (Nature Conservancy, 2014):

1. Retain water in upper parts river basin: stop deforestation, improve reforestation and develop sustainable land use and soil management.
2. Restore degraded rivers and wetlands: invest in wetlands restoration, restore natural floodplains, perform river basis management and build green corridors around cities.
3. Reduce water use by and increase resiliency of agriculture: invest in climate smart agriculture and circular farming.
4. Coastal protection: invest in building with nature and natural coastal protection.

5.2.3 Implementation of nature-based solutions in cities

More and more cities and deltas, such as the Netherlands, Singapore, but also Berlin and Paris, consider NBS an integral part of their water management plans. Common examples of NBS are: green roofs, green walls, green corridors, bioswales, retention ponds, urban wetlands, urban forestry, water squares, floating islands, community gardens and rain gardens. NBS have proven their potential for implementation in urban areas. Traditional examples are vertical vegetation (walls) and green roofs/roof gardens. Other solutions include measures to recycle and harvest water, water retention hollows to recharge groundwater and the protection of watersheds that supply urban areas.

Such interventions are commonly termed Low Impact Development (LID), Green Infrastructure (GI) or Sustainable Urban Drainage Systems (SUDS). More recently, concepts such as Water Sensitive Urban Design (WSUD), have enhanced and broadened the SUDS concept by recognising the role that NBS can play in holistically managing urban water resources (Imperial College, 2017).

NBS play an important role in improving the supply and quality of water in water scarce and stressed cities. Water becomes a central element of a complex natural process taking care of the water cycle in the city: evaporation, precipitation and infiltration. The introduction of natural vegetation plays an important role in upgrading the quantity and quality of available water.

NBS tend to be more resilient to water stress than human engineered infrastructure because it is more adaptive to the natural environment and climate shocks. We have to realise, however, that NBS alone cannot replace engineered solutions. Clever mixes between intensive and extensive NBS systems need to be combined within the limited space in urban settings. For instance, smart technologies can be used for real-time monitoring and to manage extremes by using data and weather forecast information in order to deal with periods of droughts, heavy rains or heat waves.

Large-scale implementation of NBS can boost climate resilience and attract businesses and improve economic development. In many cases NBS have many co-benefits such as biodiversity, health and increased property value. Urbanisation and climate impacts require much faster and greater coverage of NBS in cities worldwide. There is increasing interest among urban planners in NBS because they provides synergies between adaptation challenges and improved quality of the living environment. However, there is a need for an evidence base to ensure large-scale implementation of these solutions. Monitoring and evaluation of NBS is required to convince urban planners and investors of the added value.

Lloyd *et al.* (2002) describes different ways that NBS can contribute to sustainable urban water management: (1) protect and enhance natural water systems in urban developments; (2) integrate storm water treatment into the urban landscape by incorporating multiple use corridors that maximise visual and recreational amenity of developments; (3) protect water quality draining from urban development; (4) reduce run-off and peak flows from urban developments by emplacing local detention measures and minimising impervious areas; (5) develop integrated solutions for flood reduction, drought and heat mitigation. Here we introduce the concept of the Water Smart City to facilitate faster implementation of NBS in urban water management. A Water Smart City is designed to retain and infiltrate rainwater, store it when needed and only drain it when other options are insufficient. We define a six-step approach for Water Smart Cities (Figure 5.1).

1. Retention: retention of rainwater where it falls to reduce surface run-off on, e.g. green roofs, (communal) (rain) gardens, curb extensions.
2. Infiltration: improve infiltration capacity of the subsurface by reducing impervious paved surfaces by pervious pavements and creating green infiltration zones allowing rainwater to soak into the subsurface, natural treatment and subsequently also restoring ground water (for example, green space, bioswales, tree pits).
3. Water storage: rainwater harvesting systems collect rainwater from roofs and other paved surfaces for on-site use (for example, rainwater tanks, water squares). Storages or retention basins can temporarily hold back water and thus prevent floods.
4. Water treatment: to protect receiving water bodies like streams and groundwater from contamination with pollutants present in the stormwater run-off, different treatment elements may be needed. Here filter soil, dual porosity filtration, filter strips, wetlands and wet basins may provide an option. For use of stormwater for supply purposes additional treatment, like a disinfection step, may be necessary.
5. Adaptive water management: water level management by anticipating long-term weather forecast to increase storage capacity in surface water.
6. Water drainage: drain excess stormwater from pavements and roofs separately, thus disconnect from the traditional combined sewage system.

Figure 5.1. Steps to redesign urban water management (drawing by Natasha de Sena).

5.2.4 Close the urban water cycle and treat water as a resource

By closing the urban water cycle cities can reduce their dependency on dams and desalination plants to supply themselves with the water they needs in dry times. Instead of draining the water away from the city, rainwater is harvested and reused where it falls, reducing the volume rushing into stormwater retention facilities. To close the urban water cycle the following actions are relevant (Figure 5.2):

1. Reduce water use: reduce the use of water by citizens and businesses by raising awareness.
2. Improve water efficiency: improve the water efficiency of new and existing buildings by technical measures such as water-saving toilets and showers.
3. Water storage: store larger amounts of rainwater from rooftops and paved surfaces in tanks or retention ponds to prevent pluvial floods and for reuse purposes.
4. Water treatment: stormwater treatment or treatment of grey wastewater in wetlands or helophyte filters.
5. Water reuse: reuse of rainwater and grey wastewater in buildings (flushing toilets, washing machines, car washing) or for irrigation.

Urban retrofitting and climate proofing of urban infrastructure is required to prevent major damage from floods and water supply problems in the event of severe droughts. One useful framework for achieving sustainable urban water management is the Urban Harvest Approach (UHA), which promotes better urban water resource management (Wielemaker *et al.*, 2016).

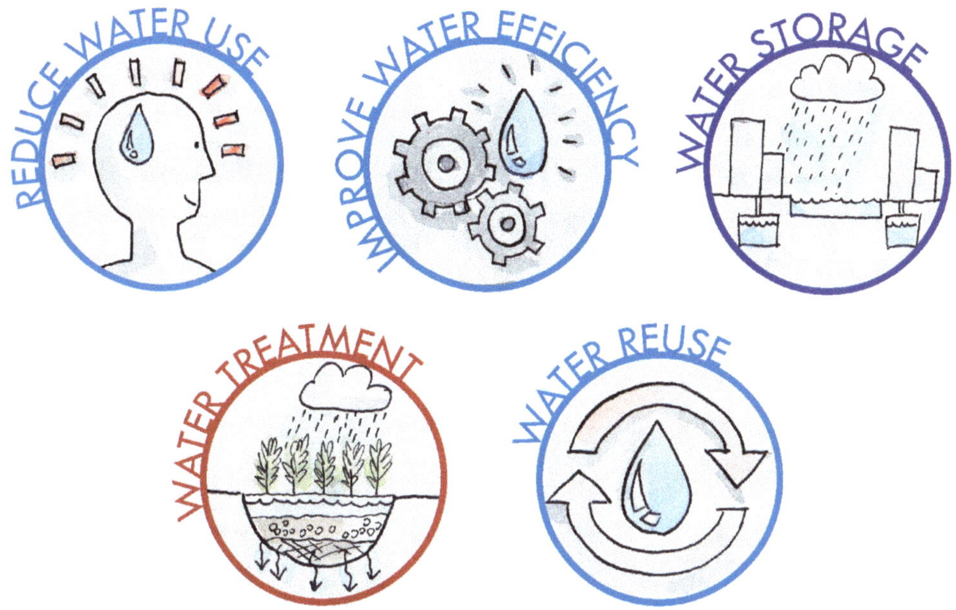

Figure 5.2. Closing the urban water cycle and treating water as a resource (drawing by Natasha de Sena).

The basic idea of UHA is inspired by the concept of urban metabolism. Urban metabolism includes 'the total sum of the technical and socioeconomic processes that occur in cities, resulting in growth, production of energy, and elimination of waste' (Kennedy *et al.*, 2007). Urban metabolism can be determined by a combination of (1) the physical needs of a city – infrastructure; (2) natural and geophysical opportunities and limitations; and (3) the socio-economic and political processes and power structures within the city (Giezen and Roemers, 2015). Urban metabolism enables the monitoring of the material flows in order to improve and achieve the sustainable development of the urban area. Urban metabolism can be used as the baseline to determine the type of metabolism in the urban area: either linear or circular (Figure 5.3).

The aim of the UHA is to identify and harvest the possible resources within the urban boundaries (Rovers, 2007) in order to establish a strategy to shift linear urban metabolism towards circularity (Leusbrock *et al.*, 2015). The UHA framework consists of several steps starting from a baseline analysis followed by demand minimisation, output minimisation, and multi-sourcing (Agudelo-Vera, 2012).

The UHA aims to improve urban water resource management by closing the cycle through the implementation of a mix of innovative technologies and harvesting local and secondary resources. An example of a secondary resource is outflow water – i.e. wastewater that is still of sufficient quality to be used. An example of a local resource is rainwater that can serve as another alternative in the water metabolism of a city. The UHA can be applied on various temporal and spatial urban scales, from household to city scale (Agudelo-Vera, 2012).

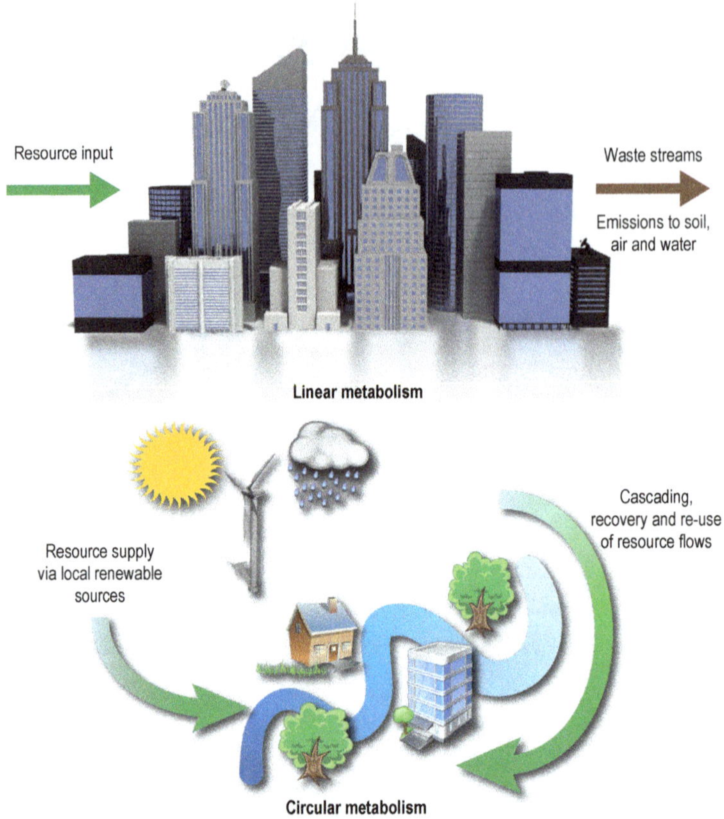

Figure 5.3. Linear vs circular city metabolism.

The first step of the UHA approach is a baseline assessment for analysis of the current water metabolism. Baseline assessment is used to identify existing urban water supply, distribution and wastewater discharges. The second step of UHA is demand minimisation, to finetune existing water infrastructure to its optimal capacity. Via demand minimisation, a strategy is established to improve current urban water metabolism through the reduction of excessive use of water. Output minimisation is the third step that aims to reduce the volume of urban water outflow from a city. To utilise residential wastewater, its quantity and quality need to be assessed to judge its potential to be transformed into reclaimed water. Furthermore, reclaimed water can be used as an alternative resource in the multi-sourcing step.

The main challenge of multi-sourcing is to identify and assess local potential water resources that can be harvested, such as rainwater, brackish water or reclaimed water. These alternative resources partly substitute existing primary resources. In that way multi-sourcing will decrease fresh water abstraction and lower the pressure onto natural water resources. Figure 5.4 gives an example of UHA at the city block level.

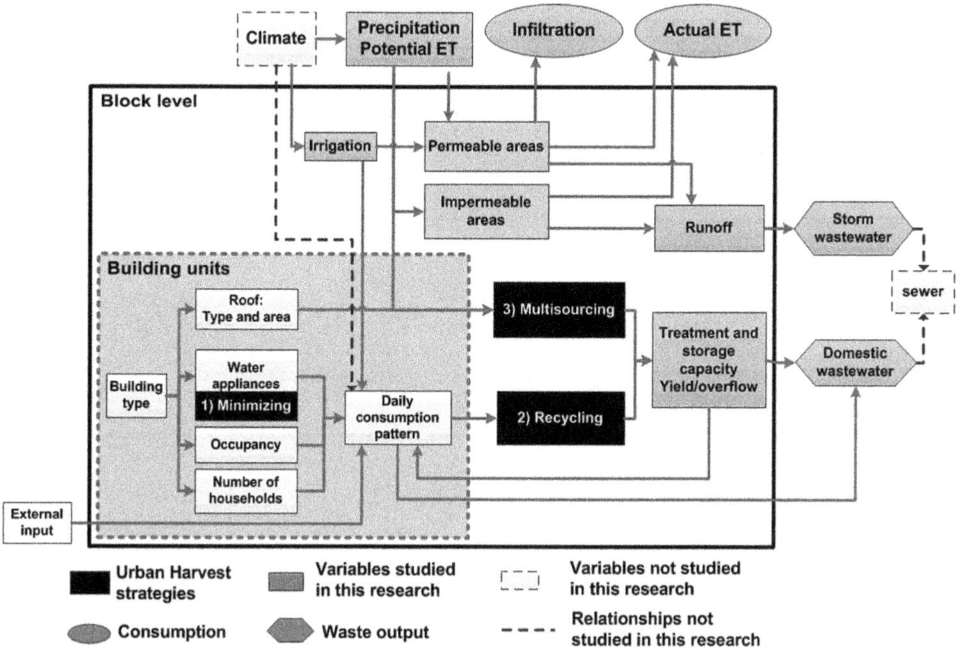

Figure 5.4. Example of UHA at the city block level (Agudelo-Vera, 2012).

5.3 The transition to Water Smart Cities with 3 case studies

5.3.1 Introduction

Brown *et al.* (2009) describe six phases of urban water management where the ultimate goal is to establish a Water Smart City. The six phases are illustrated as overlapping rings in Figure 5.5. Most cities in the developed world are in phase 3, 4 or 5, while most cities in the developing world are at phase 1 or 2. So, the question is, how can all cities reach the ultimate phase, where all the benefits of previous states are obtained together with intergenerational equity and climate resilience (movement from left to right in Figure 5.5)?

This is where the Water Smart City approach comes in, targeting cities in phase 3 (drained city), 4 (waterways city) or 5 (water cycle city), with the ambition of facilitating a fast transition to the Water Smart City. Cities in these phases are all facing the challenge of transforming already fully extended, well-developed water infrastructures with a likewise well-developed institutional set-up for its management, into systems that can take on the current and predicted pressures from both continued urbanisation and climate change (Van Hattum *et al.*, 2016).

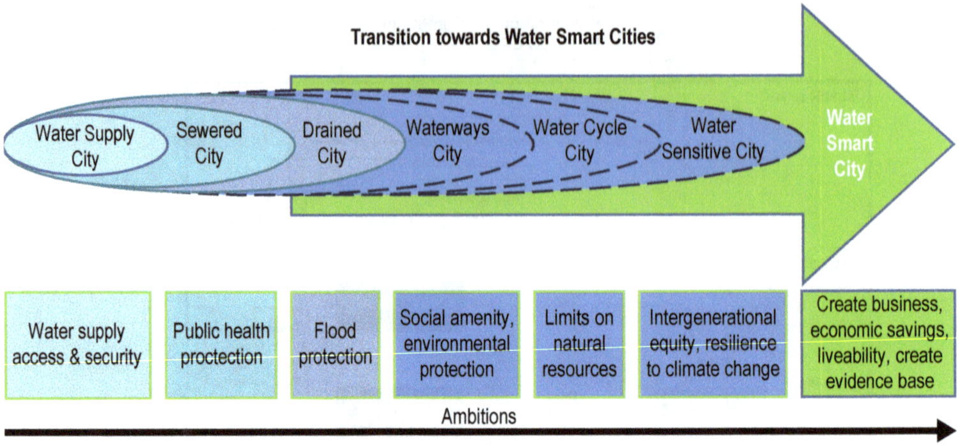

Figure 5.5. Water Smart City concept (adapted from Brown *et al.*, 2009).

What is smart about the Water Smart City concept is that it seeks to link water management to other drivers such as:

1. The need for liveable and attractive cities: the ability of a Water Smart City to brand itself and attract resourceful residents is important for the long-term stability of the city and attracting people to live there.
2. Business opportunities and attracting private finance: transition towards Water Smart Cities is a huge business opportunity for designers, product suppliers and constructors. City administrations are becoming more and more aware of the need to create liveable and resilient cities by introducing climate-proof urban design. This means that the market for these (new) businesses and innovative solutions is expected to grow rapidly in the coming years.
3. The need for an evidence base by continuously monitoring and evaluating cost and benefits and performance: the need for innovation, and to monitor and evaluate best practices in order to create an evidence base for Water Smart City solutions is an important driver. In many cities Living Labs are created to learn about the benefits of systemic water smart solutions and use the generated knowledge for upscaling to national or international level. In a Living Lab approach, academic, societal and business based knowledge is integrated to create integral solutions.

5.3.2 New waterways in Oslo

Oslo, the capital of Norway, is dealing with rapid population growth and compact city development. The sustainability of urban water management is challenged by these developments in combination with climate change. Oslo has developed a stormwater management strategy and an action plan for the period 2016-2021 (Oslo Municipality, 2016). The action plan details 18 actions for better stormwater management in Oslo, focused on developing and improving the knowledge base, preventing the consequences of stormwater runoff, having municipal projects led by example through pilot and model projects, establishing stronger collaboration between sectors and disciplines within the municipality, and providing better guidance and information on stormwater planning and solutions (Oslo Municipality, 2016).

Water smart solution

Oslo municipality is implementing a three-step strategy:

1. Catch and infiltrate water after small rainfall events.
2. Delay and retain water after moderate rainfall events.
3. Safe run-off of water to floodways after large rainfall events.

Different types of stormwater measures are part of each step (Figure 5.6).

The project New Water Ways – Towards water-sensitive and climate-adapted Nordic cities, looked at the city, district and neighbourhood scales in Oslo to develop a portfolio of urban water management combinations. Partners from the Oslo municipality, business and academia were working together to move beyond today's conventional urban water management. The case study sites in Oslo in the New Water Ways project were: Grefsen-Kjelsås, Hovinbyen, Sogn Hagekoloni, Svartdalen/Alna and Teglverket Skole.

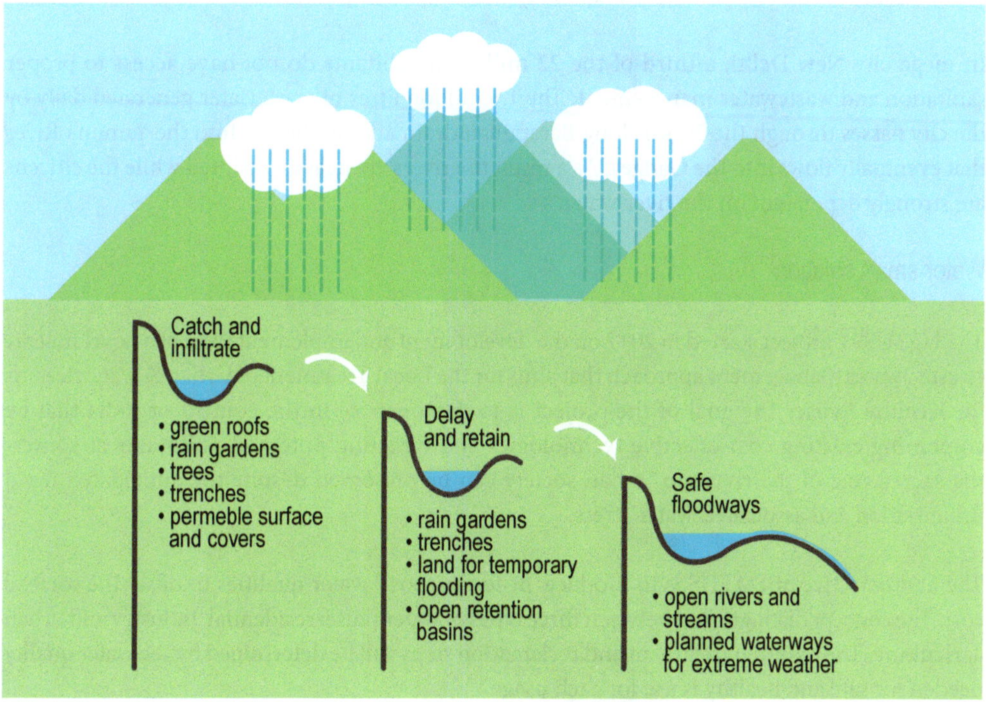

Figure 5.6. Different types of stormwater measures (Oslo Municipality, 2016).

A first step of the project was to provide a detailed assessment of the social, economic and environmental implications of sustainable urban water management in order to promote the uptake of new waterways by public and private actors. The second step was to understand the historic and current context of Norwegian and Nordic urban water management. The third step was to engage with water managers and other stakeholders in and beyond the water sector to envision a future sustainable water-sensitive and climate-adapted urban water management. An important aspect was that the project examines how citizen participation can contribute to the processes of designing, planning or implementing a multifunctional stormwater infrastructure. They applied a learning-by-doing approach for two new forms of citizen participation, namely Urban Living Labs (ULL) and crowdsourcing. These were used to test the public acceptance for water-sensitive and climate-adapted solutions, as well as to engage with citizens in the design of stormwater infrastructure in Hovinbyen. A fourth step was to develop business opportunities inherent to a water-sensitive and climate-adapted transition of urban water management (Li, 2020; Van Bezooijen, 2019; Wilbers *et al.*, 2022).

5.3.3 Local treatment of urban sewage streams for healthy reuse (LOTUS^{HR}), in New Delhi, India

Alleviating water stress in large metropolitan areas requires proper management of freshwater resources through retention, proper sanitation, and recovery of potable water from unconventional, secondary resources like wastewater and heavily polluted surface water. This, however, will result in increased costs, which are more than many societies can cope with.

In mega city New Delhi, a third of the 22 million inhabitants do not have access to proper sanitation and wastewater management. The 1.6 million litres of wastewater generated daily by the city passes through the 20-km-long Barapullah drain and discharges into the Yamuna River that eventually flows into the Ganges. As a result the rivers are heavily polluted while the citizens are strongly dependent on the river water.

Water smart solution

The LOTUS^{HR} project started in 2017 on the development and implementation of a novel holistic (waste-)water management approach that aims for the Local Treatment of Urban Sewage streams for Healthy Reuse. The goal of the project is to demonstrate in the context of India that by combining existing cost-effective technologies and targeting potential pollutants at source, the sacred role of the rivers in Indian society can be preserved despite the anticipated rapid urbanisation and associated water stress.

The first objective of LOTUS is to produce 'fit-for-purpose' water qualities to offset the societal cost. The concept distinguishes between three types of water reuse: residential, industry and urban agriculture. The required treatment and reclamation steps will be determined by the water quality needed for safe and healthy reuse for each usage.

To be able to produce safe reusable water, pollutants have to be removed from the drain water by the interconnected chain of technologies. A promising sub-technology is compact, robust and nature-based Vital Urban Filters (VUFs). This water treatment system is able to produce safe reusable, nutrient-rich water for agricultural purposes while having an added social and economic value through the growing of plants on the surface of the filter. The Vital Urban Filters are based on the principle of conventional vertical flow constructed wetlands that are commonly used in decentralised wastewater management. However the filter consists of a flat bed of hydroponic filter material planted with ornamental plants. The filter materials are highly porous, providing a large surface area for biomass and a large adsorption surface. As the influent percolates through the filter, heavy metals, pathogens and micropollutants are removed by the interaction of the plants, filter materials and bacteria, thus generating safe, reusable effluent.

5.3.4 Optimised water management in the industrial zone in Vietnamese delta, Ho Chi Minh City – ENTIRE project

The Vietnamese delta has a highly dense population and forms the connection between the resources of the main land and those of the sea. At the same time, it is extremely vulnerable and faces multiple threats. The combination of intensification of urban and economic land use, including agriculture, the related decrease in capacity to resist natural hazards, and climate change, often result in a decrease in the quality of life, poverty and substantial economic losses. Industrial, agricultural and residential sectors represent a huge drain on water resources to sustain and develop society and the economy. The sustainable use of fresh water resources is required to prevent it from becoming a limiting factor for economic development.

Water smart solution

The ENTIRE project aims to develop, model, design and implement a technological and governance framework for water and wastewater services for the sustainable development of industrial areas, relieving pressure on freshwater resources to make them more available for urban water functions and agricultural production.

To attain this objective baseline assessment of current water management and institutional and governance structures, it is necessary to design and pilot 3 existing industrial zones (IZs) in the Mekong delta, Ho Chi Minh City, Vietnam and to build and validate a generic, contextualised, decision support model for water use optimisation in industrial zones, and where relevant their associated residential areas. A detailed screening of the industries regarding their water use efficiency and generation of wastewater flows and their treatment along with actor networks relevant for governing water efficiency in three industrial zones in Vietnam is far advanced. Manufacturing processes and green area irrigation have been identified to represent the highest water demands in industrial zones.

Stricter requirements on supply water quality and increasing environmental protection fees are driving industrial zone infrastructure companies (IZICs) to consider optimisation of water use among the industrial sectors (clusters) by assessing the potential of, other than conventional, water sources and introducing closed or semi-closed water cycles. The Industrial Harvest Approach (IHA) with its strategies of (1) multi-sourcing, (2) recycling, and (3) cascading presents a potential framework for optimised industrial water supply.

5.4 Conclusion: the roadmap to Water Smart Cities

Cities around the world are rethinking how they should deal with water in their neighbourhoods, districts and regions in order to deal with the challenges of biodiversity decline and climate change. They are trying to find ways of caring for water in such a way that future generations of urban dwellers as well as ecosystems can have sufficient access to clean, safe freshwater, and so that the city with all its physical, biotic and abiotic assets can function despite the changes in climate. This requires becoming a Water Smart City, and developing an integrated approach to the urban water management of cities and their hinterlands.

Below a roadmap is defined (Figure 5.7) that describes the different steps and actions needed from the different actors in order to create a Water Smart City. These steps and actions are based on the three case studies that are presented above. Becoming a Water Smart City is a long-term process but every (re)development project should contribute to a long-term vision. City authorities need to play a leading role in this transition. They decide the goals and ambitions for their city and the measures that need to be taken for these goals to be reached. Although cities have a leading role in the transition, other actors (citizens, businesses, and research institutes) also play important roles, such as raising awareness & stressing the urgency, demanding functions, developing technologies and creating an evidence base. In this roadmap the municipality takes the lead, but the possible actions or possibilities for other actors are inserted in the different steps.

1. Identify current and future challenges and opportunities: at first the challenges and opportunities related to the water system on different levels in the urban area (neighbourhood, district, city, region) need to be identified. This can be done for the current and future situation and can be related to different ambitions and goals that the city has defined, for example in the area of health, climate, energy, safety, and poverty reduction. From this analysis a new hydro-social contract between the city and the citizens and business can be defined.
2. Define WSC vision: a Water Smart City vision needs to be defined for the city. Each city has its specific challenges and opportunities related to the natural system and its unique opportunities for liveability. Therefore, each city needs to set up a vision with the leading question 'When are we a Water Smart City?' and strive to share and embed this vision among all urban actors.
3. Explore co-creation opportunities: in this step, we are moving towards the implementation of a WSC where different opportunities and pathways can be followed. Opportunities need to be explored with those already dealing with the water management of the city, but also with health institutes, nature organisations and other societal actors that can envision the new water smart future of the city. The scenarios can be linked to already planned actions/projects, making small adjustments to a plan in order to make it more water smart, or a total new trajectory can

Figure 5.7. Roadmap towards Water Smart Cities (Van Hattum *et al.*, 2017).

be set up. Visualisation of the timeline of infrastructural projects and investigating co-creation possibilities by linking climate adaptation goals with planned infrastructural projects are important aspects. Urban (re)development offers huge opportunities to create value and synergy with Water Smart City goals.

4. Co-design solutions: in this step the best (combination of) possible measures are analysed to determine the optimal way and possible pathways (a specific project or a new trajectory) to reach the cities' ambitions. This can be best done with close collaboration on the basis of the double triple helix model (Chapter 13). Working on co-design in a network of representatives of government, citizens, business organisations, and knowledge institutes is essential as is making use of a design canvas for inclusion to get distributional, representational and recognitional equity in the proposed solutions (Chapter 12).
5. Define business cases: from the proposed combination of possible measures (scenarios), the best (most optimal) business case should be chosen. Each scenario has its benefits and disadvantages in relation to its performance (incl. flexibility), costs and value, and risks over a certain timespan. All these factors should be balanced with each other to determine the most optimal business model, which will be part of the project plan to implement the measures.
6. Implement, monitor & evaluate: there is a strong need for showcases and best practices about cost-effective urban adaptation strategies and experimentation space for co-creating new solutions. (Continuously) implement WSC measures, celebrate and learn from them and where needed adjust the process or ambitions until it is business as usual (plan-do-check-act). Cities need this knowledge for upscaling in order to become climate proof. It is important to invest in Living Labs or demonstration projects where governments, citizens, researchers and

businesses test and monitor innovative and nature-based adaptation approaches using smart technologies (sensors, satellite, citizen science). Data management and knowledge exchange between cities is important. Continuously improving the evidence base is required.

References

Agudelo-Vera, C.M., 2012. Dynamic water resource management for achieving self-sufficiency of cities of tomorrow. PhD thesis Environmental Technology, Wageningen University, Wageningen, the Netherlands. Available at: https://edepot.wur.nl/216692.

Brown, R.R. and Clarke, J.M., 2007. Transition to water sensitive urban design: the story of Melbourne, Australia. Melbourne: Facility for Advancing Water Biofiltration, Monash University, Australia.

ESPON, 2013. Climate change and territorial effects on regions and local economies. ESPON & IRPUD, TU Dortmund, Germany. Available at: https://tinyurl.com/yckxmprk.

European Commission, 2018. Drought and water crisis in Southern Africa. JRC Technical Report Publications Office of the European Union, Luxembourg, 2018. Available at: https://publications.jrc.ec.europa.eu/repository/handle/JRC111596.

European Commission, 2015. Toward a research and innovation policy agenda for nature-based and re-naturing cities. Final report of the Horizon 2020. European Commission, Publication Office, Brussels, Belgium. Available at: https://data.europa.eu/doi/10.2777/479582

European Environmental Agency, 2012. Urban adaptation to climate change in Europe. Challenges and opportunities for cities together with supportive national and European policies. EEA, Copenhagen, Denmark. Available at: https://tinyurl.com/4bu94mjc

Van Hattum, T., Blauw, M., Bergen Jensen, M. and De Bruin, K., 2016. Towards water smart cities: climate adaptation is a huge opportunity to improve the quality of life in cities. Wageningen Environmental Research rapport 2787. Wageningen University & Research, Wageningen, the Netherlands, 60 pp. Available at: https://edepot.wur.nl/407327.

Imperial College, 2017. Blue green solutions. a system approach to sustainable, resilient and cost-efficient urban development. Imperial College, London, UK. Available at: https://doi.org/10.13140/RG.2.2.30628.07046.

International Water Association (IWA), 2017. The IWA principles for water wise cities. 2nd Edition. For Urban Stakeholders to Develop a Shared Vision and Act towards Sustainable Urban Water in Resilient and Liveable Cities. International Water Association, London, UK. Available at: https://tinyurl.com/3sjbwhct.

IPCC, 2018: Global Warming of 1.5°C. An IPCC Special Report on the impacts of global warming of 1.5 °C above pre-industrial levels and related global greenhouse gas emission pathways, in the context of strengthening the global response to the threat of climate change, sustainable development, and efforts to eradicate poverty [Masson-Delmotte, V., P. Zhai, H.-O. Pörtner, D. Roberts, J. Skea, P.R. Shukla, A. Pirani, W. Moufouma-Okia, C. Péan, R. Pidcock, S. Connors, J.B.R. Matthews, Y. Chen, X. Zhou, M.I. Gomis, E. Lonnoy, T. Maycock, M. Tignor, and T. Waterfield (eds.)]. Available at: https://tinyurl.com/2p8sp3jf.

Li, H., Gao, H., Zhou, Y., Xu, C.Y., Rengifo, Z., Ortega, M. and Sælthun, N.R., 2020. Usage of SIMWE model to model urban overland flood: a case study in Oslo. Hydrology Research 51(2): 366-380. https://doi.org/10.2166/nh.2020.068

Ligtvoet W. *et al.*, 2018. The geography of future water challenges, PBL Netherlands Environmental Assessment Agency, The Hague, the Netherlands. Available at: https://tinyurl.com/2p8aszx6.

Lloyd, S.D., Wong, T.H.F. and Chesterfield, C.J., 2002. Water sensitive urban design – a stormwater management perspective. Industry Report No. 02/10. Cooperative Research Centre for Catchment Hydrology, Melbourne, Australia. Available at: https://ewater.org.au/archive/crcch/archive/pubs/pdfs/industry200210.pdf.

McDonald, R. and Shemie, D., 2014. Urban water blueprint: mapping conservation solutions to the global water challenge. Nature Conservancy, Washington DC, USA. Available at: https://tinyurl.com/ytu5s5nx.

Mekonnen, M.M. and Hoekstra, A.Y., 2016. Four billion people facing severe water scarcity. Science Advances 2(2): e1500323. https://doi.org/10.1126/sciadv.1500323

Meyer, H. and Peters, R., 2016. A Plea for putting urbanizing deltas on the new urban agenda. Delta Alliance/ Tu Delft, Delft, the Netherlands. Available at: https://tinyurl.com/bdz5695d.

Oslo Municipality, 2016. Action plan for stormwater management in the City of Oslo. Oslo Kommune, Executive summary. Oslo Municipality, Oslo, Norway. Available at: https://tinyurl.com/3fwr9yxe.

Timmermans, W., Woestenburg, M., Annema, H., Jonkhof, J.F., Shlakku, M. and Yano, S. (eds.), 2015. The rooted city: European capitals and their connection with the landscape. Blauwdruk, Wageningen, the Netherlands.

UCCRN, 2018. The future we don't want. How climate change could impact the world's greatest cities. UCCRN Technical Report. C40 Cities. Urban Climate Change Research Network. New York, NY, USA. Available at: https://tinyurl.com/ymnez6cf.

UNEP, 2016. Adaptation Finance Gap Report. UNEP, Nairobi, Kenya. Available at: https://tinyurl.com/ycknj7wn.

UNESCO, 2015. United Nations World Water Development Report 2015 – Water for a sustainable world facts and figures. Published in 2015 by the United Nations Educational, Scientific and Cultural Organization, Paris, France. Available at: https://unesdoc.unesco.org/ark:/48223/pf0000231823_eng.

United Nations World Water Assessment Programme (WWAP), 2017. The United Nations World Water Development Report 2017. Wastewater: The Untapped Resource. UNESCO, Paris, France. Available at: https://unesdoc.unesco.org/ark:/48223/pf0000247153_eng

United Nations World Water Assessment Programme (WWAP), 2018. The United Nations World Water Development Report 2018: Nature-Based Solutions for Water. UNESCO, Paris, France. Available at: https://tinyurl.com/mry4zdku.

Van Bezooijen, T., 2019. Stormwater measures and urban fabrics. A case study of Hovinbyen, Oslo. Master Thesis, Water Systems and Global Change Group, Wageningen University and Research, the Netherlands. Tool developed for the thesis. Available at: https://tinyurl.com/3exvy635.

Van Vliet, M.T.H., Wiberg, D., Leduc, S. and Riahi, K., 2016. Power-generation system vulnerability and adaptation to changes in climate and water resources. Nature Climate Change. Available at: https://www.nature.com/articles/nclimate2903.

Wilbers, G.J., De Bruin, K., Seifert-Dähnn, I., Lekkerkerk, W., Li, H., Budding-Polo Ballinas, M., 2022. Investing in urban blue-green infrastructure – assessing the costs and benefits of stormwater management in a peri-urban catchment in Oslo, Norway. Sustainability 14(3): 1934. https://doi.org/10.3390/su14031934

World Bank, 2016. High and dry: climate change, water, and the economy. World Bank, Washington, DC, USA. Available at: https://tinyurl.com/2pzw3aba.

World Bank Group, 2018. Water scarce cities: thriving in a finite world – full report. World Bank, Washington, DC, USA. Available at: https://openknowledge.worldbank.org/handle/10986/29623.

World Economic Forum, 2018. Global risk report. World Economic Forum, Geneva, Switzerland. Available at: https://www3.weforum.org/docs/WEF_GRR18_Report.pdf.

Chapter 6
Routes to a climate resilient city with climate services: the examples of Nijmegen and Lisbon

Ingrid Coninx, Wim Timmermans, Rob Swart, Luís Dias and Maarten van Ginkel

Abstract

Climate change can be a driver for cities to become frontrunners in terms of both climate mitigation (for instance, in moving towards carbon-neutrality) and climate adaptation in developing green and liveable cities. In metropolitan areas, climate threats such as floods, heat and drought, have a significant impact on nature, the economy, human health and well-being. The concentration of people, infrastructure and other assets makes cities vulnerable to climate risks as they have rarely been designed with climate change in mind. In 2009, the World Meteorological Organization (WMO) introduced the concept of 'climate services' to improve the accessibility of the vast amount of available climate data in a way useful for those who would need them. Since then, the concept has evolved from 'science-driven and user-informed' to 'user-driven and science-informed', actively supporting climate change responses by citizens, companies, economic sectors, and cities (see for instance European Commission, 2015). Climate services are tools, products and processes that can help to integrate climate information into urban policymaking and design for improving urban resilience. Because of the long-term nature of climate change, climate services force urban developers to adopt a long-term, transformative perspective in urban planning and design. This chapter elaborates on how climate change can be a driver for green urban development, by pursuing resilience with the help of climate services. Two illustrative cases are presented: Nijmegen (the Netherlands) and Lisbon (Portugal). We also illustrate the importance of the next generation of climate services to go beyond risk assessment by informing the design of response measures like nature-based solutions. The cases illustrate the modalities of climate services to trigger action: climate services have to be flexible and tailored to the specific local situation and they should be integrated with other urban challenges in order to develop innovative multifunctional urban solutions.

Keywords: water management, climate adaptation, climate services, transformation, urban solutions

10.3920/978-90-8686-935-0_6,

6.1 The metropolitan challenge of climate change

The relationship of cities with climate change is complex. On the one hand, cities contribute to climate change because of vehicle emissions, domestic heating and their economic activities; urban sources produce about 75% of all carbon dioxide emissions (UNHABITAT, 2018). But on the other hand, large numbers of inhabitants experience climate change impacts, such as heavy rains, river flooding and storm surges; social discomfort and health impacts from heat waves; declines in economic productivity from water shortages, alongside many other direct and indirect effects.

The UNFCCC Paris Agreement, the main international agreement to address climate change, reminds us of the need 'to undertake rapid reductions ... in accordance with best available science, so as to achieve a balance between anthropogenic emissions by sources and removals by sinks of greenhouse gases in the second half of this century, on the basis of equity, and in the context of sustainable development and efforts to eradicate poverty' (UNFCCC, 2015). To achieve this, cities will also have to both mitigate and adapt to climate change, posing new challenges to urban planning and development. In this chapter, we mainly focus on how cities adapt to the projected impacts of climate change, supported by climate services.

The impact of climate change on cities varies throughout the world. This is mainly caused by their diverse configurations, their different exposures to temperature rise and changes in rainfall, their locations and land use. Cities across Europe are extremely varied in terms of their cultural diversity, plurality in legislation, and landscapes. This heterogeneity results in very different exposure to actual or anticipated climate change impacts.

Despite this, drought is expected to become an ever-growing problem for many European cities. For example, even for a low-impact (RCP8.5, 10th percentile) scenario cities in southern Europe may face twice as many severe droughts in the second half of the century than experienced today. In the corresponding high-impact (RCP8.5, 90th percentile) scenario, it was found that 98% of cities across the whole of Europe are projected to exceed the historical maximum drought severity index (1951-2000), with Athens, Lisbon, Madrid, Nicosia, Sofia and Valletta considered the most vulnerable. Similarly, flooding represents an increasing threat for cities near rivers and in coastal areas, and high run-off is expected to increase in cities in the UK, Belgium, Ireland, Netherlands and Scandinavia amongst others. Furthermore, many European cities, including Athens, Nicosia, Prague, Rome, Sofia, Stockholm, Valletta and Vienna (Guerreiro *et al.* 2018) are expected to experience more frequent, prolonged and intense heat waves, with 72% showing increases of the maximum heat wave temperature of 10 °C. Also, for 72% of cities, particularly in northern Europe, an increase of 10-year high river flows is projected for a medium-impact scenario. More than 100 cities are particularly vulnerable to two or more impacts. These and other extreme climate-change driven events are capable of causing the long-term disruption of urban communities and societies leading to a decline in welfare and well-being and a deterioration of the quality of life, with the increased frequency providing the greatest concern. It is clear, therefore, that actions are required to increase the resilience of cities to prevent, absorb and overcome these impacts.

6.2 Climate services for urban action

To be able to mitigate their contribution to climate change as well as to increase their resilience, cities first need access to accurate and reliable information if they are to comprehend both the potential problems and possible solutions. Climate services are a way to provide the required climate data. The European Research and Innovation Roadmap for Climate Services report defines climate services as 'the transformation of climate-related data – together with other relevant information – into customised products such as projections, forecasts, information, trends, economic analyses, assessments (including technology assessments), counselling on best practices, development and evaluation of solutions and any other service in relation to climate that may be of use for the society at large.' (European Commission, 2015). As such, climate services in the context of the Roadmap provide data, information and knowledge that support adaptation, mitigation and disaster risk management.

From around 2000, climate services developed from an initial meteorological, observation-based focus into tools that are more predictive (Lourenço *et al.*, 2015). Yet, the current EU definition highlights the rise of user-oriented services rather than the current mostly science-driven efforts (Street, 2016). Therefore, the concept of climate services faces new challenges, and recognition of the actual demands of a diverse set of stakeholders is needed to enable a useful contribution to local decision-making.

The long-term profound changes in the climate make urban transformation key to improving the climate-resilience of cities (Carter *et al.*, 2015). For instance, whilst incremental measures – this means doing things basically the same, but a bit better, such as replacing drainage pipes with pipes of greater capacity during on-going maintenance and renewal work – can have great initial benefits, these may be insufficient to cope with long-term change. Consideration of long-term impacts encourages more transformative urban planning and design (i.e. doing things differently to account for long-term risks and opportunities). Information about climate change, its impacts and opportunities for cities' future design is therefore very much needed. Moreover, this information has to be actionable, to make sure that policymakers and other urban organisations can make use of it.

6.3 Case studies: Nijmegen and Lisbon

In this paper, we discuss the urban transformations via climate services in both Nijmegen (the Netherlands) and Lisbon (Portugal), which were, respectively, the EU Green Capitals of 2018 and 2020. Both are regarded as front-runners in the pursuit of green climate adaptation yet their planning processes were established far before the rise of climate services and certainly preceded the EU's definition. Therefore, we have analysed the kind of knowledge used in the respective planning processes. We have also explored how evidence provided by climate services has affected decisions on urban development. The case studies will provide lessons learned for other cities.

6.3.1 Nijmegen – Room for the River in Europe's 2018 Green Capital

The first case study is the Room for the River development in Nijmegen. Nijmegen is situated in an archetypical river delta landscape within the densely populated Rhine-Meuse system. Given its Roman origins, it is one of the oldest cities in the Netherlands with the oldest parts of the city found on a hill alongside the river Waal. Since the 1950s most of the so-called urban sprawl has occurred in the lower-lying areas adjacent to the river.

Plans for new residential and working areas

In the 1990s, the city's Structure Plan identified the need for a large-scale development of over 12,000 houses and additional business areas (the 'Waalsprong' or Waal jump) (Gemeente Nijmegen, 1996) to be built over a 25-year period on the clay soils of the flood-plain across the river from the old city. The proposals were subjected to Environmental Impact Assessment and designed with climate change in mind although it pre-dated the concept of climate services. However, with hindsight, it is possible to identify the adoption of the climate services approach in the design and decision-making process. A planning strategy, the Strategy of the Two Networks (S2N), partly based on rudimentary insights into the urban water cycle has been applied for sustainable, including climate-adaptive, urban development.

The challenges

The location presented two main challenges: drainage and accessibility.

The proposed development area was low-lying and comprised farmland and several small villages protected by flood defences alongside the river. Despite its location, the site did not drain into the Waal, but to the smaller Linge river nearby. However, it was appreciated that the Linge was not capable of accommodating the anticipated run-off from the Waalsprong. Furthermore, with high water levels in the Waal, the Waalsprong was exposed to seepage problems.

At the time of the proposal, the Waalsprong was served by both a road and a railway bridge. It was then realised that a further road bridge would be required but due to the cost, this was not pursued. An alternative plan for a cable car was proposed but rejected. In 2004, a bike and pedestrian-only extension to the existing structure served as a partial solution.

The strategy of the two networks (S2N)

While in other Dutch cities like Almere and Haarlemmermeer a blueprint planning strategy enabled the planning and construction of a similar number of homes in just five years, the Waalsprong proposal was considered to be too complex for that approach (Van Rossum *et al.*, 2001). Instead, the Structure Plan introduced a new sustainable planning strategy developed by

Wageningen UR (Tjallingii, 1995, 1996). The S2N planning strategy produced an organic, flexible planning approach, which smartly arranged settlement and infrastructure development along sustainable water and traffic networks.

In S2N, water and traffic networks are the two guiding forces of urban development (Figure 6.1). Along the water network green infrastructure drinking water and recreation were planned. Alongside the traffic network industries, trade and mass recreation were proposed. The outcome was a clear demarcation between relatively unpolluted, clean and tranquil areas, and the busier, more polluted zones. Between these, residential areas were proposed, ranging from low to high density settlements. However, the actual allocation of land is flexible and can follow changes in societal preferences. Water and traffic flows are managed sustainably.

Drainage problems in the Waalsprong

However, development of the Waalsprong led to drainage problems. The transformation from agricultural to urban settlement reduced infiltration of rainwater into the soil and because of this issue being anticipated and the knowledge that the Linge would not be able to accommodate increased flows, the municipality was obliged to develop an autonomous water system in the Waalsprong (Figure 6.2). Although a major challenge, the adoption of S2N principles led to a design based on three, large, interconnected basins which also improved water quality. In addition, the system included numerous measures within neighbourhoods and for each dwelling, such as permeable surfaces, green roofs and rain barrels to reduce the volume of run-off and to delay the response. These measures were supplemented by a small outlet for drainage into the Linge in the north. Similar systems are nowadays incorporated into all new developments as part of standardised climate change adaptation techniques.

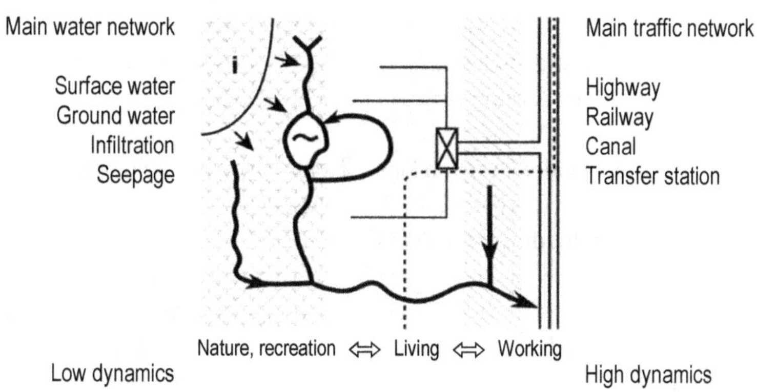

Figure 6.1. Water and traffic networks in the S2N.

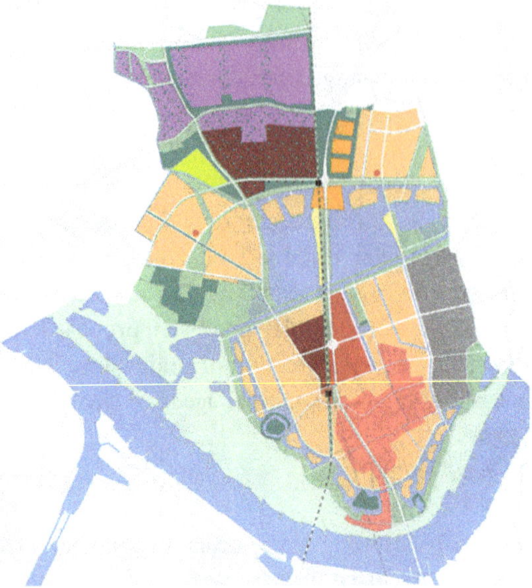

Figure 6.2. Water management system Waalsprong.

New problems: near flooding of the Waal in 1993 and 1995

In both 1993 and 1995, heavy rainfall in the Rhine and Meuse catchment in Germany and France led to severe problems in the Netherlands. In 1995, more than 140,000 inhabitants had to be evacuated from areas alongside the Waal because of the risk of dyke failure. These near-floodings and evacuations prompted a new approach to flood defence. The Waal river discharge was already rising and became more irregular yet new models using IPCC scenarios showed that the river regime of the Rhine-Meuse delta would change further. Subsequent calculations estimated that the maximum discharge of the Rhine/Waal in the Netherlands would increase from 15,000 m^3/s in 1995 to 16,000 m^3/s in 2001 and 18,000 m^3/s in 2006. These estimations can be considered as climate services on which urban development has been based.

In 2006, the National Planning Core Decision 'Space for the river' (Planologische Kernbeslissing, PKB) came into effect. It included two paradigm shifts. First, dyke-raising was no longer presented as the only solution to prevent flooding, and second, the creation or restoration of flood plains was included as a viable option. 'Space for the river' paid special attention to urban areas close to the river, and in Nijmegen it was decided that the riverbed should be widened despite the necessary loss of the village of Lent and reduction in the size of the Waalsprong. The scale and impact on communities represented a major shock not only to the most affected residents but also to the wider city. The ministry and local authorities felt the necessity to define dual goals combining flood protection and development goals for the areas along the Waal.

Waalsprong revised

Given the impact of these policy changes, completion of the Waalsprong was threatened even though construction had begun just a few years before and long-term contracts with developers had already been signed. However, the city council decided to continue to cooperate with the Ministry, retaining the dual-goal ambition for the Waalsprong and guaranteeing proper accessibility. Financial losses to developers, municipality, house owners and other stakeholders were to be compensated.

The ministry and municipality agreed that the city of Nijmegen would coordinate a shared planning process. Three integrated scenarios were considered: an overspill river north of the Waalsprong, another one through the middle of the Waalsprong, and a shift of part of the Waal dyke northwards, into the southern part of the Waalsprong. This last option was pursued.

The result of this process was a radically new urban development plan with three important elements: urban development could continue, financial losses would be compensated, and Nijmegen would receive over 90 million euros to construct an extra bridge across the Waal. In turn this prompted production of a new Waalsprong development plan (Figure 6.3 and 6.4), including a side channel that increases drainage capacity during high flows. At other times, the side-channel area serves as a cultural, recreational, leisure and water sports area, the so-called Mirror Waal. Compared with the original plans, the location of the Waalsprong centre has shifted to the southwest and now enjoys a direct connection to the old city via the new bridge. In between the Waalsprong and the river Waal, the former floodbank now represents a new city island, a stepping-stone between the old city and the new.

Long-term urban development along the Waal

Meanwhile, perceptions about flood protection and dykes have changed again. Notwithstanding the height of dykes and the overflow space for the river, dyke stability is becoming a growing issue. Future high water levels in the Waal are expected to last for longer than at present, with the risk that dykes will become waterlogged and unstable. A new, broader design, the so-called 'climate dyke' has been proposed which, through its design, provides even greater opportunities to combine flood defence, urban development and habitat creation. Similarly, the design of new urban waterfronts is also evolving to adapt to rising water levels. The city hopes that future developments can make use of state-of-the-art knowledge from climate and climate impact science via effective climate services and as a result can continue to develop innovative, sustainable urban concepts.

Climate services lessons

The case study aims to illustrate how climate services contribute to urban green transformation. However, the urban transformation of Nijmegen started long before the concept of climate services emerged.

Figure 6.3. New Waalsprong development plan, including a river bypass in the Waal.

First, the case study makes clear that in the early nineties already decision-makers were becoming aware of the effects of climate change when the Waalsprong was first planned, but there was no common ground on how climate change had to be integrated in the development. As a consequence a sustainable planning strategy (S2N), based on a sustainable design of water and traffic networks was applied.

Second, we learned that the S2N planning strategy represented a major step forward in delivering these development goals and became a very powerful planning strategy. This approach has now been (at least partly) applied in over 40 other municipalities across the Netherlands (Aalbers and Jonkhof, 2003) and now sits at the centre of the sustainable urban planning debate at the local and national scale, playing a major role in the mainstreaming of sustainable development issues

Figure 6.4. Looking to the West; the river bypass in the Waal.

in Dutch planning practice. However, climate scientists should be aware that the development of robust integrated urban planning strategies can rarely deal with climate mitigation and adaptation alone and that other societal issues have to be combined with climate knowledge to improve urban planning. This means that climate services are only one of a number of data sources that are used as a basis for urban development.

Third, Nijmegen benefits from abundant data on river behaviour in the Rhine/Waal system. Yet, time after time, scenarios and expectations changed. This uncertainty is on the one hand a major pitfall because new housing developments are planned to last for over a century. On the other hand, urban planning and design can be very creative processes, easily adapting to new drivers such as the need to widen riverbeds even at the cost of planned houses, or broader dykes because of long-term stability. However, it is crucial to do this in an open integrated debate and design process combining water safety and urban development. Climate service providers should take into account this flexible, interactive and creative process as well.

6.3.2 Lisbon: an old but dynamic city between ocean, river and hills

Lisbon is Portugal's capital and largest city. The city is located on the bank of the Tagus river close to the Atlantic coast and is the most western capital in Europe. Major expansion of the city occurred during the dictatorship (1933-1974), based on a long-term urban planning vision. Since then, the world exposition of Lisbon (Expo 98) and the urban renewal project Alta de Lisboa were the main urban interventions, which aimed to regenerate urban dereliction.

In the 1980s, the population in the city centre declined due to the poor standard of housing, itself a function of a rent freeze and the availability of better housing conditions in the suburbs. In recent years, Lisbon has grown in importance as a tourist destination resulting in some population growth accompanied by a gentrification process across the city. As the country's capital, Lisbon is the energetic political, economic and cultural centre. In order to satisfy the increasing demands of Lisbon's citizens, tourists and businesses, the city's vulnerability to climate extremes, such as heat, floods and droughts is becoming ever more important to keep the city attractive and integration of climate considerations into city planning to ready the city for the 21st century is a growing concern. Climate services can support this process.

A city vulnerable to climate extremes

Lisbon has a Mediterranean climate with mild, rainy winters and warm to hot, dry summers. As a result of climate change, the number of hot days and tropical nights is projected to increase, while the day time temperature difference between built-up areas and green spaces could reach 3 °C. Portugal and Lisbon have already experienced changes in seasonality of rainfall over recent decades, with more dry winters, but also increasing summer droughts (Câmara Municipal de Lisboa, 2021; Município de Lisboa, 2017). While annual rainfall may decrease, the intensity of rainfall in showers is projected to increase, with a resulting rise in flooding risk, notably downtown (Dias, 2016). High intensity rainfall events have already led to serious flooding problems in this area and elsewhere in 1967, 1983, 2008, 2011 and 2014. Sea level rises and storm surges may also increasingly affect the drainage system and Lisbon water front (Figure 6.5).

A hierarchical municipal planning system

Lisbon has a hierarchical planning system, with the Regional Land Use Plan for the Metropolitan Area of Lisbon (CCDR-LVT, 2012) identifying the strategic spatial framework, and the Municipal Master Development Plan (Câmara Municipal de Lisboa, 2012) setting the scene at the city level. Yet, even before climate change risks and climate services appeared on the political agenda, the municipality was pursuing an urban regeneration strategy that promoted the rehabilitation/conversion of vacant buildings and the qualitative improvement of public spaces (Santos *et al.*, 2015). In addition, the Lisbon Strategic Charter for 2010-2024 sought to address a number of other key issues including, importantly, the provision of a Green Infrastructure Plan with the aim of transforming Lisbon into an environmentally sustainable and energy-efficient city. This Plan incorporated a Climate Adaptation Strategy and a Biodiversity Local Action Plan to design and develop concrete projects to address water management and urban heat islands. The Strategy sat above Lisbon's Drainage Plan (Plano Geral de Drenagem) which provided a more traditional approach focusing on 'grey' infrastructure to alleviate flooding problems. The importance of climate risk management and the role of climate services varies between these instruments, but has generally increased in significance over time. The provision of climate information by Portuguese institutions in support of urban planning is an example of climate services well before the term was widely employed.

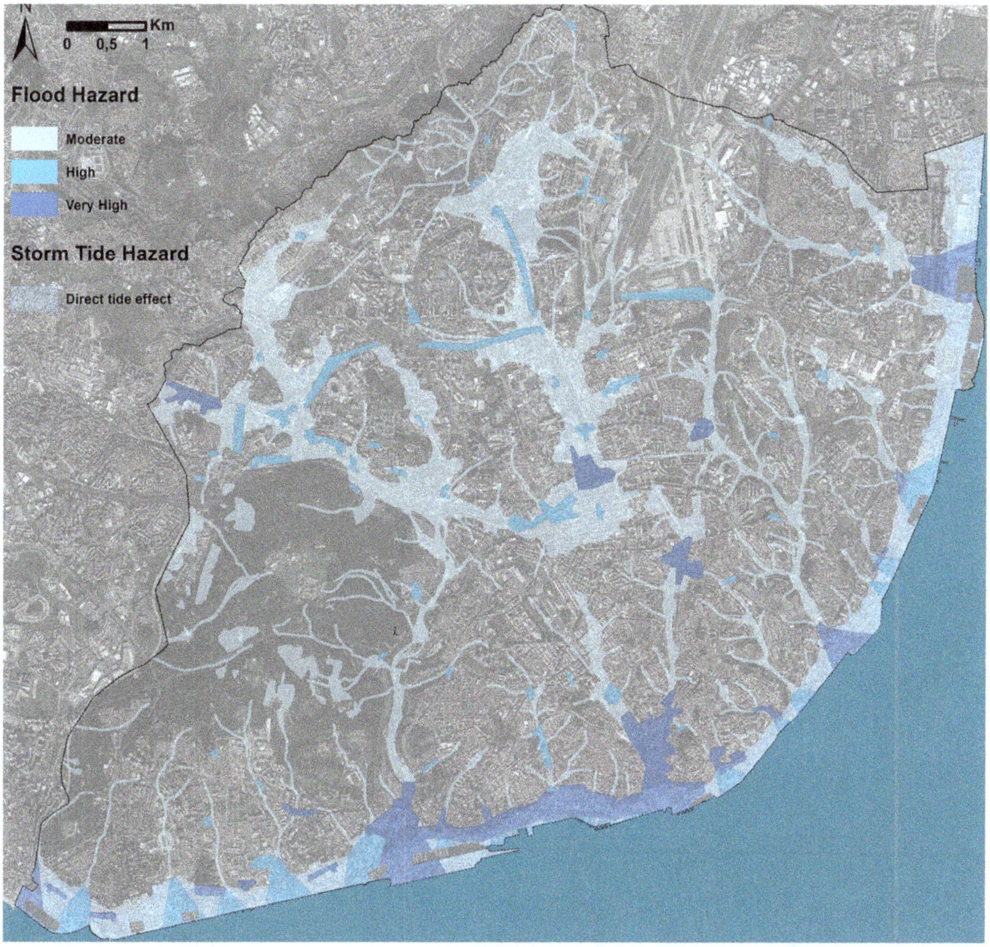

Figure 6.5. Floods and storm surges hazard in Lisbon (adapted from Câmara Municipal de Lisboa, 2012).

Greening Lisbon

Driven primarily by a desire to improve health and the quality of life, Lisbon is extending the number and area of its green spaces and the degree of connectivity between them although increasing heat and flooding risks now play a more significant role in their design and development. One example of this changing approach is the establishment of green corridors (Figure 6.6) such as the Monsanto Greenway, which connects the downtown area with the large Monsanto park area west of the city (Figure 6.7). Elsewhere, green infrastructure is provided by stimulating urban agriculture via new allotments, such as those clustered around existing allotment plots such as the Quinta da Granja (Central corridor) and Campolide (Monsanto corridor) (Figure 6.7 and Figure 6.8). It is considered that municipal involvement in these allotments secures better hygiene and health conditions and that they promote better safety in socially deprived neighbourhoods. Elsewhere, large, previously abandoned areas have been developed as 'extensive' landscapes (such as Figure 6.7 and Figure 6.9) with low requirements for maintenance and water as an alternative to the traditional, more formal parks which demand the opposite to retain their traditional appearance. As a consequence, the extensive parks are better adapted to dry, hot summers and their arid appearance is generally accepted by the citizens. Within the heavily urbanised environment, the planting of street trees is also utilised, creating environmental, social and economic synergies, such as cooling of public spaces, improving biodiversity, decreasing direct radiation which promote public space users' enjoyment and support commercial activities in surrounding areas. The Eixo Central revamping project is an example of public space improvement, combining the development of 'extensive' landscapes with plantation of street trees, where convenient (Figure 6.9).

Figure 6.6. Green corridors being implemented.

Figure 6.7. Monsanto Green Corridor farm (Campolide).

Figure 6.8. Quinta da Granja allotment plot (Municipality of Lisbon).

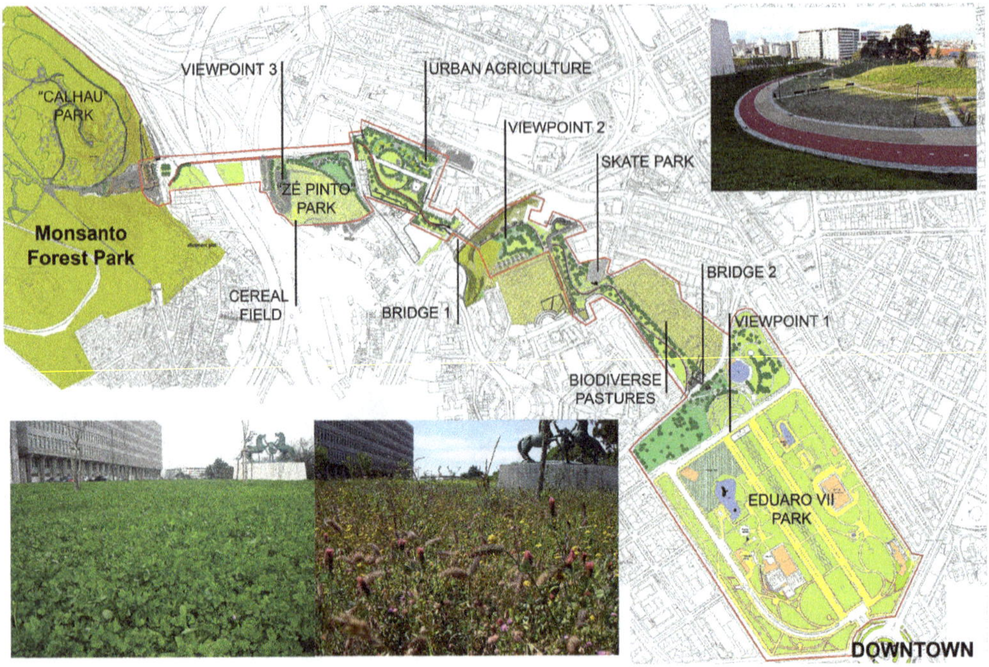

Figure 6.9. Eixo Central revamping project (Monsanto corridor) (Municipality of Lisbon).

A participatory frontrunner city with hidden climate services

Lisbon is rightly regarded as a frontrunner in terms of participatory planning. The city's Master Plan focuses not only on economic and environmental aims, but also on individual well-being and social cohesion by encouraging the use of environmentally-friendly means of transport, healthy habits and connecting people to the city. Examples of these are the development of networks of bicycle lanes, bicycle-friendly streets, ecological corridors and allotment gardens (Santos *et al.*, 2015). With the municipality behind the steering wheel, both citizens and the business community are increasingly engaged in the identification of needs and the development of innovative solutions, e.g. via regular citizens meetings in the parish councils.

A further key element is the 2.5 million euro Lisbon Participatory Budget which enables non-governmental actors to propose projects. Between 2008 and 2017, more than 6,204 projects were proposed, of which 120 were approved and implemented, levering in an additional 34 million euros from other sources. Many of the approved proposals include a green infrastructure component.

In addition, the successful BIP-ZIP programme aims to connect citizenship with entrepreneurship in priority intervention neighbourhoods. One reason for its success is that it is supported by various communication mechanisms, which provide easily accessible information, build capacity, enable consultation and encourage the submission of ideas by citizens, such as Lisboa Participa (Figure 6.10).

Figure 6.10. Participatory budget promoting in Rua Augusta (Downtown Lisbon).

Together, this programme of citizen engagement and information exchange combined with a well-organised planning system and funding streams enables the public to actively participate in shaping their environment, develop a sense of ownership and attachment to public spaces. Moreover, collaboration with scientists and highly qualified technicians (universities and public institutions) works well. Partnerships with private companies are also sought to sponsor and fund green infrastructure projects. However, traditional restraints still apply and the poor maintenance of existing green spaces, a lack of resources and/or technical knowledge and short project timelines all still hinder buy-in (Santos *et al.*, 2015).

The role of climate services

Throughout the planning process, the Municipality continues to work closely with several universities to capture their expertise in landscape architecture, land-use planning, biodiversity, ecology, but also increasingly in climate change. In this sense, the latter is often just one of many drivers of change, and indeed, it is certainly not the main driver of urban green solutions in Lisbon. However, attention is increasing. While the term 'climate services' is not yet often used for such collaborations in Lisbon, climate information is becoming increasingly integrated in city planning as an important element of a broader set of services, and is thus a good example of how climate services can work in practice. Recently, the actual term has been used in connection with the new climate scenarios portal (http://portaldoclima.pt/en/) developed by Lisbon University to support local adaptation plans in municipalities across Portugal. One can argue that the increased demand for climate information for urban planning is likely to further enhance the development and uptake of climate services in the coming years.

6.4 Conclusions

The two case studies illustrate that climate change is only one of many drivers influencing urban development. Other challenges such as urban renewal and the need for more housing, business and leisure facilities alongside a desire for greener cities and a better quality of life also spur urban development. Multi-objective solutions, which promote a mix of green (nature- or ecosystem-based) infrastructures, technological and social elements, can deliver both effective climate change adaptation and mitigation. Cities that are grappling with these issues are therefore recommended to explore the use of multi-purpose nature-based solutions when exploring ways to increase urban resilience. Many cities in Europe and elsewhere are increasingly adopting similar integrated approaches to steer their urban planning process, often combined with an increasing level of citizen participation. They have proven that this approach is successful, not only because it contributes to more efficient and effective urban planning but because active public participation and support also facilitates project implementation.

Whereas the two case studies focus on increasing resilience to projected climate change, the challenge remains to connect and integrate with climate mitigation solutions. Yet, those solutions that incorporate urban and building design can address both. Many more creative solutions are possible beyond the ones discussed above for Nijmegen and Lisbon, dependent on specific local challenges and opportunities. For example, nodal, lobe or garden city designs[1] combine built-up and natural areas in a way that ensures citizens' homes lie within walking distance of work, services and natural areas, reducing the generation of greenhouse gases whilst reducing heat and flood risks. Where possible, cities are therefore recommended to integrate mitigation and adaptation solutions.

What do the cases tell us about the application of climate services? First, climate-resilient urban development was already taking place well before the emergence of the concept of 'climate services'. At the same time, it appears that only a little information on climate change was sufficient to guide the design of solutions. No complex climate data was needed to encourage policymakers, citizens, researchers and companies to consider climate concerns in a qualitative manner. More than a millennium ago, the Lisbon Alfama was built with narrow and windy streets which encouraged cooling breezes throughout the summer. In recent history, the influence of climate change on urban planning declined but with climate risks now better understood, this should change. In Nijmegen and Lisbon, the 'climate services' that were used were projections of river flows, run-off

[1] Nodal cities refer to an urban design with connected, functional, attractive and sustainable urban centres; lobe cities are characterised by blue-green fingers penetrating deep into the urban area; garden cities to self-contained communities surrounded by 'greenbelts', containing proportionate areas of housing, industry, and agriculture.

and information on the cooling and water retention capabilities of green spaces. These climate services were used to achieve an urban design that would not only address social, economic and environmental goals, but also make those goals sustainable over the longer climate change timeframe. Climate services can vary from assessing urban climate stresses, to simple rules of thumb on how to make cities more climate-resilient, to detailed quantitative analyses supporting specific urban renewal projects.

At this moment, we know that urban policymakers need climate services, not only to raise the awareness of politicians and citizens, but also to select and plan long-term solutions, to implement them in the most efficient and effective way as well as to monitor their impact. Yet at present, they are mainly used to raise awareness and inform the planning process only in general terms. In the future, more advanced climate services would need to be embedded within the actual design and implementation of solutions. The application of climate services in urban settings is beneficial and introduces a new dimension to urban planning. More specifically, policymakers are encouraged to plan for a longer time horizon and use climate services to assess their plans in terms of climate resilience. Climate services encourage a more preventive spatial planning approach in cities that consider long-term challenges, with in many situations a larger role for nature-based solutions.

Furthermore, it appears that climate services, if developed jointly with stakeholders, can encourage collaboration between the city government, citizens, businesses and other stakeholders. The cases also show that climate services are only one of the many information services used in the urban planning process. To become relevant for the planning process, climate data will have to be combined with social, economic and/or infrastructure information to allow for meaningful stress testing and response planning. It is important that climate services are tailored to the local context. Furthermore, the interpretation of climate service data may not always be straightforward and can be interpreted differently by other disciplines, which means its use will have to go hand in hand with effective, participatory dialogue with relevant stakeholders.

We can consider access to climate data as the first generation of climate services, and information on climate impacts (e.g. risks of floods of droughts, heat health effects) as the second. The two case studies suggest that climate services have to further evolve towards a third generation in which climate data is connected and integrated with other, technical and socio-economic, data to allow assessment of the current and future vulnerability of neighbourhoods and socio-economic groups. The role of scientists in urban resilience is to clarify the information needs of the urban planners and to gather data and information in a climate service that is tailored to their specific needs. Furthermore, scientists can play a role in designing solutions for different urban challenges and to assess the proposed solutions in terms of resilience in a fourth generation that would also incorporate services supporting the development of both climate mitigation and adaptation solutions in urban design.

The way forward for urban resilience is by integrating climate services. When cities aim to deal with climate change, they can rely on a wide range of climate services to not only inform subsequent risk assessments (stress tests) but also influence the design and implementation of urban renewal and expansion programmes. The projected increases in heat, flood and drought risks, as well as the transition of the energy system to mitigate climate change, are forcing city governments to take a long-term perspective on planning. The growing evidence base that is emerging from climate services studies can support this, but only jointly with other services since it represents only one of the many aspects to consider in urban development. The cases illustrate that climate services have to continuously be adapted to respond to evolving knowledge and to trigger action. They have to be flexible and tailored to the specific local situation and they should integrate other urban challenges as well in order to provide multifunctional innovative urban solutions.

References

Aalbers, C. and Jonkhof, J., 2003. Planning met principes. S2N de strategie van de twee netwerken herzien. AENEAS. Boxtel, the Netherlands.

Câmara Municipal de Lisboa, 2012. Regulamento do Plano Director Muncipal de Lisboa. Available at: https://tinyurl.com/bddt3j3h.

Câmara Municipal de Lisboa, 2021. Plano de Ação Local para a biodiversidade em Lisboa, Relatório intercalar de execução 2020, 1° Semestre 2021. Available at: https://tinyurl.com/yc6ar5rj.

Carter, J.G, Cavan, G., Connelly, A., Guy, S., Handley, J. and Kazmierczak, A., 2015, Climate change and the city: building capacity for urban adaptation. Progress in Planning 95: 1-66.

CCDR-LVT, 2012. Regional land use plan for the metropolitan area of Lisbon. Regional Coordination and Development Committee for Lisbon and the Tagus Valley. Available at: https://tinyurl.com/2p89zndv.

Dias, L.F., 2016. As alterações climáticas, as inundações e a cidade: contributos para o estudo da resiliência urbana em situações de chuvas torrenciais. Tese de Doutoramento. Universidade de Lisboa, Faculdade de Arquitetura, Lisbon, Portugal. Available at: http://hdl.handle.net/10400.5/12584.

European Commission, 2015. A European research and innovation roadmap for climate services. European Commission, Directorate-General for Research and Innovation, Brussels, Belgium. Available at: https://data.europa.eu/doi/10.2777/750202

Gemeente Nijmegen, 1996. Structuurplan 'het land over de Waal', gemeente Nijmegen juli/augustus 1996, Nijmegen, Nederland.

Guerreiro, S.B., Dawson, R.J., Kilsby, C., Lewis, E. and Ford, A., 2018. Future heat-wave, droughts and floods in 571 European Cities. Environmental Research Letters 13: 034009. https://doi.org/10.1088/1748-9326/aaaad3

Lourenço, T., Swart, R., Goosen, H. and Street, R., 2015. The rise of demand-driven climate services. Nature Climate Change 6: 13-14. https://doi.org/10.1038/nclimate2836

Município de Lisboa, 2017. Projeto da estratégia municipal de adaptação às alterações climáticas de Lisboa. Assembleia Municipal de Lisboa, Lisbon, Portugal. Available at: https://www.am-lisboa.pt/301000/1/008022,000380/index.htm.

Santos, A., Branquinho, C., Gonçalves, P. and Santos Reis, M., 2015. Lisbon, Portugal – case study city portrait; part of a GREEN SURGE study on urban green infrastructure planning and governance in 20 European cities. Fundação da Faculdade de Ciências de Lisboa (FFCUL), Portugal.

Street, R., 2016. Towards a leading role on climate services in Europe: a research and innovation roadmap. Climate Services 1: 2-5.

Tjallingii, S.P., 1995. Ecopolis, strategies for ecologically sound urban development. Backhuys Publishers, Leiden, the Netherlands.

Tjallingii, S.P., 1996. Ecological conditions. IBN scietific contributions 2. Wageningen, the Netherlands.

UNFCCC, 2015. Paris Agreement. UNFCCC, Bonn, Germany. Available at: https://unfccc.int/process-and-meetings/the-paris-agreement/the-paris-agreement.

UNHABITAT, 2018, Urban themes – climate change. Available at: https://unhabitat.org/urban-themes/climate-change/.

Van Rossum, H., Van Wijk, F. and Baljon, L., 2001. Verkennigstocht naar VINEX-land. Nai Uitgevers. Rotterdam, the Netherlands.

PART 3

FOOD IN THE SYMBIOTIC CITY

Chapter 7
Governance towards nature-based City Region Food Systems

Marian Stuiver, Sabine O'Hara, Lotte Roosendaal and Bertram de Rooij

Abstract

In the wake of the Milan Urban Food Policy Pact, cities around the world have been taking the lead in designing their own City Region Food System (CRFS). A City Region Food System aims to foster the development of resilient and sustainable food systems which encompass an urban centre, a peri-urban ring, and the rural areas surrounding a city. This chapter argues that a CRFS must also focus on nature and the context it provides rather than on the social and environmental indicators typically used to describe a city and its context. This includes the nature context of the rural-urban region where a city is located. A CRFS needs to include a wide range of food system actors including those who give expression to the nature context of the region. It must also focus on cooperation, decision-making, learning, and distributive leadership among the various CRFS actors. We discuss examples of emerging CRFSs from Amsterdam, Rotterdam and Washington DC to illustrate these principles.

Keywords: CRFS, environmental drivers, socio-economic drivers, systems approach, social inclusion, good governance, decision-making, learning, distributive leadership

7.1 Introduction

Cities are increasingly taking the lead in the City Region Food System (CRFS) transformation which is taking place around the world (Edwards and Mercer, 2010; FAO *et al.*, 2021). A CRFS perspective is committed to developing a resilient food system that is embedded within its urban centre, its peri-urban ring and the rural surroundings of the city. A CRFS perspective has the potential to advance an urban food resilience strategy which seeks to mitigate food security risks associated with external shocks (Figure 7.1) such as national and international supply chain disruptions, natural disasters, and climate change impacts (IPCC, 2022). All these shocks can lead to food shortages and rising food prices (FAO, 2019).

A growing number of urban dwellers support the development of new urban food strategies that recognise that cheap food is quite expensive when one considers the social and environmental costs of a highly centralised, globalised food system. These urban dwellers value healthy, sustainable, safe and fair food alternatives (O'Hara and Stagl, 2001). CRFS offer a strategy to respond to these

10.3920/978-90-8686-935-0_7,

Figure 7.1. Climate change and the food system (courtesy of Bertram de Rooij 2020).

consumers who have a preference for buying local food (Blake *et al.*, 2010; Carolan, 2017) and seek urban food alternatives with a smaller carbon food print, shorter and more stable supply chains, and more sustainable production methods (FAO *et al.*, 2018).

This CRFS perspective is reflected in the new city focus of the United Nations which integrates all facets of the sustainable development agenda (UN-Habitat III, 2017). An urban food systems focus was also included in other national and international strategies, such as the 2018 Farm Bill of the United States and the Urban Food Agenda of the FAO (Cabannes and Marocchino, 2018; FAO, 2019; FAO *et al.*, 2021). The latter has launched projects in the city-regions of Colombo (Sri Lanka), Lusaka and Kitwe (Zambia), Medellín (Colombia), Quito (Ecuador), Toronto (Canada) and Utrecht (the Netherlands). Each of the CRFS initiatives seeks to strengthen rural-urban linkages and identify bottlenecks and opportunities for the successful launch of a CRFS. For example, the FAO, RUAF and IWMI have been working since 2019 in the city-regions of Kigali (Rwanda), Antananarivo (Madagascar), Colombo (Sri Lanka), Quito (Ecuador) and Tamale (Ghana) to incorporate climate and pandemic risk management strategies into regional food systems plans (FAO *et al.*, 2018).

Yet while its popularity is growing, the City Region Food Systems (CRFSs) perspective is by no means uniform (Figure 7.2). At least three CRFS scenario's emerge in the big puzzle (Van der Schans and Wiskerke, 2012; Wiskerke, 2020). They are: (1) large scale rural – urban metropolitan foodsheds which include a peri-urban ring around the city (Kranendonk *et al.*, 2018); (2) neighbourhood based city urban food systems with urban agriculture in community gardens

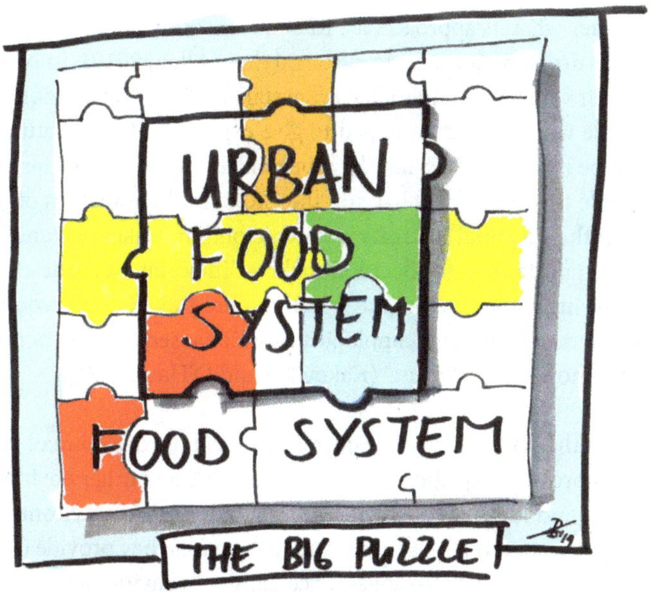

Figure 7.2. The big puzzle (courtesy of Bertram de Rooij 2020).

or roof top gardens (Van der Schans and Wiskerke, 2012); and (3) vertical farming and capital-intensive greenhouse and indoor farming systems in the urban core and/or the peri-urban ring (Butturini and Marcelis, 2019). These three configurations of the food production components of CRFS can serve as a structuring element, while also recognising that consumption and lifestyle issues may provide other relevant structuring elements.

A CRFS encourages integrated urban policies and action perspectives across social and cultural characteristics, economic sectors, and (academic) disciplines. It focuses on food from a systems perspective. This implies that it takes the socio-economic drivers of food production and consumption into account (such as the role of actors in regional economic development) as well as its environmental drivers (such as nature-based solutions). The goals of CRFS are to improve a range of outcomes, including public health, ecosystem restoration, and climate resilience to name just a few (Blay-Palmer *et al.*, 2018; FAO *et al.*, 2018; van der Gaast *et al.*, 2020).

A CRFS therefore offers a holistic perspective on food and nutrition security. At its best, it will consider how the urban food system interacts with other systems including ecological, economic, social, and political ones. CRFSs therefore provide an excellent opportunity to analyse how different elements of the food system interact with each other and with natural systems that form its context and impact its outcomes. We propose explicitly including the region's natural characteristics and the context it provides rather than just focusing on the social and environmental indicators typically used to describe a city and its context.

We argue that no matter what its approach, a CRFS must include the nature context of its rural-urban region to really develop pathways to sustainability (Allen, 2010). In other words, nature characteristics and their value must guide all relevant stakeholders. Furthermore, special attention must be given to those who are best equipped to give expression to the nature context of their CRFS. This includes the impact a CRFS may have beyond its region and the impact resources and emissions from outside the region may have on the CRFS. In their analysis of the impact of the hydrological cycle on the potential to create economic output, Kakovitch and O'Hara find that nature provides important lessons for decision makers. They state that the efficiency increases needed to realise growing economic output '... will require the efficient processing capacity of nature rather than relying solely on technological solutions; second, the non-use of resources may be as valuable or more than their use' (Kakovitch and O'Hara, 2021).

This example illustrates that a CRFS must consider the efficient use of resources such as land, water, and energy; it must improve soil quality, water and air quality and other environmental assets; it must reduce emissions and ideally reduce their accumulation in the environment to reduce the pressures emissions put on nature's ecosystems and the services they provide (see Chapter 9 for a more in-depth discussion); it must also consider consumer behaviour and seek to move toward a more sustainable and healthy diet that reduces the burden on natural resources.

Examining the linkages between resource use, emissions and the quality of the nature context of a region is particularly relevant for cities, where population density is high, and the urban core is inextricably linked to its peri-urban and rural surroundings. Cities depend on their surrounding areas to ensure resources are provided for the population-rich urban core; and waste and emissions from the core are typically deposited and processed in the surrounding area. In other words, the peri-urban and rural surroundings of a city provide both resourcing and 'sink capacities' (see Chapter 9 for a definition of sink capacities).

Urban sprawl places significant pressures on the nature context of cities. For example, cities may be competing with their surrounding area for freshwater resources that are essential for food production. The fact is that waste produced in a city is typically transported to landfills and waste-to-energy facilities in peri-urban and rural areas can negatively impact the soil, water and air quality of the communities receiving the waste (De Rooij *et al.*, 2020). At the same time, the food demand of cities can provide incentives for food production, processing, packaging and transportation to surrounding areas. These interdependencies between cities and their peri-urban and rural neighbours must be acknowledged in shaping a sustainable and resilient CRFS.

A successful CRFS must be nature-based (Figure 7.3) and inclusive of all the nature perceptions and values of the stakeholders involved. A pressing question in Section 7.2 is therefore how to design a CRFS so that its physical/environmental and social/cultural characteristics form a systems perspective that departs from the siloed approach that has been shaped by the discrete social, ecological and environmental characteristics of a particular city and its surrounding region. Section 7.3 argues that it is crucial to develop good governance structures (van Bers C *et al.*, 2016)

Figure 7.3. Nature-based Food Systems (courtesy of Bertram de Rooij 2020).

that reflect the multitude of perspectives of actors that form the integrated whole of a CRFS. In order to achieve this transformation, it is essential to deliberately steer the decision-making and knowledge-production processes in a systems direction akin to responsible leadership scenarios. Section 7.4 presents the conclusions.

7.2 A systems approach of City Region Food Systems

Adopting a CRFS implies that urban actors and their peri-urban and rural partners reconfigure the entire food system of a city and metropolitan region across economic sectors, scales, and institutions. In order to advance such a cross-cutting structure and identify its aims, the currently dominant structuring elements that are at the root of the social, economic and environmental injustices of the current food system must be scrutinised (Wigboldus *et al.*, 2020). Current food systems are frequently driven by economic interests. A CRFS approach must advance a more integrated approach in which social and environmental interests gain weight in decision-making processes.

Over the last decade, scholars and policymakers have increasingly acknowledged the benefits of an integrative systems perspective as a holistic approach to addressing food insecurity and other challenges like climate change, poverty and disparities. A 2013 report by the Food and Agriculture Organization (FAO, 2013) states 'Food systems encompass the entire range of activities involved in the production, processing, marketing, consumption and disposal of goods that originate from agriculture, forestry or fisheries, including the inputs needed and the outputs generated at each of these steps. Food systems also involve the people and institutions that initiate or inhibit change in the systems as well as the socio-political, economic and technological environment in which these activities take place'.

Cities with their high land-use density, space constraints and ecosystems deprivations can be particularly effective in illustrating the important linkages between the social and environmental dimensions of a sustainable and resilient food system, and its specific physical/spatial context that we refer to as its nature context. For example, urban food production can introduce permeable surfaces into the black-topped urban scape of cities; food-producing green roofs can absorb storm water run-off (Li *et al.*, 2014), and thus reduce pressure on urban storm water systems (Agudelo-Vera *et al.*, 2011); producing food in dense urban neighbourhoods requires attention to integrative pest management practices (Wittmer *et al.*, 2010); and urban food production can mitigate heat islands (Sustainable Urban Delta, 2021). In addition, urban food production in community gardens and native plant nurseries can also have strong social implications, such as increasing social cohesion, improving liveability and aesthetics, re-connecting city dwellers with fresh and healthy foods, supporting general well-being, and strengthening sustainability literacy (Veolia Institute, 2019). Some of these food-environment and food-people connections are shaped by the distinct location of a city as priorities will differ between an arid location and one prone to flooding; and between a location with a short growing period requiring six months of heating, and one where year-around outdoor growing is possible.

The multi-dimensionality of a CRFS also calls attention to the fact that it must address all three core areas of economics, namely resource allocation, food production, and food consumption. In addition, an effective CRFS must move beyond the focus of standard economics and follow the principles of restorative economics, outlined in Chapter 9 (see also O'Hara (2015).

Our CRFS narrative (see Chapter 3) for Washington DC and its eight Wards illustrates the power a food system can have to connect seemingly disparate spheres such as food production, diet and health, culture, the economy, natural resources, and technology. A systems approach thus shifts the focus from food as a source of calories to food as a system that interacts with multiple other systems and therefore leads to outcomes that include not only food and nutrition security, but also livelihoods and restored ecosystem services, which can in turn have positive implications for the food system itself. The connectivity of the food system and its elements is shown in Figure 7.4.

We address the systems linkages of the food system to other systems that are commonly treated as separate entities including the connection between: (1) food and environmental drivers and outcomes; (2) food and diets including their cultural expressions; and (3) socio-economic drivers and outcomes of food.

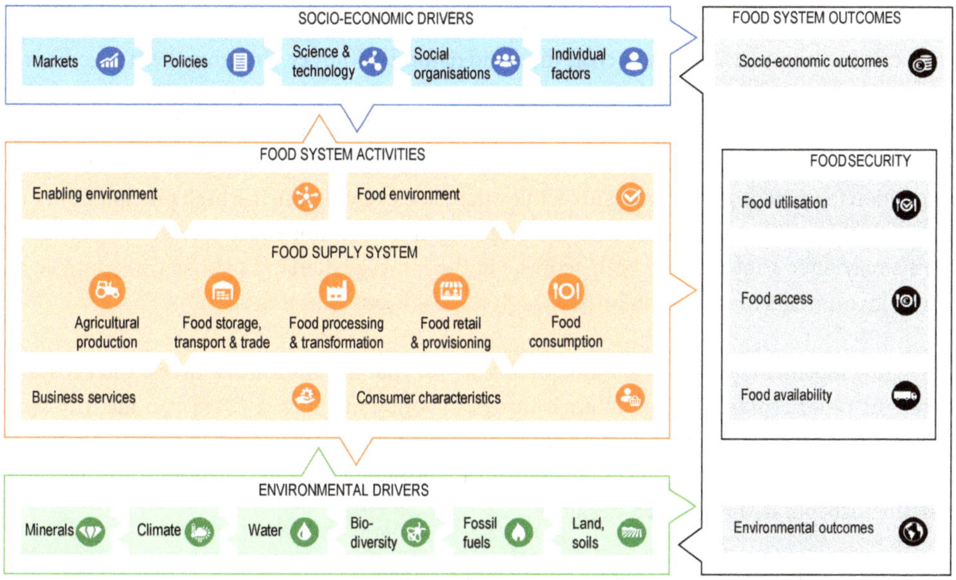

Figure 7.4. The food systems approach (van Berkum *et al.*, 2018).

7.2.1 Food systems and environmental drivers and outcomes

The current food system comes with many environmental challenges such as high freshwater use, declining soil quality, nutrient run-off, and a large energy footprint associated with the transportation needs of a highly centralised food system. Given its long food supply chains and the perishable nature of food, the current food system is also characterised by enormous food waste losses of close to 30% of the overall food supply. In addition to the loss of food itself, food waste also represents an enormous loss of natural resources embedded in the wasted food (Dubbeling *et al.*, 2016). This implies not only the loss of water, nutrients, and energy but also the loss of habitats and biodiversity. As human settlements and food production needs continue to expand, they will encroach more and more on wildlife habitats resulting in the accelerating rate of species extinction, and pandemic risks (O'Hara and Toussaint, 2021; Wallace, 2016).

A regionally embedded food system offers opportunities for climate adaptation and ecological resilience. As green/blue infrastructure elements are incorporated into urban and peri-urban food systems, the win-win potential is tremendous. Soil enhancements can improve soil quality and carbon sequestration; soil-less food production methods can reduce water use by 90% or more and reduce water run-off, eutrophication, and groundwater pollution; green cover can absorb heat, improve oxygen levels and reduce carbon dioxide; nutrient recycling can reduce nutrient losses and reduce pollution levels. These practices, along with the shorter supply chains of a CRFS, can save some natural resources and reduce pollution pressures on others.

These examples also illustrate that CRFSs can connect different spheres that are typically addressed in separate institutional settings, such as food and natural resources. Bringing them together may also point to the fact that local food production may not be the best option. The best solutions may instead depend on the specific climate conditions and energy demands of a year-round growing environment. In large cities where fresh water is scarce, local food production may mean competition for drinking water supplies, while other cities may experience high precipitation rates, a high risk of flooding or of producing food on polluted soil. Indoor and soil-less food production systems may offer a solution on both scores, yet their energy demand may be considerable and upfront investments may be prohibitively high for traditional funding streams.

By bringing together typically disparate spheres of food, nature, natural resources, and ecosystem services or other representations of environmental context factors, a CRFS can identify viable solutions based on the specific interconnections evident in a given city and region. This can add tremendous value by creating the conditions for identifying win-win collaborations that can be explicitly incorporated in a CRFS design.

7.2.2 Food systems and food security, nutrition and health

Food Security, Nutrition Security and Health are closely linked. For instance, the lack of access to fresh and unprocessed food characterises a diet that results in illness rather than health (Yeoman, 2018). Our examples from Washington DC (Table 9.1 and 9.2 in Chapter 9) illustrate that there are neighbourhoods where citizens lack access to fresh unprocessed food. These same neighbourhoods are characterised by high incidents of obesity, diabetes, and hypertension as well as low household incomes and a high percentage of non-Hispanic black populations. The resulting food injustice that leaves black and brown populations more food-insecure than white and Asian populations has been amplified during the COVID pandemic (Chapter 9; O'Hara and Toussaint, 2021).

An effective CRFS must therefore consider the individual perspective (Figure 7.5) on Food Security, Nutrition Security and Health, as well as the systemic discrimination based on age, race, gender, ethnicity and socio-economic status. This includes discriminatory practices with respect to the quality of food, dietary practices and access to clean water, sanitation, health services and safe housing. This implies that not only food access, but entire food environments must be transformed including (1) economic environments like market access, transportation, and food prices, (2) social environments like cultural or spatial disparities, and (3) personal food environments, such as dietary preferences, opportunities, and barriers to participation.

7.2.3 Food systems and socio-economic drivers and outcomes

A CRFS approach can act as a catalyst to transition to diverse food environments that include local food venues where urban dwellers can connect with each other and with their surrounding region through food. The challenge is to develop food environments that are based on regional produce, while being culturally sensitive to the different populations within a diverse city.

Figure 7.5. The individual perspective (courtesy of Bertram de Rooij 2020).

The socio-economic benefits of food can be amplified through a CRFS that connects the urban core to its peri-urban and rural surroundings and envisions a diverse food economy of highly perishable, nutrient-dense, high-revenue crops in the urban core; less perishable, high-demand vegetables in the peri-urban ring; and grains and root vegetables in the surrounding rural areas where land values are lower, and space is more readily available. These distinct, spatially aligned food system components can also offer linkages to other economic sectors including hospitality, food processing, technology, energy, soil enhancement, landscaping, recreation, and more.

Bringing more and better-appointed grocery stores to a food insecure neighbourhood may be a starting point. However, there are other models that may be more effective in connecting food and cultural diversity than a full-service grocery store. Examples of such alternative models were mentioned during the 'Five Pillars' project in Washington DC, which engaged residents from Ward 7 (Deanwood) and Ward 8 (Congress Heights) in a story-writing process about a sustainable future for their neighbourhoods (O'Hara, 2018). The Five Pillars approach views health, education, social and cultural amenities, environmental quality and information and transportation access as the five pillars that undergird a sustainable economic development future. Food featured prominently as a catalyst for better health outcomes, more attractive social and cultural amenities and improved environmental quality. The residents also stressed the importance of food as an expression of the culturally diverse neighbourhoods.

While it may be challenging for urban areas to provide equitable access to nutritious food to all their residents, the agricultural sector of the surrounding areas also struggles. The ageing population of farmers and the continued loss of farmland to urbanisation, put growing pressures on a resource-strapped farm sector. An economy rooted in a localised and regionalised food system can strengthen both the urban core and its surrounding rural areas. Mark Winne (2019) describes the positive economic impact of creating food-based local economies even in unlikely mid-sized cities that are not known for their culinary prowess.

7.3 Good governance of City Region Food Systems

To be effective in its inclusivity, a CRFS needs to be based on a systems perspective that goes beyond socio-economic factors and includes nature-conscious, environmental drivers and outcomes. Such a nature-conscious CRFS needs good governance. Good governance ideally expresses the collective values associated with nature and its care beyond sustaining the natural resources it delivers. True to its commitment to collective values, a central principle of good governance is its commitment to engaging a wide range of diverse actors from the public, private, and non-profit sectors. By improving the collaboration of all relevant stakeholders who play a role in the CRFS, all can engage in identifying governing principles that can move the region toward a nature-based CRFS (Gerritsen *et al.*, 2013).

Andrée *et al.* (2019) argue that '... in the realm of food systems, governance spans from the internal decision-making structures of companies, non-profit organisations, and informal associations, to the more formalised policies and procedures of state governments and indigenous authorities, and then to multilateral trade and environmental agreements.' In other words, policies form only one aspect of governance. Good governance must reflect a range of diverse decision processes that involve the broadest possible representation of diverse actors.

Focusing on a CRFS implies the need for governance at the regional level. Yet, regional governance structures that connect the rural, peri-urban and urban are rare. Existing structures may not match the administrative and jurisdictional levels and scales necessary to implement a CRFS (Dubbeling *et al.*, 2016). Efforts to implement a CRFS approach can therefore trigger a debate about the level of decision-making needed to support the transition toward a CRFS. This will have to include considerations of the linkages between new governance structures and existing ones at the city and regional level (Figure 7.6). We propose to foster good governance by focussing on two principles: (1) the engagement of a wide range of actors and (2) decision-making, learning and leadership.

7.3.1 Engagement of a wide range of food system actors

A wide range of stakeholders must be engaged in envisioning, designing, and implementing a nature-based CRFS. For instance, a regional food network can exchange future scenarios with (social) entrepreneurs, professionals, government representatives and citizens. Such networks of stakeholders can incorporate the goals of their municipalities, as well as those of citizens groups and private sector businesses. They can inspire and connect, strengthen and accelerate food initiatives that support nature-based goals and solutions.

Figure 7.6. Urban-rural linkages (courtesy of Bertram de Rooij 2020).

An example of such a stakeholder network is the cooperation Midden Delfland in the Netherlands, where citizens and farmers in the region of Rotterdam, Den Haag and Delft aim to restore the peat meadow lands, while at the same time organising and providing regional food produced by agricultural business in the region. Examples include farms that take care of landscapes and nature; a new *bokashi* business, based on the fermentation of roadside plant clippings and raw manure into high-quality fertiliser for public green spaces and a cooperative farming initiative where citizens collectively own and rent out land to farmers (Westerink *et al.*, 2018; Westerink and van Rooij, 2018; WUR, 2018).

This is just one example that shows the range of stakeholders a CRFS can involve while also engaging partners at varying scales from regional, to national and even global players. Stakeholders may also come from a wide range of sectors including food producers, food processors, researchers, the nutrition and health sector, technology and AI, green infrastructure, the knowledge industry, and of course consumers, including households, food banks, schools, assisted living and health care facilities and a range of other private, non-profit and public sector organisations. Some of these actors represent formal and credentialled knowledge fields and perspectives, while others represent the more informal and local perspectives of non-credentialled experts. Cutting across sector-specific, individual and organisational perspectives are the perspectives associated with different cultural and socio-economic backgrounds.

Cities can take intentional steps toward establishing decision-making structures that acknowledge both urban and rural interests and consider the various linkage points of urban, peri-urban and rural food systems within a region. These connections are not easily forged especially across urban and regional spaces where existing administrative boundaries may stand in the way of rural-urban collaborations. Agreeing on a CRFS, can therefore legitimise collaboration across administrative boundaries and foster meaningful rural-urban strategies. Fontein and Stuiver (2011) describe the ideal process of establishing an effective CRFS as exhibiting four levels of cooperation: strategic, tactical, operational, and reflexive. For instance, the municipality of Rotterdam operates on a strategic level and invites other private and non-profit sector organisations to participate in a strategic envisioning process. On the tactical level it is important to link food production in the urban scape to other policy themes such as health, urban planning, economic development, utilities (especially energy and water) and education. This coordination across well-established municipal and regional governance structures requires active officials willing to forge collaborations. For example, the Municipality of Rotterdam is actively seeking collaborations between urban food, housing, and health insurers. In doing so, Rotterdam is elevating urban agriculture to be an active partner of other policy areas, which have not been accustomed to working with the food and agriculture sector. On an operational level, local and regional initiatives such as organising festivals, regional markets, and pop-up restaurants, can forge meaningful collaborations. Municipalities can facilitate these kinds of operational collaborations in various ways from granting subsidies, to making room in zoning regulations, to playing the role of broker and facilitator. The reflexive level is an important element of the cooperation process since it makes room for stakeholders to meet and participate in learning networks outside of their own municipality or organisation. Municipalities are well positioned to take the lead in this reflexive learning process. In Rotterdam, for example, civil servants participate in Communities of Practice focused on nature-based solutions, green circles, and urban food production to learn from other organisations what measures they are considering or have implemented to move towards a nature-based CRFS (Fontein and Stuiver, 2011).

Ensuring the broadest possible representation in developing and implementing a successful CRFS will require social inclusion, equity, and participatory planning at every level (Cabannes and Marocchino, 2018; Feltrinelli, 2015; IPES-Food, 2017). In addition to developing wide-ranging participation strategies, more power must be given to community-based organisations, small food sector representatives, social movements and individual consumers who represent the less powerful fringe of the food system. These typically excluded perspectives are especially indispensable to building more just and sustainable decentralised food systems that are anchored in the nature characteristics of their regions.

However, engaging local and regional stakeholders can be a challenge. Even when an inclusive approach is widely supported, the channels for effective stakeholder participation may not be in place and stakeholder participation may therefore be limited (Feltrinelli, 2015; Harper *et al.*, 2009; Termeer *et al.*, 2018). And even when key actors are brought together, power differences and conflicting interests can seriously hinder effective participation and governance (Feltrinelli, 2015; Harper *et al.*, 2009). Addressing these challenges to ensure a truly inclusive and collaborative stakeholder process may require discourse strategies and other deliberative approaches that can level the playing field among divergent stakeholders (O'Hara, 2017). This work of levelling the playing field may extend to both region internal and external stakeholders (Harper *et al.*, 2009). Power imbalances among the actors who represent that full scope of the food system and therefore shape a CRFS must be addressed as part of the transition to a sustainable CRFS (Duncan *et al.*, 2019). This means that power relations, structural inequalities and power imbalances within the food system need to be analysed and addressed. In doing so, questions should go beyond who has power and who does not, and include explorations of how power is exercised, what influences existing power structures, how politics may or may not legitimise certain forms of power, how power can be leveraged, and how agency can be extended to those with the least amount of influence (Elzen *et al.*, 2019; Roosendaal *et al.*, 2020).

Working with a wide range of stakeholders with different interests and power relations can also pose a challenge in terms of continuity. Transitioning to a CRFS is not a short-term proposition. It may require deliberate actions to establish and maintain engagement, energy, and trust over the long run. It also requires considerable skill in working with a wide range of constituencies and membership structures (Harper *et al.*, 2009). With the introduction of a CRFS, existing institutions will have to broaden their scope, and engage with new networks of collaboration to deal with the multiple values of the CRFS (Figure 7.7). City governments, for example, have not focused much on agriculture and food production in the past. In the case of Washington DC,

Figure 7.7. Multiple values (courtesy of Bertram de Rooij 2020).

which is a city state, more than a dozen DC agencies have some connection to food. Yet there is no department of agriculture at the municipal level. Cooperation between different governance levels and stakeholders will be critical to ensuring that land-use pressures, which stand in the way of region-wide food security solutions, are adequately addressed. According to the recent report of the Greater Washington Board of Trade (Board of Trade, 2018) and the Resilient DC Plan (Resilient DC, 2019) there are many opportunities to link regional initiatives to hyper-local initiatives in the eight Wards of Washington DC.

Similarly, food and nutrition are not traditionally seen as the responsibility of urban planners Figure 7.8). Yet they can play a pivotal role in supporting food and nutrition security for urban dwellers, for example by allocating space for urban agriculture or food outlets that sell fresh fruits and vegetables. In the city of Rotterdam, for example, Henneman documented 20 years of initiatives and growing cooperation between food- and nature-focused organisations (Henneman, 2021). The municipality of Rotterdam had initiated Henneman's study of collaborations as part of the Rotterdam Living Lab within the European Edible Cities Network. Thirty-four collaborators were documented and provided numerous examples of mutually beneficial collaborations between stakeholders. Yet despite these positive examples, the continuity of the Rotterdam Food Policy Council and other coordinating agencies and actors is constantly under threat (Henneman, 2021).

Another example is the local food agenda called 'Food and Amsterdam' (Gemeente Amsterdam, 2014), which seeks to promote local food production and consumption, healthy food options, education, sustainability and innovation through the collaboration of partners from the private, public, and non-profit sector. The collaboration aims to increase regionally produced food from

Figure 7.8. Planning food (courtesy of Bertram de Rooij 2020).

5.5% (Gemeente Amsterdam, 2014) to 20% (Kranendonk *et al.*, 2018), an aspirational goal that was established by partners in the Metropolitan Region Amsterdam. One of the shared objectives of the collaborators was that their food production goals needed to be achieved in a sustainable manner. In 2014 the Amsterdam Economic Board, proposed the creation of a Food Council, similar to those adopted by frontrunner cities like Toronto and Boston. The Council was established in 2017. However, in the Metropolitan Region Amsterdam, initiatives have not been consistent, with new actors joining and leaving constantly. As a result, effective collaborations between organisations pursuing ecological goals and those pursuing food-related goals face ongoing challenges.

For stakeholder engagement to take place in an organised and dependable manner, continuity is essential. Successful examples include Food Policy Councils (FPC), and Urban Food Councils (Feltrinelli, 2015; Gupta *et al.*, 2018; Harper *et al.*, 2009). Harper *et al.* (2009) defines a FPC as 'both forums for food issues and platforms for coordinated action. FPCs consist of groups of representatives and stakeholders from many sectors of the food system'. FPCs therefore reflect the trans-disciplinarity of food systems and function as collaborative structures that cut across organisations, sectors, stakeholder groups, and scales (Termeer *et al.*, 2018). In addition to FPCs, cities have also formed food boards, forums, networks, and partnerships to exemplify a variety of formal and informal arrangements around food and food systems. While their name and function may vary, a common thread is their commitment to bringing together actors from across the food system, with the goal of advancing the implementation of a more sustainable and resilient food system in their respective communities and regions.

Ideally, governance arrangements must facilitate both vertical and horizontal linkages (Dubbeling *et al.*, 2016; FAO *et al.*, 2021; Gupta *et al.*, 2018; IPES-Food, 2017; UN-Habitat III, 2017). Vertical linkages ensure that the governance structure links with regional, national, and global policies and frameworks, but this also means that you need the right people on board to maintain and strengthen links with these vertical levels. Horizontal linkages ensure that the governance structure is well-connected with other entities across the full spectrum of food systems related players. Both vertical and horizontal linkages require the right people to link with the right stakeholders representing the respective linkages(Roosendaal *et al.*, 2020).

7.3.2 Decision-making, knowledge and leadership

As governance arrangements in an urban region bring together a wide range of actors, perspectives and interests, decision-making can become increasingly complex (Figure 7.9). Different actors may be drawn out of their comfort zone and challenged to adopt unfamiliar and even uncomfortable perspectives and approaches. This requires quality data, a learning mindset, ongoing reflection, and distributive leadership.

Data and knowledge (Dubbeling *et al.*, 2016; FAO, 2019; Harper *et al.*, 2009) play a key role in any decision-making processes. Yet they are especially important when diverse stakeholders have to be brought to the table. Data can help provide a joint starting point for dialogue, especially when there are likely disagreements on key trends and issues (Foresight4Food, 2022). In addition,

Figure 7.9. How to reach (courtesy of Bertram de Rooij 2020).

exchanging knowledge and data can create a level playing field that allows all actors to engage, form an opinion and make informed decisions. Data is also needed to monitor progress and adapt strategies as new insights emerge that have implications for a region's CRFS and its successful implementation (Brouwer *et al.*, 2015).

Knowledge exchange between local experts and credentialed experts is critical in order to align and co-create urban food innovations and learn how to scale them successfully into a CRFS. In addition, the knowledge exchange and reflexive learning between cities and metro regions can support cities in their own visioning process and journey towards creating a CRFS. Cities typically find it helpful to connect with other like-minded municipalities who are committed to a sustainable urban food system. A growing number of cities have signed the Milan Urban Food Policy Pact or are engaged in the C40 Network to advance their own plans toward a CRFS.

Knowledge governance is the purposeful production and dissemination of knowledge, which delivers new insights and solutions that enable actors to achieve social and policy change. A knowledge area that is indispensable to developing a CRFS is the connections between food systems and nature. One way to quantify these linkages is to focus on the natural resource needs of a CRFS, such as its land and water use, for example. A footprint analysis can serve as a useful tool with which to assess the natural resource use of a CRFS, by calculating the per unit (i.e. per person, city or country) land surface needed to support the unit's consumption (see for example

(Sukkel *et al.*, 2010; Wackernagel *et al.*, 2006; Wascher *et al.*, 2016)). The Metropolitan Food Planner (Wascher *et al.*, 2015) for instance is a research- and planning tool that studies the land and water required to support a CRFS. Another tool, the ecological footprint is a highly flexible methodology that can be used to calculate the impact of different types of land use in terms of their varying water, energy, and nutrient needs. The tools also differ in scale and some applications may proof challenging especially as different aspects of the food system, for example, may vary in terms of their local versus global implications.

An important example of the impact of local knowledge systems on good governance is visible in the case of the so-called circular farming systems in the Netherlands (Sonneveld, 2004). Farmers, citizens, and scientists have collaborated last decades to develop new local forms of knowledge and data on soils, manure and landscaping in order to develop nature inclusive farming systems (Stuiver, 2008; Wiskerke *et al.*, 2003). The recognition of the importance of farmers' knowledge (Van der Ploeg, 2003) has led to new novelties in nature inclusive farming, new rules and regulations for the agricultural sector, and governance structures to enforce circularity in the food and agriculture sector in general (Gerritsen *et al.*, 2010; Stuiver, 2009) and urban food and agriculture in particular (Browning, 2018).

Distributed leadership is key to advancing new knowledge frameworks and guiding decision-makers across the diverse sectors and stakeholders involved in moving toward a CRFS. Distributed leadership means, first of all, acknowledging that all stakeholders and participants in the CRFS have unique knowledge, skills, and experience to contribute to the design, implementation and decision processes involved in creating a CRFS. This also implies that much is to be gained from joining forces around a common goal to effectively use the knowledge resources available. Distributed leadership plays a key role in bringing the diverse actors of hyperlocal innovations together to strengthen the CRFS in their own locale, and to assess and improve their hyper-local contributions. One model for the successful coordination of distributed knowledge systems is learning ecosystems, which have also been called communities of practice (Wenger, 1999), and field laboratories (Stuiver *et al.*, 2003; Termeer *et al.*, 2013). Evidence Based Food Planning (Kranendonk *et al.*, 2018) and Multi Stakeholder Planning (Brouwer *et al.*, 2015), where diverse actors collaborate to advance knowledge development and innovation. Inclusive governance platforms can create the necessary space to store the local knowledge advanced in these learning ecosystems and bring smaller initiatives and local champions together to inspire coordinated action across the CRFS.

7.4 Conclusions

This chapter presented examples of Rotterdam, Washington DC and the Metropolitan Region Amsterdam to illustrate the characteristics of viable nature-based City Region Food Systems (CRFS). They show the important linkages between food and nature, such as food production focused on ecosystem restoration, water management, green infrastructure and new nature-based technologies, but also cultural food practices that are sensitive to nature, and food experiences of local consumers who are looking for socially and environmentally sustainable outcomes.

We argue that the complexities of a CRFS require the engagement of a wide range of stakeholders, from hyper-local neighbourhood representatives to regional governance structures across private, public, and non-profit sectors. Yet, bringing diverse stakeholders to the table is not sufficient. The social inclusion of all stakeholders and actors is equally important. Food is a right and not a privilege. Everyone should have access to quality food regardless of their social or economic position, class, race, gender, or creed (FAO, 2017). Such a right-based perspective requires a more active role of the public sector to ensure that all are represented especially in times of crises. Leaving food production and distribution to market forces alone can result in heightened risks, especially in highly centralised and global markets. The associated risks of vulnerable supply chains, and worker shortages due to impeded mobility, are especially concerning when the disruptions impact a product which is both a necessity and perishable. The most vulnerable will be the first to suffer from disrupted harvests, fragile transportation systems, and disrupted distribution chains; they will also be least able to adapt to new workplace conditions, and rising food and energy prices (O'Hara and Toussaint, 2021; Reardon *et al.*, 2020). Growing attention and demand for locally produced food, shorter food chains, and an active role of local governments (FAO, 2020; Tefft, 2020) is a step in the right direction and may bode well for the continued growth of the CRFS movement.

For nature-based CRFSs to advance, it is crucial that all urban stakeholders collaborate towards an improved understanding of the connections between food production, nature, ecosystems restoration, and food security at the local and regional landscape level. A wide range of perspectives is needed to ensure that social, economic, ecological and environmental interests from different actors at different levels and scales are taken into account in the decision-making processes. This broad representation is key to arriving at policies and actions within the CRFS that are aligned with existing sector-specific, and geographic scales at the local, regional, and national level. A CRFS approach therefore needs to affirm distributed leadership, and collective decision-making in inclusive governance platforms. In this way urban innovations that are hyperlocal can offer great potential to contribute to a regional solution that can advance a nature-based City Region Food System.

References

Agudelo-Vera, C.M., Mels, A.R., Keesman, K.J. and Rijnaarts, H.H.M., 2011. Resource management as a key factor for sustainable urban planning. Journal of Environmental Management 92: 2295-2303. https://doi.org/10.1016/j.jenvman.2011.05.016

Allen, A., 2010. A PERIscope on the PERI-urban, pathways to sustainability: agendas for a New Politics of Environment, Development and Social Justice, Sussex, UK. STEPS Centre.

Andrée, P., Clark, J.K., Levkoe, C.Z. and Lowitt, K., 2019. Civil society and social movements in food system governance (1st ed.). Routledge, London, UK. https://doi.org/10.4324/9780429503597

Blake, M.K., Mellor, J. and Crane, L., 2010. Buying Local food: shopping practices, place, and consumption networks in defining food as 'local'. Annals of the Association of American Geographers 100: 409-426. https://doi.org/10.1080/00045601003595545

Blay-Palmer, A., Santini, G., Dubbeling, M., Renting, H., Taguchi, M. and Giordano, T., 2018. Validating the city region food system approach: enacting inclusive, transformational city region food systems. Sustainability 10: 1680.

Board of Trade, 2018. 2019-2021 strategic plan, Greater Washington Board of Trade, Washington DC, USA.

Brouwer, H., Woodhill, A.J., Hemmati, M., Verhoosel, K.S. and Van Vugt, S., 2015. The MSP guide: how to design and facilitate multi-stakeholder partnerships. Wageningen University & Research, Wageningen, the Netherlands.

Browning, G., 2018. Circular urban agriculture in The Hague: where we are and where we could go. TU Delft, Leiden University, the Netherlands.

Butturini, M. and Marcelis, L.F.M., 2019. Vertical farming in Europe: present status and outlook. In: Kozai, T., Niu, G. and Takagaki, M. (eds.) Plant factory. Elsevier, New York, NY, USA.

Cabannes, Y. and Marocchino, C., 2018. Integrating food into urban planning. UCL Press, London, UK. https://doi.org/10.2307/j.ctv513dv1

Carolan, M., 2017. More-than-active food citizens: a longitudinal and comparative study of alternative and conventional eaters. Rural Sociology 82: 197-225. https://doi.org/10.1111/ruso.12120

De Rooij, L.L., Verweij, P. and Agricola, H.J., 2020. Feeding cities and migration: urban food systems in a spatial environmental perspective. No. 3002, Wageningen Environmental Research, Wageningen, the Netherlands. https://doi.org/10.18174/520018

Dubbeling, M., Bucatariu, C., Santini, G., Vogt, C. and Eisenbeiβ, K., 2016. City region food systems and food waste management: linking urban and rural areas for sustainable and resilient development, GIZ, FAO, RUAF Foundation. Available at: https://www.fao.org/3/i6233e/i6233e.pdf.

Duncan, J., Levkoe, C.Z. and Moragues-Faus, A., 2019. Envisioning new horizons for the political economy of sustainable food systems. IDS Bulletin 50: 37-56. https://doi.org/10.19088/1968-2019.117

Edwards, F. and Mercer, D., 2010. Meals in metropolis: mapping the urban foodscape in Melbourne, Australia. Local Environment 15: 153-168. https://doi.org/10.1080/13549830903527662

Elzen, B., Bosselaar, J., Brouwer, H., Bulten, E., Kraan, M., De Steenhuijsen Piters, B., De Roo, N. and Vernooij, A., 2019. Power and politics in food systems: a quick scan. Food Systems Governance Mechanisms' Food systems Community of Practice meeting, Utrecht, the Netherlands.

FAO, 2013. The state of food and agriculture. Food and Agriculture Organization of the United Nations, Rome, Italy. Available at: https://www.fao.org/3/i3300e/i3300e00.pdf.

FAO, 2017. Governance of food and nutrition security – factors for viability and sustainability: case studies from seven Latin American countries, Food and Agriculture Organization of the United Nations, Santiago, Chili.

FAO, 2019. FAO framework for the Urban Food Agenda. Food and Agriculture Organization of the United Nations, Rome, Italy. Available at: https://doi.org/10.4060/ca3151en.

FAO, 2020. Urban food systems and COVD-19: the role of cities and local governments in responding to the emergency, Food and Agriculture Organization of the United Nations, Rome, Italy.

FAO, RUAF and MUFPP, 2021. The Milan urban food policy pact monitoring framework – a practical handbook for implementation. Food and Agriculture Organization of the United Nations, Rome, Italy.

FAO, RUAF and WLU, 2018. City region food system tools/., Food and Agriculture Organization of the United Nations, RUAF Foundation, Wilfrid Laurier University: Centre for Sustainable Food Systems. Available at: https://www.fao.org/3/i9255en/I9255EN.pdf.

Feltrinelli, 2015. Milan urban food policy pact: selected good practices from cities. Milan, Italy.

Fontein, R.J. and Stuiver, M., 2011. Naar duurzame regionale voedselsystemen. Wageningen Universtity & Research, Wageningen, the Netherlands.

Foresight4Food, 2022. A framework for understanding foresight and scenario analysis. available at: https://tinyurl.com/3sbdj9ay.

Gemeente Amsterdam, 2014. Voedsel en Amsterdam: een voedselvisie en agenda voor de stad. Amsterdam, the Netherlands. Available at: https://tinyurl.com/ywxpchtx.

Gerritsen, A.L., Stuiver, M. and Termeer, C.J.A.M., 2013. Knowledge governance: an exploration of principles, impact, and barriers. Science and Public Policy 40: 604-615. https://doi.org/10.1093/scipol/sct012

Gerritsen, A.L., Stuiver, M., Termeer, C.J.A.M. and Werkman, R., 2010. Knowledge governance for sustainable rural development; exploring potentials, characteristics and barriers in the northern Frisian woodlands, ERSCP-EMSU Conference on Knowledge Collaboration & Learning for Sustainable Innovation, Delft, the Netherlands.

Gupta, C., Campbell, D., Munden-Dixon, K., Sowerwine, J., Capps, S., Feenstra, G. and Van Soelen Kim, J., 2018. Food policy councils and local governments: creating effective collaboration for food systems change. Journal of Agriculture, Food Systems, and Community Development 8: 11-28. https://doi.org/10.5304/jafscd.2018.08B.006

Harper, A., Shattuck, A., Holt-Giménez, E., Alkon, A. and Lambrick, F., 2009. Food policy councils : lessons learned, food first: Institute for Food and Development Policy, Oakland, CA, USA.

Henneman, R., 2021. Naar een Rotterdams netwerk van groene (Voedsel)Initiatieven. Available at: https://tinyurl.com/usxr4u48.

IPCC, 2022. Climate Change 2022: Impacts, adaptation, and vulnerability contribution of Working Group II to the Sixth Assessment Report of the Intergovernmental Panel on Climate Change. Cambridge University Press, Cambridge, UK.

IPES-Food, 2017. IPES-Food. 2017. What makes urban food policy happen? Insights from five case studies, International Panel of Experts on Sustainable Food Systems.

Kakovitch, T.S. and O'Hara, S., 2021. Water and the circular economy: learning from nature. Sustainability 13: 12597. https://doi.org/10.3390/su132212597

Kranendonk, R., Wascher, D., Agricola, H., Jeurissen, L., van Bossum, J., Rademakers, K.W., ten Brug, L. and Wertheim-Heck, S., 2018. Evidence-based food system design (EFSD). Amsterdam Institute for Advanced Metropolitan Solutions (AMS), Amsterdam, the Netherlands.

Li, D., Bou-Zeid, E. and Oppenheimer, M., 2014. The effectiveness of cool and green roofs as urban heat island mitigation strategies. Environmental Research Letters 9: 055002. https://doi.org/10.1088/1748-9326/9/5/055002

O'Hara, S. and Stagl, S., 2001. Global food markets and their local alternatives: a socio-ecological economic perspective. Population and Environment 22: 533-554. https://doi.org/10.1023/A:1010795305097

O'Hara, S. and Toussaint, E.C., 2021. Food access in crisis: food security and COVID-19. Ecological Economics 180: 106859. https://doi.org/10.1016/j.ecolecon.2020.106859

O'Hara, S., 2015. Food security: the urban food hub solution. Solutions 6: 42-52.

O'Hara, S., 2017. The urban food hubs solution: building capacity in urban communities. Metropolitan Universities 28: 69-93.

O'Hara, S., 2018. The Five pillars of economic development: a story of a sustainable future for ward 7 and 8 in Washington DC. College of Agriculture, Urban Sustainability and Environmental Sciences, University of the District of Columbia, Washington DC, USA.

Reardon, T., Bellemare, M. and Zilberman, D., 2020. How COVID-19 may disrupt food supply chains in developing countries, International Food Policy Research Institute. Available at: https://tinyurl.com/2n4pu6f3.

Resilient DC, 2019. Resilient DC a strategy to thrive in the face of change, Resilient DC 100 Resilient Cities, Washington DC, USA.

Roosendaal, L., Herens, M., De Roo, N., Stuiver, M., Pittore, K., Soma, K. and Hetterscheid, B., 2020. City region food system governance: guiding principles and lessons learned from case studies around the world. Wageningen Centre for Development Innovation, Wageningen, the Netherlands.

Sonneveld, M.P.W., 2004. Impressions of interactions: land as a dynamic result of co-production between man and nature. Wageningen, the Netherlands.

Stuiver, M., 2008. Regime change and storylines. A sociological analysis of manure practices in contemporary Dutch dairy farming. Wageningen University and Research, Wageningen, the Netherlands.

Stuiver, M., 2009. Storylines and their travels, practices of knowledge production in the opening-up of the Dutch modern manure regime. Lambert Academic Publishing, Chisnau, Republic of Moldova.

Stuiver, M., Van Der Ploeg, J.D. and Leeuwis, C., 2003. The VEL and VANLA environmental co-operatives as field laboratories. NJAS – Wageningen Journal of Life Sciences 51: 27-39. https://doi.org/10.1016/S1573-5214(03)80025-8

Sukkel, W., Stilma, E.S.C. and Jansma, J.E., 2010. Verkenning van de milieueffecten van lokale productie en distributie van voedsel in Almere: energieverbruik, emissie van broeikasgassen en voedselkilometers, Praktijkonderzoek Plant & Omgeving B.V., Lelystad, the Netherlands.

Sustainable Urban Delta, 2021. Growing food to cool down the city: this is how urban agriculture improves climate resilience. Available at: https://sustainableurbandelta.com/food-and-climate/.

Tefft, J., 2020. How local government are strengthening the resilience of agrifood systems to COVID-19. Available at: https://tinyurl.com/2p86zcek.

Termeer, C.J.A.M., Drimie, S., Ingram, J., Pereira, L. and Whittingham, M.J., 2018. A diagnostic framework for food system governance arrangements: the case of South Africa. NJAS – Wageningen Journal of Life Sciences 84: 85-93. https://doi.org/10.1016/j.njas.2017.08.001

Termeer, C.J.A.M., Stuiver, M., Gerritsen, A. and Huntjens, P., 2013. Integrating self-governance in heavily regulated policy fields: insights from a Dutch farmers' cooperative. Journal of Environmental Policy & Planning 15: 285-302. https://doi.org/10.1080/1523908X.2013.778670

UN-Habitat III, 2017. The new urban agenda. United Nations, New York, NY, USA.

Van Berkum, S., Dengerink, J. and Ruben, R., 2018. The food systems approach: sustainable solutions for a sufficient supply of healthy food. Wageningen Economic Research, The Hague, the Netherlands.

Van Bers C, Pahl-Wostl C, Eakin H, Ericksen P, Lenaerts L, Förch W, Korhonen-Kurki K, Methner N, Jones L, Vasileiou I and Eriksen, S., 2016. Transformations in governance towards resilient food systems. No. 190, CGIAR Research Program on Climate Change, Agriculture and Food Security (CCAFS), Copenhagen, Denmark. Available at: https://cgspace.cgiar.org/handle/10568/78293.

Van der Gaast, K., Van Leeuwen, E. and Wertheim-Heck, S., 2020. City-region food systems and second tier cities: from garden cities to garden regions. Sustainability 12: 2532.

Van der Ploeg, J.D., 2003. The virtual farmer: past, present and future of the Dutch peasantry. Uitgeverij Van Gorcum, Assen, the Netherlands.

Van der Schans, J.W. and Wiskerke, J.S., 2012. Chapter 21 Urban agriculture in developed economies. In: Viljoen, A. and Wiskerke, J.S. (eds.) Sustainable food planning: evolving theory and practice. Wageningen Academic Publishers, Wageningen, the Netherlands, pp. 243-258. https://doi.org/10.3920/978-90-8686-187-3_21

Veolia Institute, 2019. Urban agriculture: another way to feed cities. The Veolia Institute Review – Facts Report. Available at: https://tinyurl.com/34uc936p.

Wackernagel, M., Kitzes, J., Moran, D., Goldfinger, S. and Thomas, M., 2006. The ecological footprint of cities and regions: comparing resource availability with resource demand. Environment and Urbanization 18: 103-112. https://doi.org/10.1177/0956247806063978

Wallace, R., 2016. Big farms make big flu: dispatches on influenza, agribusiness, and the nature of science. NYU Press, New York, NY, USA.

Wascher, D., van Eupen, M., Corsi, S., Sali, G. and Zasada, I., 2016. Metropolitan foodsheds as spatial references for a landscape-based assessment of regional food supply. In: Roggema, R. (ed.) Agriculture in an urbanizing society volume one: Proceedings of the sixth AESOP conference on sustainable food planning. Cambridge Scholars, Cambridge, UK, pp. 31-58.

Wascher, D., Zasada, I. and Sali, G., 2015. Tools for metropolitan food planning – a new view on the food security of cities. In: Deakin, M., Diamantini, D. and Borrelli, N. (eds.) The governance of city food systems. Fondazione Giangiacomo Feltrinelli, pp. 68-97.

Wenger, E., 1999. Communities of practice: learning, meaning, and identity. Cambridge University Press, Cambridge, UK.

Westerink, J., Plomp, M., Ottburg, F., Zanen, M. and Schrijver, R., 2018. Boeren voor Natuur: de ultieme natuurinclusieve landbouw? Lessen van vier pilotbedrijven en relevantie voor beleid. No. 2858, Wageningen Environmental Research, Wageningen, the Netherlands. https://doi.org/10.18174/434934

Westerink, J. and Van Rooij, S.A.M., 2018. De bokashi-keten als natuurlijke verbinding tussen land en stad. Available at: https://tinyurl.com/msjfv7cs.

Wigboldus, S., Brouwers, J. and Snel, H., 2020. How a strategic scoping canvas can facilitate collaboration between partners in sustainability transitions. Sustainability 12: 168.

Winne, M., 2019. Food town, USA: seven unlikely cities that are changing the way we eat. Island Press, New York, NY, USA.

Wiskerke, J.S., 2020. Achieving sustainable urban agriculture (1st ed.). Burleigh Dodds Science Publishing, Cambridge, UK. https://doi.org/10.1201/9780429275869

Wiskerke, J.S.C., Bock, B.B., Stuiver, M. and Renting, H., 2003. Environmental co-operatives as a new mode of rural governance. NJAS – Wageningen Journal of Life Sciences 51: 9-25. https://doi.org/10.1016/S1573-5214(03)80024-6

Wittmer, I.K., Bader, H.P., Scheidegger, R., Singer, H., Lück, A., Hanke, I., Carlsson, C. and Stamm, C., 2010. Significance of urban and agricultural land use for biocide and pesticide dynamics in surface waters. Water Research 44: 2850-2862. https://doi.org/10.1016/j.watres.2010.01.030

WUR, 2018. Midden-Delfland werkt samen aan vernieuwing. Available at: https://weblog.wur.nl/uitgelicht/innovatie-in-midden-delfland/.

Yeoman, B., 2018. The hidden resilience of 'Food Desert' neighborhoods, Sapiens. Available at: https://tinyurl.com/msr4skbx.

Chapter 8
Regional resourcefulness for food systems: the case of phosphorus in the metropolitan region of Amsterdam

Wim van Dijk, Jan Eelco Jansma, Jan Broeze and Jim Groot

Abstract

This chapter focuses on the issue of regional resourcefulness for food systems. It considers urban areas to be vital linkages for food production systems that need to close the loop in order to prevent the depletion of natural resources. Urban areas generate large 'sources' of residual material and waste flows that contain nutrients and organic matter that currently only partly return to the food system. These flows thus offer perspectives on 'circular valorisation', either recycling or replacing other (partly fossil) resources. Urban areas therefore represent attractive starting points in a transition towards circularity and to connect rural and urban flows. In this chapter we look at the transition needed in the use of phosphorus (P), essential for life and an integral part of our food system. P is a finite resource, obtained by mining. Reserves are limited; estimates of global reserves vary between 100 and several hundred years. This chapter discusses two case studies and focuses on urban-rural flows of P. The first case explores the potential of a circular food system at the regional scale of Almere. Almere is a Dutch city, part of the Metropolitan Region of Amsterdam (MRA). It was selected because of its location adjacent to a large agricultural hinterland, the Flevo Polder. The central question of this case study is: what does a circular City Region Food System with regional food production look like from the perspective of P-flows? How much land is needed to feed the urban area and to what extent can the urban P-flow be recirculated to agriculture? The second case study explores an actual urban P-flow: what are the actual quantities of the P-flows and to what degree is the P-loop already closed? These questions are addressed in a study that encompassed the entire MRA region. The chapter concludes that a circular City Region Food System is technically feasible. However, a fundamental shift is required in how the food and waste systems are organised: from farm to fork and from disposal to dung. Hence, it ends with recommendations on how to enhance circular urban food production that does not deplete but preserves P.

Keywords: phosphorus, circularity, self-sufficiency, city region, food systems

10.3920/978-90-8686-935-0_8,

8.1 Introduction

A growing group of scholars see City Region Food Systems as hotspots in the shift from a linear to a circular economy (e.g. Blay-Palmer *et al.*, 2018; Morgan, 2009; and Wiskerke, 2015). The potential impact of this transition is larger than ever now that a majority of the world's population lives, consumes, wastes and excretes in urban areas. Here, large 'sources' of residual material and waste flows contain nutrients and organic matter that only partly return to the food chain. These flows offer perspectives on 'circular valorisation', either recycling or replacing other (partly fossil) resources. Urban areas therefore represent attractive starting points in a transition towards circularity. However, this transition can't be achieved without cooperation between the city and its rural counterpart, where food production and agriculture are often important sectors. For example, the Dutch Ministry of Agriculture launched its vision on the future of Dutch agri-food production (LNV, 2018), which placed circularity of the agricultural system high on the national policy agenda. The prevalent question is, however, how to connect both 'ends' of the food system, the urban areas and the food and agricultural sector. In other words, how can we connect the rural and urban flows within the City Region Food Systems leading to a circular food system?

In this chapter we specifically look at the transition needed to connect both ends of the food system from the perspective of phosphorus (P). P is essential for life and an integral part of our food system (Cordell *et al.*, 2011), yet it is a finite resource, obtained by mining. Reserves are limited and whilst estimates of global reserves vary between 100 and several hundred years, it is believed that most current producers will have depleted their stocks within a century (Cooper *et al.*, 2011). Extractability also plays an important role in (future) P availability; the more quickly the available resources to extract P become depleted, the more the costs and the amount of waste will rise as more challenging resources have to be exploited (Cordell and White, 2014). There is also a geo-political component; 70-90% of global reserves are confined to one single country: Morocco (Cooper *et al.*, 2011).

Today, around 90% of the mined P is used in our food system, mostly as fertiliser in crop production (Neset and Cordell, 2011). Only 20% of this ends up in consumed food, the remaining 80% is lost due to a range of inefficiencies: loss, spill, and leakage throughout the system, from mining to consumption (Cordell and White, 2014). In addition, almost all consumed P eventually leaves the human body as faeces and urine (Cordell and White, 2014) before finding its way into the sewage system. In the Netherlands, about 85% of the P that enters the sewage system ends up as sludge (CBS, 2018), which is subsequently incinerated; the P-content is not currently recycled and consequently this phosphorus is lost as a resource.

8.2 The challenge: a circular food production system with re-use of P

It is widely recognised that there will be no one-size-fits-all solution to mitigate the anticipated future scarcity of phosphorus. Cordell and White (2014), referring to earlier work by Cordell, suggest an integrated approach combining measures on both the demand and supply side across all stages of the food system. They also propose that optimal measures may vary from country to country. For the Netherlands, it was suggested that measures should focus on improving efficiency, recovery and re-use because, despite its lack of mineral resources, the dense human and livestock populations provide a surplus of P in a range of manure and waste sources. Subsequently, a growing number of projects have begun to explore how both the efficiency and rate of re-use and recovery can be improved. Examples of the latter include the re-use of composted organic waste and the production of P fertiliser products from wastewater treatment plants, e.g. struvite. However, the impact of these initiatives is marginal compared to the total loss of P from the overall food system. Recovery and re-use of P on a large scale is only feasible if the current linear food system shifts towards a circular system which reclaims nutrients from the waste sector (solid waste, sewage) to be reused in agriculture as well; ideally this can result in a circular system that recovers more phosphorus (Figure 8.1).

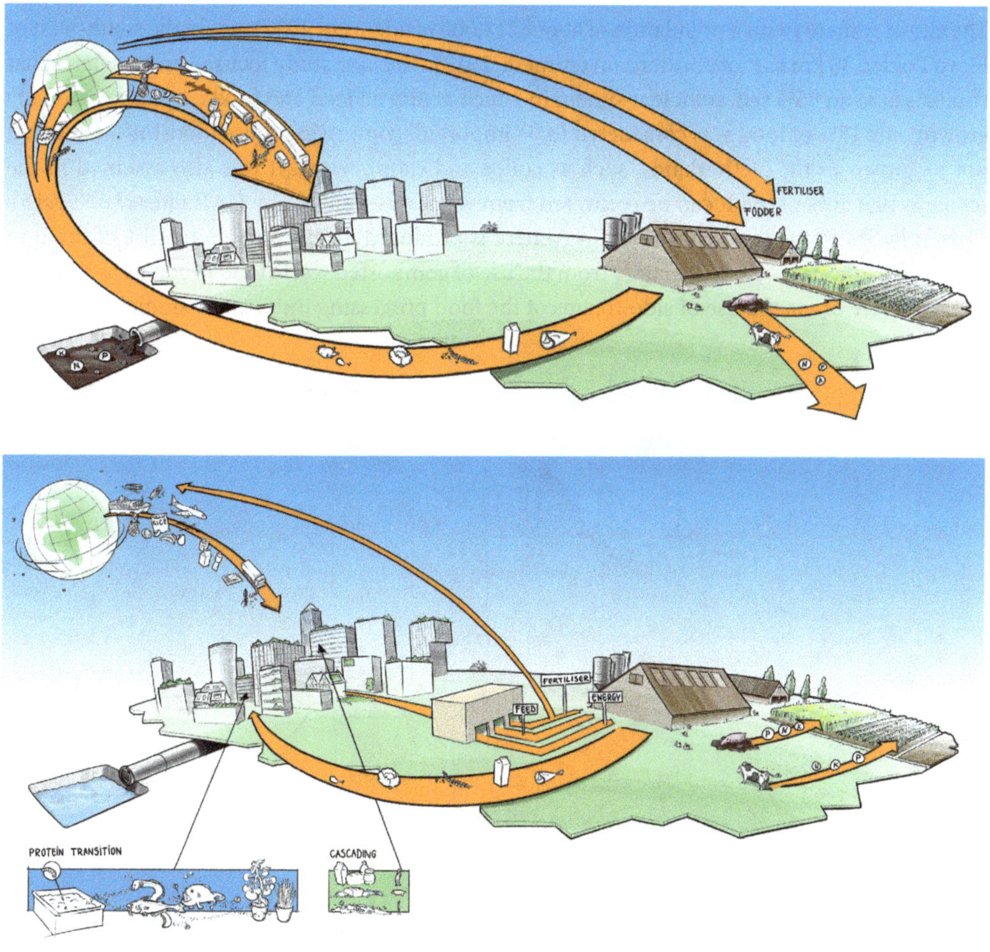

Figure 8.1. An outline of the current disconnection between agriculture and the city (above) and an outline of a future regionally circular food system (below). Picture by JAM Visual thinking.

This chapter discusses two examples of P-flows in the City Region Food System of the Metropolitan Region of Amsterdam. The first is the Almere case study that explores the potential of a circular food system in the urban area of Almere and its surroundings. Almere is a Dutch city, part of the Metropolitan Region of Amsterdam (MRA) (Figure 8.2) and was selected because of its location adjacent to a large agricultural hinterland, the Flevo Polder. The central question of this case study is: what does a City Region Food System look like from the perspective of P-flows? How much land is needed to feed an urban area and to what extent can the urban P-flow be recirculated to agriculture? The second case study explores the actual urban P-flow: what are the actual quantities of the P-flows and to what degree is the P-loop already closed? These questions are addressed in a case study that encompassed the entire MRA (Figure 8.2).

8.3 Almere case study

The city of Almere (with a population of about 218,000 residents in 2022), lies in the south-west of Flevo Polder, 30 km east of Amsterdam (Almere, 2013). The case study focused on a scenario that comprised of an 85% self-sufficient system in which nearly all feed and food crops are regionally grown; only 15% of food was considered to be imported, comprising those foodstuffs that could not be grown in the Netherlands, such as coffee and exotic fruits. It was also assumed in the scenario that 90% of the P can be recovered from waste as estimated by De Ruijter *et al.* (2015). Currently, 5% of the food consumed in Almere is produced locally and about 25% of the P in residual streams is recovered, mainly from the use of composted waste food products as fertiliser in agriculture or via the use of sidestreams of the food processing industry as animal feed.

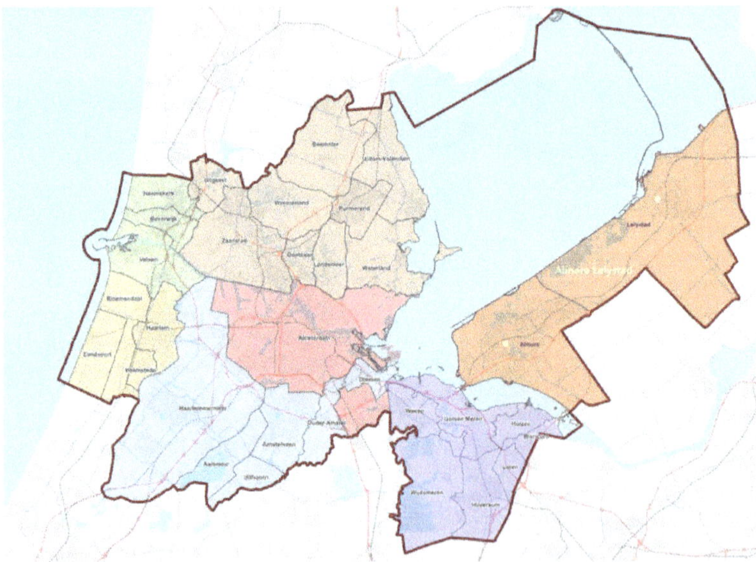

Figure 8.2. Visualisation of the metropolitan region of Amsterdam (source: www.metropoolregioamsterdam.nl). Almere is situated right middle.

8.3.1 Methods

The starting point for the study was food intake in Almere, from which point all subsequent P-flows were calculated (Figure 8.3). The food intake of different food product groups, differentiated according to gender and age group, was derived from a national food intake survey carried out in 2007-2010 (Van Rossum *et al.*, 2011). By combining this evidence with the population structure of the city (Almere, 2013), the city's total food intake could be calculated. Each food product group from the national food intake survey was connected to model food products (e.g. bread as a model food product for cereal intake). Subsequently, these model food products were linked to primary products produced by farms (e.g. bread is linked to wheat). Additionally, we assumed that 30% of wastage (food produced but not consumed) occurs throughout the food chain. This was based on figures that reveal a wastage of 18% in households (Van Westerhoven, 2013) and 13% in the processing industry (LEI factsheet, undated). Following this, we calculated the required amount of each food product and derived the area demand of the food and feed crops, based on crop yields and animal production data from farms in the Flevo Polder. The P-flows are quantified by multiplying the number of products by the standard P-concentrations in the products. To account for the inevitable leaching of soil P, values were set at the level that corresponds to the environmental P-goals for surface water (0.05-0.15 mg P/l). Further details with regard to the methodology are found in Van Dijk *et al.* (2017).

8.3.2 Results

P-flows in the food system

The P-flows in the Almere regional food system are shown in Figure 8.4. We distinguish three subsystems: agriculture, city (processing industry, retail, catering, and households) and waste. The products produced in the agriculture subsystem, including feed ingredients for concentrates, are exported to the city where they are processed and consumed. The return flow back to agriculture

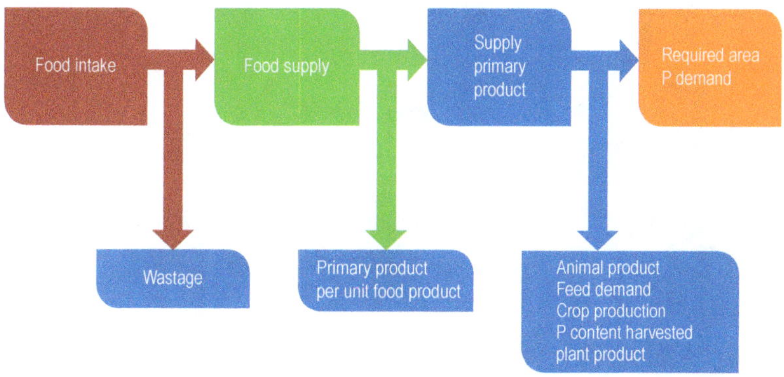

Figure 8.3. Schematic outline of the methodology of the calculation of the area demand for food production and quantification of the P-flows.

mainly comprises processed feeds (concentrates). Additionally, there's a small flow of P in dead animals to waste and, in contrast, a loss due to leaching. Manure produced by the animals is an internal flow within the agriculture subsystem. Almost all the P in primary products that is processed in food products and consumed, ends up in the waste sector (waste food, human excretion, and waste from abattoirs). There is a small export of P in non-consumed by-products like whey and organs and a small import of P through exotic products.

If a high P recycling rate from the waste sector can be assumed, this self-sufficient scenario offers a nearly-closed P-circle (Figure 8.4). The total inflow (280 tons) into the waste sector consists of 55% solid waste and 45% waste water. Recycled P from the waste sector (261 ton P) contributes about 30% of the total P agricultural demand. The remaining 70% is provided by manure produced by the livestock in the system (65%, internal flow) and imported manure/fertiliser (5%). A relatively small import of P in manure and fertilisers (35 tons) is still necessary to balance the loss of P in this scenario. A loss of P still remains because: (1) the recovery of P from waste does not exceed a maximum of 90%, (2) the export of P out of the region exceeds the import, and (3) because of leaching losses.

Area demand for food production

Key to the Almere case study is a food system that seeks to retain as much P as possible circulating within the region including, fundamentally, its supporting agricultural (processing) industry. The case study estimated that based on the productivity of the clay soils of its hinterland, 27,000 ha of crops would be needed to feed the 195,191 inhabitants of Almere. Nearly 90% of this area is required to feed livestock, of which 52% is accounted for by dairy produce. The 27,000 ha of land corresponds

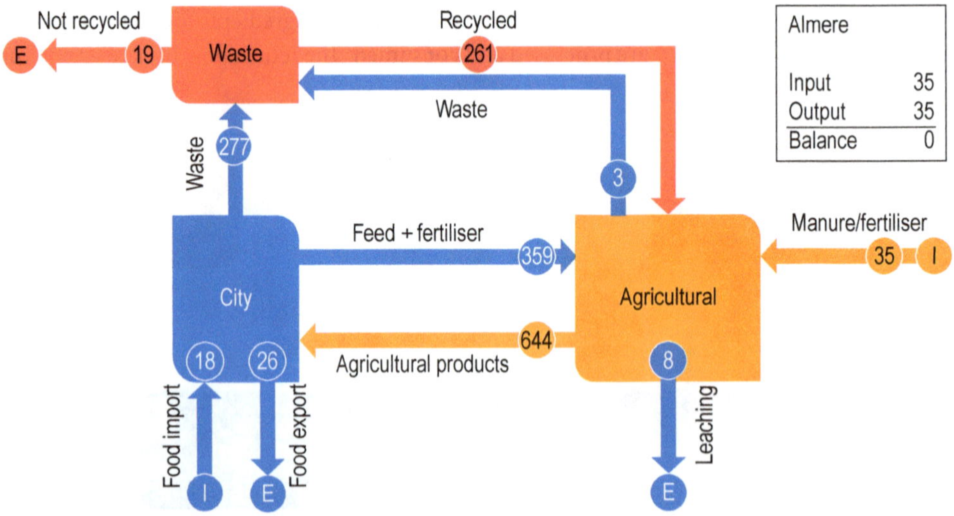

Figure 8.4. P-flows (×1000 kg P) of the regional food system of Almere in the self-sufficient scenario with maximal recycling of waste P (via GFT, digestate, fertilisers, food or feed biomass). E: export and I: import of P.

to a 'food footprint' of 1,400 m^2 per inhabitant per year. Figure 8.5 depicts the distribution and extent of crops within a self-sufficient food system for Almere. In fact, to feed itself, the city must utilise agricultural land from the adjacent municipalities of Zeewolde, Lelystad and Dronten.

8.4 Metropolitan Region of Amsterdam case study

The Metropolitan Region of Amsterdam (MRA) is an informal partnership between Amsterdam municipality, the provinces of Noord-Holland and Flevoland, and 32 municipalities (including Almere), which are all located within a 30 to 40 km radius around the city of Amsterdam (Figure 8.2). The partner organisations share a vision that embraces economics, transport, living spaces, employment and recreation[2].

The MRA is densely populated (2.48 million inhabitants across 2,580 km^2 in 2022) with a high concentration of food-related producers/industry (business to business and business to consumers), with significant input and output streams (including wastes/losses).

[2] http://www.metropoolregioamsterdam.nl/over-mra/deelnemers.

Figure 8.5. Overview of the required 27,000 ha of crops to supply 85% of the food needs of the 195,191 inhabitants of Almere. The crops are positioned in Almere's hinterland, the Flevo Polder, which is an area with highly productive soils.

In order to estimate the P contents it is necessary to identify, specify and quantify relevant streams. Here, we have focussed on those biogenic wastes defined in the EU waste directives connected to the food system and produced during food processing and consumption[2]. Table 8.1 shows that food waste is found within 'Vegetal waste', 'Animal and mixed food waste', 'Sludges and liquid wastes from waste treatment' and 'Household and similar wastes'.

Only the household waste data for the MRA are derived from local evidence; other volumes are derived from national data sources. The industrial volume was based on the number of companies per industry segment for MRA compared with the number on a national level. For the MRA, the number of companies in the region 'COROP Groot Amsterdam' was used (a geographic area that largely overlaps with MRA). Since no specific data on sizes of the companies in the MRA and at the national level are available, it was assumed that the average waste streams per company are equal to national averages. The volumes for both sludge waste streams are estimated by converting the national volumes in MRA volumes via the share of the number of inhabitants in the MRA to the number of inhabitants in the Netherlands. The national volumes are derived from CBS (2018).

Vegetal wastes (food industry) and mixed food wastes (from households and catering that is collected separately) are largely utilised as animal feed or for the production of compost for agriculture. Therefore, it is assumed that the P in these residual streams is reused in the food system. Any food waste not collected separately, ends up in the category 'Household and similar waste' and is usually incinerated. In particular, the municipality of Amsterdam does not separate food waste and non-biogenic waste at the household level.

Given these circumstances, improving the removal of P from sludges is likely to represent the best opportunity to improve the overall effectiveness of the system. Arising primarily from the wastewater of private households and industrial facilities, this typically comprises the waste products of food preparation and processing and, to a lesser extent, pulp and paper (Eurostat, 2010). The sludge is generated through wastewater treatment in both municipal and industrial wastewater plants but can also comprise the product of anaerobic digestion of animal and vegetable

Table 8.1. Estimated biogenic waste volumes (in 1000 kg) produced during food processing and consumption in Metropolitan Region of Amsterdam (MRA) in 2012.

Quantity of waste per type in MRA	(Food) industry	Waste (water) treatment	Catering industry	Households	Total
Vegetal wastes	253,562	2,667	163,303	36,233	455,765
Animal and mixed food waste	33,855	631	10,631	102,235	147,352
Sludges[1]	8,609	127,884	12,100	6	148,600
Household and similar wastes[2]	21,460	4,402	327,795	331,827	685,485

[1] Sum of categories 'common sludges' and 'sludges and liquid wastes from waste treatment'.

[2] Mix of biogenic and non-biogenic waste (due to non-separated collection).

waste from waste treatment plants. In the Netherlands, these sludges are mainly incinerated in combination with household waste; when sludges are incinerated separately (mono-incineration) the phosphorus concentration of the ash will be sufficiently high for efficient P recovery.

In short, sludges and currently non-recycled food wastes represent potential sources for increasing the rate of P recycling.

As explained above, a number of assumptions have been made to allow data from a number of sources to be utilised. National or provincial data were used to estimate the P-flow in the sludges, vegetal, animal and mixed food wastes, whilst local data was adapted to derive appropriate values for residual streams.

The amount of P in sewage sludge from communal wastewater treatment in MRA is based on the total amount of P in sewage sludge in the provinces Flevoland and Noord-Holland related to national P production data in 2012 (CBS, 2018). As the number of inhabitants in MRA is 72.5% of the total number of inhabitants in Flevoland and Noord-Holland, the total amount of P-flow in communal sewage sludges is estimated at 1,309 tons. Across the Netherlands, the total amount of P in residues discharged to industrial sewage systems sludge equates to 865 tons. Assuming that 79% of this P ends up in the sludge after wastewater treatment (the average value for communal wastewater plants in Flevoland and Noord-Holland), this results in a P content of 683 tons nationally. For the MRA alone, this represents around 92 tons. The total amount of P in communal as well as industrial sludges in MRA is estimated at about 1,400 tons.

The vegetal waste from the food industry, excluding catering, applies to sidestreams that occur during food processing. Drawing on the Almere case study (above), an average P concentration of around 0.1% would provide around 230 tons of P from an overall total of 233,000 tons of waste.

The P-content of mixed food waste from consumption (catering and households) is estimated differently. Based on FAO's food balance sheets, food waste figures and phosphorus content of food products, Conijn and Van Dijk (2018) estimated food intake and food waste for consumers (including P-content) at the national level. They found that the amount of P in food waste (consumption level, at home and out of home) was estimated at 187 gr/cap/a. Multiplied by the amount of inhabitants in the MRA in 2012 (2.35 mln (CBS)), this gives about 440 tons of P in mixed food waste (or the 'Animal and mixed food wastes', 'Vegetal wastes' (catering industry) and 'Household and similar wastes'). The latter refers to mixed food waste not collected separately, e.g. the municipality of Amsterdam. We estimated that for the MRA alone, 55% is collected separately.

Animal waste refers mainly to slaughter residues and we assumed that this was produced almost entirely by the food industry and would, therefore, more or less equal the figure of 34,000 ton in Table 8.1. Again, based on the Almere case study, the average P concentration in this material is estimated to be about 0.4% resulting in a P-content of about 140 tons.

Table 8.2 gives an overview of the estimated P-flows in the MRA; currently, only about 20% of the total P (470+1,740=2,210 tons) in residual streams of the food system is re-used. The majority of sludges are incinerated so allowing P to leave the food system.

The relationship between the availability of P in residual streams within the MRA (Table 8.2) and the amounts allowed to be applied on agricultural land in the MRA is shown in Table 8.3. Based on land-use data (area grassland and arable land, www.cbs.nl) and maximum allowed P fertilisation levels (www.rvo.nl), it can be calculated that about 2,840 tons P is allowed to be applied on agricultural land across the MRA. Figures from the Environmental Data Compendium (www.clo.nl) indicate that in the western clay region and the Flevo Polder around 2,130 tons (or about 75% of the allowed P application quota) is accounted for by the application of animal manure. Of this, 1,060 tons P is produced within the MRA region (www.ois.nl) and 1,070 tons P is imported from elsewhere; the shortage of manure in the MRA is explained by its relatively modest animal husbandry sector compared with its neighbours. The reason that the P fertilisation quota is not completely covered by manure application could be because on clay soils it is not always possible to apply manure to all crops due to the risks of harming the soil structure in spring.

The remaining quota not used for manure is about 710 tons P. In order to compensate the P crop offtake through harvested products and to prevent soil P depletion, most farmers will fully utilise their allowed P-quota. This is commonly achieved with the use of compost and/or mineral fertiliser additional to animal manure. There is no reason why, however, the latter could not be substituted with P recycling products from the waste sector, particularly the sludges, e.g. struvite or recycled P from incineration ashes (see discussion for possibilities).

Table 8.2. Estimated P-content of waste streams occurring at food processing and consumption in the Metropolitan Region of Amsterdam.

	Recycled P (1000 kg)	Non-recycled P (1000 kg) – incinerated-
Vegetal wastes	230	
Mixed food waste	240	200
Animal waste		140
Sludges (total industrial and households)		1,400
Total urban area	470	1,740

Table 8.3. Estimated amount of phosphorus used in agriculture in Metropolitan Region of Amsterdam (MRA).

	1000 kg P
Allowed application quota	2,840
Manure produced in MRA	1,060
Manure produced elsewhere	1,070
Remaining P application quota	710

In order to close the P-loop at the MRA level the imported manure P from outside the region (1,070 tons P) could be substituted by recycled P products from within the region. This would mean that including the remaining quota of 710 tons, there's room for about 1,800 tons recycled P, while the availability from residual streams is about 2,200 tons (Table 8.3). The excess amount of about 400 tons, basically stemming from MRA food imports, would be available for use outside the region and contribute to circularity at the national level. However, the gain of regional P re-use leads to a decreased use of manure in the MRA and thus an increase in the national manure surplus. Therefore, recycled P can inherently also be considered as an export of P recycling products from MRA or the Netherlands. This will also contribute to circularity but on a larger-scale level.

8.5 Analysis and discussion

This chapter compared the potential re-use of P in a circular food system in the Almere region with the actual rates of P re-use across the MRA. The Almere case study shows that the maximal production of food in the region combined with maximal P recycling from waste streams results in an almost circular system. In such a system, a high rate of P recovery and recycling from food chain residual streams is essential. However, the actual situation (in the MRA) shows that, currently, only 20% of the organic residual streams of the food chain is re-used. This is provided mainly by the re-use of vegetal sidestreams of the food industry and organic household food scraps. Therefore, significant steps have to be made in order to increase the rate of P recovery to secure a truly circular food system.

Major P-flows not currently returned in the food system are food waste that is not collected separately from other waste streams (industry and households), waste water (sludge) and slaughter residues. Adequate waste collection logistics are necessary to support a separate collection of food waste. In Almere, nearly 100% of household food waste is collected separately, so allowing for composting to take place and a re-usage in the food system. Nationally, however, only about 55% of food waste is collected separately (www.milieucentraal.nl). Furthermore, major P-losses occur in wastewater treatment and in the meat industry (from slaughter residues). In order to increase P recycling significantly, it will be necessary to devise measures which will allow P to be recycled from both residual streams.

In terms of wastewater, most of the P, about 85%, ends up in sewage sludges. Re-use of P in the food system is possible by direct application in agriculture or through chemical recovery. In the Netherlands the use of sewage sludge in agriculture is not allowed due to the amount of contamination, especially by copper and zinc. Furthermore, there are health risks associated with contaminants, microorganisms and micropollutants (including remains of medicines). In addition, due to the relatively high P concentration in sewage sludge, the nitrogen, potassium and organic matter supply per unit P supply is relatively low when compared with commonly used animal manures in agriculture (Regelink *et al.*, 2017). This imbalance between P and other components ensures that sewage sludge is a less attractive fertiliser in agriculture than manure. Because of this, wastewater treatment plants currently focus on the chemical recovery of P from sewage sludge (e.g. by the production of struvite) or acid extraction from incineration ashes. Struvite production is achieved by adding magnesium, resulting in the formation of magnesium-ammonium-phosphate salt that can be used as a fertiliser in agriculture. Depending on the applied technology, 5-50% of the P entering the sewage system can be recovered by this method (Remy and Jossa, 2015; STOWA, 2011, 2013). A consequence of this process is that the subsequent recovery of P from incineration ashes of the remaining sludge is correspondingly low. Moreover, the addition of magnesium in the production of struvite may lead to a surplus of this element in agriculture especially when used in combination with manure.

The recovery of P from incinerated sludge (ashes) is possible by acid extraction, allowing its use in mineral P fertiliser production. With this method, it is estimated that 70-95% of P can be recycled from the sludge. Currently, a covenant has been agreed between two Dutch sludge processors, HVC and SNB, and the fertiliser company Ecophos that allows the use of ashes for P fertiliser production. If this covenant is realised, it is likely that 50% of the P in sewage sludges in the Netherlands could be recycled. Although P recovery from incinerated sludge ashes may be high, the incineration process leads to the loss of nitrogen and carbon. It must also be emphasised that P recovery from incineration ashes cannot be well combined with struvite production as the latter leads to a decreased P concentration in the ashes, making the P extraction less efficient.

The above-mentioned measures to recover P from the sewage system are based on the current sanitation and water systems; an improved P recovery may also require the adoption of new sanitation and water systems (e.g. source separation of urine, faeces and water within sewage system and decrease of water use) in order to decrease the costs and environmental impact of P recovery (e.g. energy demand). Adjusted sanitation could also contribute to the recovery of other nutrients as nitrogen and potassium.

A significant amount of P is also currently lost through slaughter residues, parts of which are incinerated for safety reasons. As with sewage sludge, this P could also potentially be recovered from the incineration ashes. However, for this purpose, mono-incineration (not used as co-firing fuel) is required as a sufficiently high P concentration in the ashes is necessary for efficient P recovery. Currently, its use as a co-firing fuel in power plants is common, resulting in ashes with too low P concentration for an efficient P extraction. Other components of slaughter residues are processed into pet food and this P eventually ends up in the excrement of our domestic animals. This partly ends up in the residual waste stream or in the environment, making efficient recovery of P difficult. On the other hand, the demand for pet food and its supply from animal (and plant) residues represents a modest re-use, although in the end the nutrients are not recovered.

Apart from improved technology for nutrient recovery from sidestreams, circularity will also require changes in legislation that guarantee the quality and the safety of the recycled products. The latter applies to contaminants like heavy metals, micro-pollutants and micro-organisms. Currently, EU legislation is being developed with regard to recycled fertiliser products from waste streams. It is expected that a number of products of chemical recycling such as struvite and extracted P from incinerated sewage sludge will be allowed for use as fertiliser (Huygens, 2019).

In this chapter we have focused on the re-use of P from urban waste streams by agriculture. To mitigate the future scarcity of P, Cordell and White (2014) suggest an integrated approach across the food system including the re-use from sidestreams; however, we have not addressed the impact of the human diet. The Almere case study shows that when 85% of the food is produced regionally, the 'food footprint' of the current diet, when taking the high productivity of the hinterland into account, is 1,400 m^2 per inhabitant. For the 17 million residents of the Netherlands, Conijn and Van Dijk (2018) estimated that an area of about 2.4 million ha would be needed. However, the current Dutch agricultural area only extends to about 1.9 million ha which means, with the current diet and cropping systems, there is insufficient land to feed the existing population. For example, for Almere the present diet requires about 90% of the needed 27,000 ha agriculture land to be used for animal feed production. Yet Terluin *et al.* (2013) and Van Kernebeek *et al.* (2015) calculated the food footprint would fall to 400-1000 m^2 per inhabitant if the diet was changed to incorporate less red meat and cereal products, replacing them with more chicken meat and potatoes. Even at the upper limit of this food footprint range, self-sufficiency for the whole of the Netherlands would be possible. A reduction in cropping area needed for food production would obviously mean less P to fertilise the crops; and, in the case of Almere, not only would less land (about 7,800 ha) be needed to feed the local population but a decrease of the food footprint from 1,400 to just 1000 m^2 per inhabitant would save 200-250 tons P annually. However, in a more or less circular system most of the P necessary for crop production will originate, obviously, from internal sources (manure, residual streams from industry, retail and households). The external P-input with fertiliser is relatively small (simply representing compensation for losses), around 35 ton P for Almere (Figure 8.4). Therefore, the decrease in the food footprint will also decrease the input of fertilisers and manure from outside the region by about 10 ton P.

City Region Food Systems may have positive side effects including a reduction in long-distance food transport and its related nuisances such as air pollution, traffic, etc., but it will also affect the regional infrastructure, and the organisation and economics of agriculture. For example, farmers in the Almere region may, in the case of self-sufficiency, have to substitute their current cash crops (potatoes, carrots and onions) with low-cash animal feed crops, which are normally imported from elsewhere.

Finally, a maximal recycling of P does not necessarily require a regional food system, as long as the recycled P products are used elsewhere in a sustainable way. Research is needed to explore at what level (local, regional, national or global) the production and consumption of food can best be optimised in order to reduce the dependency on mined P. The key is to recover and re-use as much P as possible from all the urban food waste streams, in addition to an increased P efficiency throughout the food system, from mining to consumption.

8.6 Conclusions and recommendations

This study started with the quest to connect rural and urban flows, to establish a circular City Region Food System, and looked at the MRA region. The Almere case demonstrates that connecting the urban and rural flows of the essential and finite mineral P is technically feasible. Fundamental to this feasibility is a regionally more self-sufficient food system merged with a maximum recovery and re-use of P from the waste streams. However, the MRA case study also shows that the gap between this more self-sufficient food system and the reality of food production at present is large.

To bridge this gap, a fundamental shift is required in how the food and waste flows are organised at present and can be organised within future City Region Food Systems: regional resourcefulness is necessary, from farm to fork and from disposal to dung. The shift has different yet critical angles. It is about a reduction in the consumption of animal proteins which will reduce the acreage of feed crops and thus the amount of P fertilisers needed. The shift also involves a reduction in the leakage of P from the food chain, and should be combined with the maximum re-use of P from food waste streams. In order to achieve this, significant transitions need to be carried out across the waste sector. Improvement is possible via more efficient P recovery in the current waste treatment systems. Moreover, the adaptation of wastewater collection and treatment may also be necessary in order to recover other major nutrients such as nitrogen and carbon. The shift also implies that agriculture will have to be able (legally, economically and healthily) to use (organic) fertilisers from the diverse waste streams, including from sewage sludge. The level (local, regional, national or global) at which this shift in the food and waste systems should be facilitated remains an important issue to be addressed.

References

Almere, 2013. Sociale atlas 2013 Almere. Municipality of Almere, the Netherlands.

Blay-Palmer A., Santini, G., Dubbeling, M., Renting, H., Taguchi, M. and Giordano, T., 2018. Validating the city region food system approach: enacting inclusive, transformational city region food systems. Sustainability 10: 1680. https://doi.org/10.3390/su10051680

CBS, 2018. Milieu-zuivering stedelijk afvalwater. Available at: http://statline.cbs.nl/Statweb/.

Conijn, J.G. and Van Dijk, W. 2018. Food related land requirement and phosphorus balance of the Netherlands; Wageningen Research, Report WPR-757. 18 pp.

Cooper, J., Lombardi, R., Boardman, D. and Carliell-Marquet, C., 2011. The future distribution and production of global phosphate rock reserves. Resources, Conservation and Recycling 57 (2011) 78-86.

Cordell, D., Rosemarin, A., Schröder, J.J. and Smit, A.L., 2011. Towards global phosphorus security: a systems framework for phosphorus recovery and reuse options. Chemosphere 84 (2011), 747-758.

Cordell, D. and White, S., 2014. Life's bottleneck: sustaining the world's phosphorus for a food secure future. Annual Review of Environment and Resources 39:161-188.

De Ruijter, F.J., Van Dijk, W., Van Middelkoop, J.C. and Van Reuler, H., 2015. Phosphorus recycling from the waste sector. Wageningen University, Plant Research International report 641, Wageningen, the Netherlands.

Eurostat, 2010. Guidance on classification of waste according to EWC-Stat categories Supplement to the Manual for the Implementation of the Regulation (EC) No 2150/2002 on Waste Statistics. Available at: https://tinyurl.com/39bmw4mf.

Huygens D., Saveyn, H.G.M., Tonini, D., Eder, P. and Delgado Sancho, L., 2019. Technical proposals for selected new fertilising materials under the Fertilising Products Regulation (Regulation (EU) 2019/1009) – process and quality criteria, and assessment of environmental and market impacts for precipitated phosphate salts & derivates, thermal oxidation materials & derivates and pyrolysis & gasification materials, EUR 29841 EN, Publications Office of the European Union, Luxembourg. Available at: https://doi.org/10.2760/186684.

LNV, 2018. Landbouw, natuur en voedsel: waardevol en verbonden: Nederland als koploper in kringlooplandbouw. Ministerie van Landbouw, Natuur en Voedselkwaliteit, Den Haag, Nederland.

Morgan, K., 2009. Feeding the city: the challenge of urban food planning. International Planning Studies 14: 341-348.

Neset, T.S.S. and Cordell, D., 2011. Global phosphorus scarcity: identifying synergies for a sustainable future. Journal of the Science of Food and Agriculture 92: 2-6. https://doi.org/10.1002/jsfa.4650

Regelink, I.C., Ehlert, P.A.I. and Römkens, P.F.A.M., 2017. Perspectieven voor de afzet van (fosfaatveramd) zuiveringsslib naar de landbouw. Wageningen, Wageningen Environmental Research, Rapport 2819; 76 pp.

Remy, C. and Jossa, P., 2015. Life cycle assessment of selected processes for P recovery from sewage sludge, sludge liquor, or ash. P-Rex deliverable 9.2, Kompetenzzentrum Wasser Berlin, Germany.

Stichting Toegepast Onderzoek Afvalwater (STOWA), 2011. Fosfaatterugwinning in communale afvalwaterzuiveringsinstallaties. Stichting Toegepast Onderzoek Afvalwater, rapport nr. 2011-24. Available at: https://tinyurl.com/2p92fj2b.

Stichting Toegepast Onderzoek Afvalwater (STOWA), 2013. Fosforhoudende producten uit de communale afvalwaterketen. Stichting Toegepast Onderzoek Afvalwater, rapport nr. 2013-32. Available at: https://tinyurl.com/36cmrt5k.

Terluin, I.J., Kamphuis, B.M., Oudendag, D.A. and Van Leeuwen, M.G.A., 2013. Voedselvoorziening in Nederland onder buitengewone crisisomstandigheden. LEI-rapport 2013-012. LEI, Den Haag, Nederland.

Van Dijk, W., Jansma, J.E., Sukkel, W. Van Reuler, H., Vermeulen T. and Visser, A.J., 2017. Closing the life cycle of phosphorus in an urban food system: the case Almere (NL). WPR-report 725. Wageningen University and Research, Wageningen, the Netherlands, 44 pp.

Van Kernebeek H.R.J., Oosting, S.J., Van Ittersum, M.K., Bikker P. and De Boer, I.J.M., 2015. Saving land to feed a growing population: consequences for consumption of crop and livestock products. The International Journal of Life Cycle Assessment 21: 677-687. https://doi.org/10.1007/s11367-015-0923-6

Van Rossum, C.T.M., Fransen, H.P., Verkaik-Kloosterman, J., Buurma-Rethans, E.J.M. and Ocké, M.C., 2011. Dutch National Food Consumption Survey 2007-2010. Diet of children and adults aged 7 to 69 years. RIVM report number 350050006/2011. RIVM, Bilthoven, the Netherlands.

Van Westerhoven, M., 2013. Bepaling voedselverliezen in huishoudelijk afval in Nederland. Vervolgmeting 2013. CREM rapport nr. H13. CREM, Amsterdam, the Netherlands.

Wiskerke, J.S.C., 2015. Urban food systems. In: De Zeeuw, H. and Drechsel, P. (eds.) Cities and agriculture: developing resilient urban food systems. Routledge, London, UK, pp. 1-25.

Chapter 9
Restorative economics – food hubs as catalysts of a new urban economy

Sabine O'Hara and Marian Stuiver

Abstract

The common conception of the economy is a throughput model where inputs are converted into useful goods and services. Emissions and waste are the unfortunate by-products of this conversion process. Inputs and emission and waste by-products seemingly come from nowhere and disappear into nowhere. In other words, the common conception is that the economy has no context. As a result, the unintended side effects of economic activity disappear from the purview of economics. Economists refer to these unaccounted side effects of economic activity as 'externalities'. The deficiencies of this contextless model have long been recognised by ecological and social economists. This chapter outlines a new economic model developed by the lead author called restorative economics. Like the doughnut model developed by Kate Raworth (2012), it places the economy in its social and environmental context and makes long neglected social and environmental impacts of economic activity visible. The restorative economics model, however, makes a further distinction regarding the context of economic activity that makes the model more operational. It distinguishes between sources and sinks. Sinks are the counterpoints to the resources that feed the front end of the economic process. Sinks deal with the tail end of the process. The restorative economics model therefore places value not only on the resources used to produce goods and services and on these outputs of goods and services themselves, but also on the many social and environmental sink factors that absorb emissions, process waste, reduce stress, and alleviate social tension. Restorative economics pays attention to both the resources and the sinks needed to sustain economic activity in the long run. We illustrate the value of these sinks by reviewing several local and regional initiatives from Washington DC, namely the Urban Food Hubs model, the Five Pillars of Economic Development, the Food Connects vision and the ambitious agenda of the Sustainable DC plan. We conclude the chapter by reviewing policies that support a restorative economy.

Keywords: restorative economics, emissions, immissions, social and environmental side effects, sink functions, externalities

9.1 Introduction

Our economy is vulnerable. This is true at every level, from global to national, regional, and local scales. The global COVID pandemic heightened our awareness of these vulnerabilities especially related to our food system. Stay-at-home orders and travel restrictions made it impossible for

10.3920/978-90-8686-935-0_9, © Wageningen Academic Publishers 2022

seasonal farm workers to get to fields that needed harvesting. Workers in meat-packing plants who stood shoulder to shoulder in cold and wet conditions fell ill at alarming rates. Restaurants, and the tourism sector came to a virtual standstill due to strict coronavirus measures. Lay-offs across the hospitality industry placed enormous stress on families and communities. Despite the lower CO_2 emissions associated with reduced traffic during the height of the pandemic, world temperatures continued to rise, and hurricanes, tornadoes, wildfires, and floods continued to hit communities around the world. This indicates that long-term climate change may be unfolding faster than anticipated despite lower emissions in the short term (IPCC, 2021).

While these and other vulnerabilities of the throughput model of the economy were always present, they remained largely invisible on a planet of 3 billion people which we had reached in 1960. Yet by 2000 we had a world population of 6 billion, and only 20 years later in 2020 7.8 billion. Vulnerabilities become more acute on a full planet as compared to an emptier one. The vulnerabilities of supply chain disruptions, climate change, and other shock events are further heightened in population-dense urban communities. Cities are home to the majority of our global population and our global workforce. The density of cities puts pressure on resources to produce the large number of consumer goods and services needed, and it also creates enormous emission and waste streams. Moreover, cities bring into focus social disparities between rich and poor, old and young, long-time residents and newcomers, differences in race, ethnicity, and language. Yet while social disparities may be ever present in cities and metropolitan areas, nature has all but disappeared from view. Since more than 60% of the world's population now lives in cities, the majority of the world's population has become increasingly disconnected from nature.

Economists may well have ushered in this growing disconnect with the contextless model of the throughput economy. This model has *sources* or resources that feed the production process of goods and services for consumer markets. Historically, economists distinguished between three categories of resources: labour, land, and capital. Labour stands for the human inputs into the economy in the form of muscle power and mental ingenuity; land stands for all non-human-made or natural resources; and capital stands for all human-made resources like tools, materials etc. Yet already in the 1950s 'land' disappeared. In his famous production function, the Nobel Prize winning economist Robert Solow uses only two input factors, labour and capital, plus a rate of technology to produce an economy's total output of goods or services. The model therefore assumes that technology makes it possible to replace non-human-made natural resource with human-made ones (Solow, 1956). In other words, labour and capital can fully substitute nature. This assumption is especially questionable for the agricultural sector where arable land and the power of photosynthesis continue to play an indispensable role despite the advances of controlled environment agriculture.

Given the firm belief in the power of substitution, it is not surprising that another feature of nature never made it into the economic model in the first place, namely 'sinks'. Just as resources, or 'sources' for short, are indispensable for the continued production of useful goods and services, so too *sinks* are indispensable for the absorption and processing of the emission and waste by-products of economic activity. Oceans, for example, are sinks that absorb the carbon emissions that are the result of transporting perishable food items across long distances and storing them in cold chains that require energy; plants are sinks that can absorb nitrogen and prevent the nitrate contamination of aquifers and the eutrophication of rivers and lakes; permeable surfaces are sinks that absorb stormwater run-off from roads and roofs; sunflowers are sinks that absorb arsenic, and Indian mustard and Indian grass can remove contaminants from soils and water. These qualities, called phytoremediation, are examples of sink capacities provided by the diverse ecosystems of our planet. Losing these sink capacities is as costly to food production, for example, as losing valuable resources like phosphorous, nitrogen, and potassium fertilisers or water.

And the costs of lost sink capacities extend not only to environmental sinks but to human sinks as well. Our bodies are sinks that absorb vitamins and nutrients, but also saturated fats, sugar and chemicals that can cause diabetes, heart disease and cancer; and our human communities are the ecosystems that support the capacities of our bodies to reduce illness and stress. The increased mental health and suicide rates during the recent COVID lock downs are just one example of the negative effects of isolation and lost social connections.

This chapter introduces an economic model called restorative economics which was developed by the lead author (O'Hara, 2001, 2016, 2021). The restorative economic model addresses the deficits of the existing mainstream economic model and identifies four principles for restoring the degraded social and environmental context systems that underpin any thriving economy. The concept of restorative economics therefore addresses the deficits of the standard economic model that is typically used to determine urban food policies and adds several critical dimensions. They are: (1) redefining efficiency as the efficient use of a range of assets beyond labour and capital; (2) reducing emissions and waste through improved processes and closed loops; (3) strengthening sinks through improved ecosystem services and social support networks; (4) redefining production as restorative by meeting principles 1, 2 and 3.

By defining the production of food plants, for example, as not only dependent on (1) inputs like nitrogen, phosphorus, potassium, water and labour, but also on (2) emissions like nitrate leachate, CO_2 emissions, food contaminants, and food waste, and (3) sink services like heat absorbing green cover, permeable surfaces, physical and mental health conditions associated with food access, exercise, and vibrant social connections, the model offers more than a conceptual framework like the one offered by the doughnut economy model, for example (Raworth, 2012). It takes a model that is familiar to policy analysts and economists as its starting points and adds long-neglected social and environmental context systems that underpin all economic activity. Like the doughnut model, the restorative economics model points out that an economy which constantly overshoots its social and environmental boundaries will destroy the very systems that support it. Our restorative economic model offers actionable intervention points at three interconnected components that redefine economic activity as restorative based on the four principles outlined above.

To illustrate our model, we review several local and regional initiatives from Washington DC, namely the Urban Food Hubs of the University of DC (O'Hara, 2017), the Five Pillars of Economic Development project (O'Hara, 2018), and the Food Connects vision (Stuiver and O'Hara, 2021).

We review these initiatives through the lens of the four principles of the restorative economics model (see above). Lastly, the chapter reviews policies that can shift the underlying valuation framework of the current mainstream economic model towards a framework that reflects our restorative economic model and incorporates the true value of economic activity.

9.2 Revisiting externalities

The negative social and environmental externalities of economic activity are well documented as ecological, and institutional economists have long warned that economic markets are ill suited to reflect the value of ecological and social context systems (Costanza *et al.*, 1997; Daly, 1996; Ostrom, 1990; O'Hara, 1997, 2001). As a result, these context systems are constantly overused and undervalued. The United Nations report on Environment and Development was one of the first reports to call for a more sustainable approach to economic development that pursues a triple bottom line of profits, people, and planet (UNCED, 1987). The implication is that economic activity must balance economic, social and environmental goals and account for the negative social and environmental impacts of economic activity.

Well before the UNED report, which became known as the Brundtland report, the report of the Club of Rome (Meadows *et al.*, 1972) had called attention to the declining resources of our planet, especially non-renewables like fossil fuels. They cautioned that the rate at which non-renewable resources were used put future economic success at jeopardy, and the rate at which renewable resources were harvested was also reason for concern. Groundwater and timber, for example, are in principle renewable. However, the rate at which groundwater is pumped to the surface and used for irrigation far exceeds the rate at which it can be replenished through precipitation that percolates through the soil and into the aquifer. Likewise, the rate at which forests are cut outpaces the rate at which they grow back. This makes even renewable resources like timber and groundwater non-renewable.

Only recently the focus has also been on sinks in addition to resources. The United Nations Sustainable Development Goals (SDGs) illustrate this expanded focus (UNSDG, 2015). The seventeen (17) SDGs focus on more than the sources that feed the economy, by also addressing the critical environmental and social implications associated with the tail end of the economic process, such as emissions, waste, and growing disparities. For example, goals 1, 2, and 3, no poverty, zero hunger, and good health and well-being; goal 6, clean water and sanitation; goal 10, reduced inequality; goal 11, sustainable cities and communities; goal 12, responsible consumption; goal 13, climate action; and goals 14 and 15, life below water and life on land; all wrestle to some degree with the need to address the negative by-products and leftovers of our economic activities. These leftovers must be processed and absorbed somewhere. If deliberate efforts to absorb and process them fail, the burdens of failed sink capacities will be felt as substantial and unequal costs to society.

The contributions of sinks are enormous. For example, an average 55% of global anthropogenic CO_2 emissions are taken up by carbon sinks on land (especially forests) and in oceans (especially plankton). Unfortunately, data about the effectiveness of carbon sinks is difficult to obtain since both emission measurements and the measurement of carbon reservoirs fluctuate widely. It is therefore difficult to assess how quickly we are losing environmental sink capacities or what the costs of these losses are. What we do know is that overused carbon sinks lead to increased CO_2 accumulations. The climate vulnerabilities of the CO_2 accumulations are felt around the world in the form of heat waves, torrential rains, hurricanes, mudslides, wildfires and other disasters. Durant and colleagues (Durant *et al.*, 2011), for example, suggest that this lack of knowledge is itself costly. They estimate that even 5% of undetected global CO_2 emissions may have a marginal economic impact of approximately US $20 billion per year by 2050.

Soils too have sink capacities that can be overused. When one considers that a 0.3 metre layer of topsoil forms the outer skin of our 12,742 km diameter planet, it is astonishing how little we care for our topsoils. This paper-thin layer of topsoil constitutes the arable land that allows us to grow our food. When soils are overused and organic material is not sufficiently replenished, the results are erosion, reduced crop production, nutrient run-off, and ground- and surface water pollution (Pimentel, 2006). Overused freshwater reservoirs result in more vulnerable drinking water supplies and less water for crop irrigation. Plastic waste the size of continents that does not break down in a reasonable timeframe burdens landfills and oceans, while micro plastics bio-accumulate throughout the food chain and plastic waste is now reminiscent of the impact of pesticides first reported in Rachel Carson's Silent Spring (Carson, 1962). The burdens we add regularly to the sink capacities of our planet are astonishing and yet the costs associated with losing planetary and regional sink capacities go unaccounted for. Typically, planetary sink capacities are made up of smaller local and regional sinks so that local overuse translates into a loss of planetary sink capacities. Some regions do better than others in preserving their soils, lakes, forests, grasslands, and oceans. Cities are notoriously bad at it. Yet compensating for the loss in sink capacities in one place by adding capacity in another is not necessarily a simple proposition. Sink capacities reflect differences in temperature, precipitation, and historical conditions that result in different processing capacities in different locations. The nitrate contamination of the groundwater, for example, is not only a reflection of nitrogen fertilisation levels, but also reflects soil types, precipitation patterns, green cover of the soil and so much more. A simple bait and switch of 'we pollute and you preserve' is therefore rarely possible (O'Hara, 1984).

The burden of exhausted sink capacities is also distributed unequally. Human bodies are sinks. Humans literally absorb the pollutants that have accumulated in rivers, soils, the air, and in our food. Human bodies also absorb the emotional burdens of being overworked, isolated, and financially stressed. Some of these burdens are not simply absorbed by any one individual, but by social networks like family units, or communities. When we work long hours and don't get enough rest, we don't just suffer as individuals. Our families and friends suffer too. Just as exhausted soils and oceans carry the burden to other members of their ecosystem, humans too carry their physical and mental exhaustion to others in their social ecosystems. The other members of the community take care of those who are sick, and nurture those who are physically, mentally, and emotionally exhausted so that they can regain their ability to work tomorrow and the day after. And just as ecosystems can collapse, humans may act out their collapse in post-traumatic stress syndromes, violence, and suicide when the collective burden gets too high for the existing social sink capacities. Vulnerable populations have few options to shift their burdens and suffer disproportionately from the loss of physical, mental, and social sink capacities. Feminist scholars and ecological economists have long argued that the poor and marginalised carry a disproportional share of the burden of overused and undervalued sink capacities (Perkins, 2007; O'Hara, 2010).

Sink capacities demand systems thinking. The term refers to the capacity of the whole system of environmental and human communities to absorb the unwanted physical and emotional by-products of economic activity. The Intergovernmental Panel on Climate Change (IPCC) found that large-scale discontinuities in the Earth's processing capacities, which ensure the stability of regulatory cycles that impact hydrological, carbon, and climate systems, are nearing a tipping point (Lenton *et al.*, 2019, PIK, 2017). The ecological economist William Rees writes: 'A cascade of data shows that the human enterprise is in ecological overshoot, consuming nature's goods and services faster than ecosystems can regenerate and dumping (often toxic) wastes beyond nature's processing/recycling capacity. In short, we are currently 'financing' economic growth by liquidating the biophysical systems upon which humanity ultimately depends.' (Rees, 2020).

At the root of this overuse and degradation of invisible sinks lie long-held economic valuation biases. Standard economic theory holds that consumer goods are visible and rendered valuable while the restorative work of nature and human support systems, especially of women, is invisible and rendered valueless (O'Hara, 2014; Perkins, 2007). Likewise, inputs used for the production of marketable goods and services have value, while inputs that are conserved or used for the production of non-market goods have no value. These valuation biases undermine the provision of sink services and the uneven burdens resulting from their loss. As Case and Deaton (2020) argue, death is not simply the result of physical deprivation but encompasses a 'death by despair' as some no longer have the means to sustain themselves and lack the support necessary to deal with the burdens of illness, poverty and isolation. Mary Mellor writes: 'If payment is not made in economic terms someone will pay in other ways: they will die before their time, sleep on the street, be nursed by a relative, go without shoes, and walk miles to the well.' (1994:3)

The so-called market failures of standard economics create costs that remain outside of the valuation framework of standard economics (Gowdy and O'Hara, 1995; Kakovitch and O'Hara, 2021). Yet sinks remain altogether invisible and are used up, degraded and lost without much awareness of the valuable processing capacities they provide. This raises serious questions about the conceptual framework of standard economics which would argue that all that is needed is a correct price that internalises the externalised costs back into the standard economic model. We suggest that instead of internalising externalised costs, our economic activities themselves must be embedded into the physical/environmental and social/cultural context systems that support them. The price of food can therefore not simply be based on its market price; it may not even be based on a market price that has been increased to reflect the cost of reduced groundwater contamination and higher nutrient density; it may instead be indicated to create diverse local and regional food systems that improve water quality, biodiversity, safe and meaningful working conditions, and diverse food cultures. In other words, food systems that support and strengthen the sink capacities provided by healthy and thriving ecosystems and social networks.

The recognition that both resources and sinks are needed to sustain a productive and resilient economy is not anti-economic. Instead, it questions the kind of economic activity that no longer creates value but has devolved into what Mariana Mazzucato calls 'market taking' rather than market making (2020). To shift from taking to making, markets must promote individual and public interests. The work of the late Nobel Laureate Elinor Ostrom (1990) provides guideposts for developing new economic models based on collaboration, trust, and public responsibility. Ostrom's research revealed that economic actors, when given the opportunity to build norms of reciprocity and trust, can successfully develop innovative institutional frameworks that are capable of correcting the failures of markets. Mitigating risks, preserving natural resources, and reversing unsustainable practices can reduce the burdens on social and environmental sinks. Yet this requires deliberate measures to address the productivity and health of sink capacities provided by the social and environmental context systems that sustain all economic activity.

9.3 Making contexts visible: the model of restorative economics

As we have argued in the previous section of this chapter, accounting for the true value of economic activity, the value of resource inputs used and goods and services produced needs to be captured as well as the emissions and waste products generated, and the burdens they place on social and environmental sink capacities. Ecological economists have long drawn attention to the deficiencies of the standard economic model (Georgescu-Roegen, 1984; Daly and Cobb, 1989; Gowdy and O'Hara, 1997; O'Hara, 2016). An economic model that qualifies as restorative economics must be able to capture all processes involved and must be capable of capturing the burden placed on the sink capacities provided by nature and human communities (O'Hara, 2010, 2014, 2016). This demands an expanded economic model.

The restorative economic model pioneered by the lead author of this chapter is depicted in Figure 9.1. The figure is a graphical representation that builds on the schematic representation of a production function that is familiar to economists and expands it to account for both emissions and sink capacities. Each quadrant of the model is a schematic depiction of the processes associated with one systems component of the overall process.

The first quadrant in Figure 9.1 depicts the common formulation of a production function whereby the quantity of useful goods and services produced (Q) is a function of the inputs used. All inputs are summarised as an input vector I so that output Q is a function of the resource inputs I or Q = f(I) (Gowdy and O'Hara, 1995). The input vector (I) reflects the technology, labour, material, and energy resources as well as the management practices and intellectual capacity necessary to generate the output of useful goods and services (Q). As technology changes, the

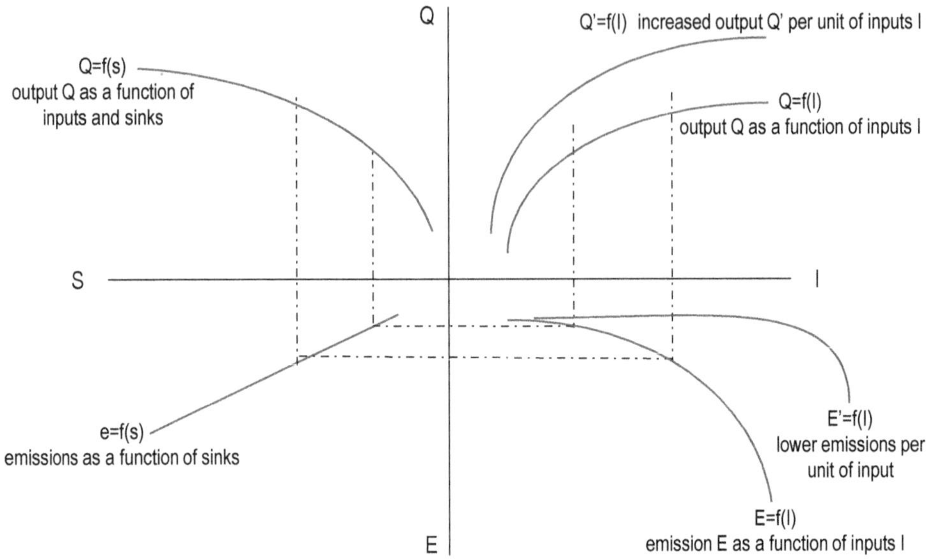

Figure 9.1. Restorative economics model.

relationship between inputs and outputs will change with the objective of increasing efficiency, defined as producing more output with the same or fewer inputs. This efficiency increase can be shown as a shift of the production function up and away from the I axis.

The function underneath in the second quadrant of Figure 9.1 depicts the emission function corresponding to the production process depicted in the first quadrant. Here the use of inputs captured by the vector I are shown as generating emissions E whereby higher levels of inputs used generate higher levels of emissions and waste causing negative externalities. E is therefore a function of the same input vector I used in the production process whereby E=f(I). As technology changes, the relationship between inputs and emissions can change as well, yet the objective of technological change is not solely to increase output but also to reduce emissions and waste. In some cases, technological change may reduce both inputs and emissions; in other cases, it may lead to a shift in the type of emissions generated but not necessarily to their overall reduction. Legislation is typically the vehicle through which emission-reducing innovations are initiated with the goal of shifting the emission function E closer towards the I axis, which indicates a reduction in emission levels.

Quadrant three of Figure 9.1 describes the complex assimilation, absorption, and support processes associated with social and environmental sink capacities. The German language offers a useful illustration of the role sinks play. It distinguishes between 'Emission' and 'Immission'. Emission refers to the pollutants that are emitted into the environment and human communities. Immission refers to the pollutants in the environment, that is in the water, air, soil, ecosystems, and human bodies. Different ecosystems and communities react differently to these emission burdens. The environmental, physical, biological, and emotional processing capacities associated with the transformation of Emissions into Immissions are a reflection of the multifaceted conditions and complex interactions in environmental and social context systems we have defined as sink capacities S. Strong sink capacities can reduce the negative effects of emissions E, while weak sink capacities can exacerbate them.

This allows us to reconceptualise the production of useful goods and service Q as a function of inputs, emissions, and sinks. A restorative production function Q is the result of sink capacities S that are the vector of complex sink capacities or Q=f(S). A process of output generation Q that qualifies as restorative will maintain or improve the vector of sink capacities (S); a process that will reduce or undermine them is unsustainable and even destructive. This restorative economics model must be understood as an integrated system and not simply as three isolated systems components.

The two-dimensional representation of the complex processes represented in the model recognises that the different spheres involved are all interconnected. For example, producing vegetables requires labour, water, fertiliser, soil minerals or nutrient rich water. Most of the focus in agriculture has been on producing higher crop yields with the same or fewer inputs and especially with less labour. Agriculture has been wildly successful on that front and has made it possible to free up labour for many economic sectors outside of agriculture. During the past 50 years the focus has also been on reducing nutrient run-off from fertilisers, and since Rachel Carson's Silent Spring, also on reducing pesticide use (Carson, 1962). Pesticides have been a concern from a human health and ecological perspective, as pollinators and other useful insects and microorganisms have been threatened. Efforts to reduce the use of chemical emissions have also ushered in a shift from managing crops with chemicals to genetic modification. The change in emissions associated with this shift are not altogether clear yet, but if history is a guide, we will only be concerned when it becomes impossible to ignore any potentially harmful side effects of new crop production practices.

Only within the past ten years has attention slowly begun to shift from efficiency increases and emission reductions to the economic contributions of ecosystem services (Costanza *et al.*, 1997; O'Hara, 2010, 2016). Soils can sequester carbon and reduce CO_2 emissions; the increased capacity of soils to absorb water can mitigate flooding; maintaining biodiverse ecosystems provides habitats for pollinators that make crop production possible. The accelerating loss of ecosystem services has confronted us with the need to maintain sink capacities if we want to continue to maintain crop yields. The logical question then is how to produce food in a way that maintains healthy ecosystems in the first place and that avoids nutrient run-off, groundwater pollution, and the loss of stormwater and heat absorption capacities.

When actors shift the focus from inputs to sinks, in accordance with Figure 9.1, it will dramatically change the perceived value of economic benefits and costs. Seemingly costly (in the present economic model) decentralised, smaller-scale urban food production methods that increase permeable surfaces in the urban scape, sequester carbon, preserve biodiversity, mitigate flooding and heat effects and improve overall sink capacities may become outright cheap and highly valuable when their restorative side effects are considered. Similarly, housing developments that introduce expensive common area spaces such as shared gardens and kitchens may be considered outright cheap and valuable when considering their positive impact on environmental sinks like permeable surfaces and green cover, and social sinks like aesthetics, social interactions, and the health of residents.

These examples illustrate the opportunities and challenges the restorative economics model presents. The system we call the economy is itself complex. Yet it is nested within the system 'society' and the larger system we call the 'environment'. These nested systems are interconnected and function according to different timespans, spatial dimensions, and success measures. We propose four generally applicable principles of a restorative economics model:

1. Redefine efficiency as more than labour efficiency, as the efficient use of environmental and social assets.
2. Reduce emissions and waste by closing loops as much as possible and reducing environmental and social burdens on environmental and human sinks.
3. Strengthen sink capacities through improved ecosystems services and social support networks that strengthen both social equity and ecological resilience.
4. Redefine the production of goods and services as restorative and dependent on the ability to meet principles 1, 2 and 3 and strengthen rather than undermine environmental and social contexts systems.

In addition to the four generally applicable principles, local and regional principles may be necessary to meet the specific characteristics of environmental and social contexts associated with a particular region or locality.

9.4 Making resources and sinks visible: the case of Washington DC

Like many cities around the world, Washington DC, the capital of the United States of America, is a tale of two cities. Its workforce is highly educated, and an impressive array of social and cultural amenities including world-class museums, a diverse and creative restaurant scene, theatres, retail, large parks and two rivers are available to some of DC's residents. Yet there are also pockets of poverty, high unemployment, food insecurity, short life expectancies, and high rates of crime.

Administratively, the city is divided into eight Wards. It is also a carefully designed city with the United States Capitol forming the centre from which four quadrants emanate – northwest, southwest, northeast, and southeast. One of the rivers, the Anacostia, is the dividing line of the city, and Washingtonians refer to the Wards west of the river, which are home to well-educated, high-income households and plenty of amenities, and those east of the river, which are home to low-income residents and lack amenities. Ward 1 forms the city centre and is the most densely populated Ward. Wards 2 and 3 are west of the river and in the north of the city, while Wards 7 and 8 are east of the river and in the city's south. The remaining Wards 4, 5, and 6 have neighbourhoods on both sides of the east-west divide. The demographic data summarised in Table 9.1 speaks for itself and illustrates the socio-economic and racial divides of Washington DC.

Table 9.1. Demographic and socio-economic data for Washington DC by Ward (O'Hara, 2018).

Demographic, socio-economic & health data	Ward 1	Ward 2	Ward 3	Ward 4	Ward 5	Ward 6	Ward 7	Ward 8
Total population	82,859	77,645	83,152	83,066	82,049	84,290	73,290	81,133
Black (non-Hispanic)	33%	10%	6%	59%	77%	43%	95%	94%
White (non-Hispanic)	40%	70%	78%	20%	15%	47%	2%	3%
Hispanic	21%	9%	8%	19%	6%	5%	2%	2%
Asian	5%	10%	8%	2%	2%	5%	0.3%	0.5%
Household income	$113,972	$209,147	$257,224	$123,353	$82,425	$140,853	$56,759	$45,239
Unemployment	5.1%	3.8%	3.7%	9.8%	14%	6.2%	19%	22%

In 2012 Washington DC launched its Sustainable DC plan spearheaded by the then mayor, Vincent Grey (DC Government, 2012). The objectives of Sustainable DC remain largely unchanged and DC's current Mayor, Muriel Bowser, describes the initiative as her vision '...to make the District of Columbia the healthiest, greenest, most liveable city for all District residents' (DC Government, 2021).

Sustainable DC is structured along 13 topics that are translated into 36 goals and 167 actions. The topics and goals are as wide ranging as the SDGs. They cover government accountability, equity, the built environment, climate mitigation, the economy, education, energy, food, health, nature, transportation, waste and water. Three topics are of particular interest for the restorative economic model: first, food, which has the goals of 'expanding urban agriculture', 'increasing access to healthy food', 'growing the food economy', and 'reducing food waste'. Second, the topic nature outlines the goals of 'protecting and expanding aquatic wildlife and habitat', 'protecting and expanding land wildlife and habitat', and 'improving residents' access to nature'. Third, the goals of the topic water are 'making waterways fishable and swimmable', 'reducing the amount of stormwater runoff', 'reducing the amount of potable water used', and 'ensuring safe drinking water'. Different entities within the District of Columbia government are responsible for implementing these different action items associated with each goal and reporting progress towards them.

While there has been progress towards a number of the Sustainable DC goals over the past nine years since their launch (DC Government, 2021), persistent social disparities remain. Household incomes and unemployment rates are key economic indicators that speak to the inclusion or exclusion of residents from the economic success story of Washington DC, which ranks among the U.S. cities with the highest household income. Progress towards the health goals of 'enabling active lifestyles' and 'improving community health', for example, can be more readily achieved when food access is plentiful and income levels are high, as illustrated by the low diabetes rates in Wards 2 and 3; yet it is challenging to reach these goals when food access is limited and household incomes are low, as illustrated by the health outcomes in Ward 7 and 8. Similarly, positive health outcomes are more likely when walking and biking opportunities are plentiful, while neighbourhoods with a large number of blighted buildings tend to have fewer safe outdoor recreation settings. High crime rates also impact life expectancies and can reflect neglected infrastructure investments as well as the stress factors associated with unstable living conditions and higher vulnerabilities.

These disparities indicate that the capacity to process environmental and social stressors does not receive equal attention across different neighbourhoods with different socio-economic and demographic characteristics. Some neighbourhoods show signs that their capacity to process physical and social stressors is overwhelmed. Table 9.2 shows some of the social and environmental context conditions for Washington DC by Ward. To use the language of restorative economics: the ability to deal with social and environmental burdens is unevenly distributed across the 8 Wards of the city. And improvements may be slow. For example, neighbourhoods where food access has been limited for decades may need cooking classes and nutrition counselling so that improved food access can be translated into improved eating habits and health outcomes. Recognising the complex connections between lost sink capacities and negative outcomes is an important step in the right direction, yet finding the right strategies and intervention points to address them may be challenging. At the very least, it requires long-term commitments, strong public engagement, and building trust.

Three examples may illustrate initiatives that could advance progress towards the goals articulated in the Sustainable DC Plan and address persistent disparities in social and environmental processing capacities. The first example is the Urban Food Hubs of the University of the District of Columbia. As described in Chapter 3, the Urban Food Hubs model has four integrated components: Food Production, Food Preparation, Food Distribution and closing the loop through Waste and Water Management. Ultimately, the Hubs seek to contribute to a decentralised network of local food systems. To date, four Hubs are in various stages of implementation, and each offers opportunities for improved food access, nutrition education, business development and job creation across a range of sectors including food production, the culinary sector, retail, education, green infrastructure management, heat island mitigation, park and garden design, and soil enhancement. The Food Hubs are well aligned with the principles of restorative economics.

Table 9.2. Social and environment context indicators for Washington DC by Ward (O'Hara, 2018).

Indicators	Ward 1	Ward 2	Ward 3	Ward 4	Ward 5	Ward 6	Ward 7	Ward 8
Life expectancy	78 yrs	86 yrs	85 yrs	78 yrs	75 yrs	77 yrs	73 yrs	70 yrs
Infant mortality	4	3	4	11	10	10	7	10
Diabetes deaths	3	3	4	15	21	12	21	20
Violent crimes per 1000	13	9	2	10	13	12	18	19
Grocery stores per 1000	7	7	9	5	5	4	2	2
Full-service restaurant	364	777	304	104	167	144	26	43
Residents exercising	84%	86%	92%	80%	72%	85%	68%	69%
Blighted buildings	24	39	4	38	86	16	119	123

The Hubs contribute to land-use efficiency and labour productivity through their capacity to produce food and create jobs; they reduce emission burdens by expanding local production and reducing waste and water use; and they strengthen social sink capacities by improving the quality of life and resilience of neighbourhoods which are lacking on all counts. Moreover, three of the four Hubs are located in neighbourhoods (in Ward 5, 7 and 8) that have low socio-economic, environmental, and health outcomes but are well positioned to meet the principles of a restorative economics we have identified. Since the Food Hubs design can be adapted to reflect specific local neighbourhood conditions, they can be replicated in other cities to address deficits in specific social and environmental components of the restorative economic model.

The second example is the so-called Five Pillars project which is also described in Chapter 8. The project convened two focus groups in Ward 7 (Deanwood) and Ward 8 (Congress Heights) to engage residents in writing a story about a sustainable future for their neighbourhoods (O'Hara, 2018) The story is impressive and reflects the articulate neighbourhood-based vision of a restorative economy including a local food system of locally produced, farm-to-table restaurants, and a focus on healthy lifestyles; off-the-grid water filtration and energy generation, health, wellness and amenities-based jobs; and a proud African American history of entrepreneurship and innovation. The particular strength of the Five Pillars project lies in its commitment to engaging the local expertise of residents in re-defining their local economy. Their vision of creating local jobs, reducing negative social and environmental impacts, reducing local vulnerabilities, and strengthening social and environmental ecosystems is consistent with the principles of a restorative economy.

The third example is our own vision of a regional food economy that utilises the physical/environmental and social/cultural assets of the DC metropolitan region. This vision is based on the two previous initiatives, the Urban Food Hubs and the Five Pillars of Economic Development, both of which are anchored in local engagement and expertise. This larger, regional vision paints a future of a food-based economy that provides meaningful work and a living wage to a diverse workforce, affirms a circular economy of linked sectors where the waste product of one is the resource of another, and strengthens the environmental, social, and economic assets of the entire DC metro region.

The case of Washington DC illustrates that disparities in the social and environmental burdens of economic activity are plainly visible. While markets may be efficient allocation mechanisms that capture the individual pursuits of interest-maximising consumers and profit-maximising producers, they fail to account for social and environmental resource and sink capacities that are in the long-term best interest of the city. The disparities between haves and have-nots and the loss of social and environmental sink capacities are examples of these market failures. Their costs are not borne by those who benefit most, but by those who benefit least. To mitigate these distortions is both a moral imperative and a prerequisite for the long-term sustainability of the economy itself.

9.5 Toward restorative urban food policies

Today's dominant economic model in urban food policies is anti-restorative. Rather than strengthening the four principles of restorative economics, it undermines efficiency by leading to higher and higher market concentration, high emission rates exacerbated by long global supply chains, and neglected sink capacities. Policies that seek to protect markets are therefore not enough; they must instead facilitate real competition and enforce restorative outcomes. Put simply, ethical goals like social justice and the protection of sink capacities require deliberate action in the field of urban planning. In fact, Adams Smith, the father of modern economics, stressed the importance of ethical goals that provide a fence for economic markets. In his famous book, *An Inquiry into the Nature and Causes of the Wealth of Nations*, Smith laid the foundation for free market capitalism by defining the economic marketplace as the interaction between countless consumers and producers pursuing their individual satisfaction- and profit-maximising interests as if 'led by an invisible hand' (Smith, 1776). Yet in his book *The Theory of Moral Sentiments*, which Smith considered his most important work, he argued that the true power of markets is their ability to achieve socially desirable outcomes. Yet they are only achieved when markets operate within a moral framework of empathy and compassion (Smith, 1759). In other words, economies flounder without a moral compass that ensures socially optimal outcomes. Policies must therefore establish ethical principles that address social disparities and environmental exploitation while preserving the agency and efficiency of decentralised markets (Toussaint and O'Hara, 2020). Economic activity that undermines sink capacities either by adding emissions and waste, or by destroying the ecosystem services and social networks that strengthen sink capacities, cannot be considered valuable or efficient but has a negative value for society. Economic value is therefore determined by the health and vitality of the context systems that sustain social and environmental sink capacities (O'Hara, 2016).

So, what do urban policies that support the four principles of restorative economics look like? The Five Pillars project and the Sustainable DC agenda provide some guidance. The first principle of a restorative economy focuses on efficiency. Yet, rather than defining efficiency based on economic indicators like reduced costs and increased production, efficiency is defined as increasing local and regional assets. The Five Pillars project defines these assets as good education, health, strong social and cultural amenities, environmental quality and access to information and connectivity (O'Hara, 2018). These Five Pillars of economic development shift the narrow cost-based definition of efficiency to an asset-based one that engages communities in designing their own future and measuring progress towards it. The asset-based approach also requires a broader definition of agency. The expectation that new entrepreneurs in community-based businesses can pull themselves up by their own bootstraps may not be realistic. More promising models include cooperatives and employee-owned businesses that pool assets and resources.

Secondly, urban policies must focus on reducing emissions and waste to stop adding to the problem. The SDGs as well as many of the goals and actions outlined in the Sustainable DC plan provide examples including reducing food waste, implementing zero carbon targets, making rivers fishable and swimmable again, and creating loop-closing and circular-economy networks. Reducing air, water and soil contamination especially in low-income communities can also reduce health disparities, yet more explicitly equity-oriented policies will be needed to reduce the long-standing physical, mental, and emotional impact of social and environmental injustice. Implementing these policies may not be popular since poverty has become big business and cities are replete with well-meaning public, private and non-profit sector service providers.

Urban policies that address the third principle of restorative economics and improve the sink capacities of nature and of human communities may be the least developed and understood. The 'Food Connects' example and other examples of City Regional Food Systems illustrate both the complexities and leverage points of restoring social and environmental sinks (Stuiver and O'Hara, 2021). Learning from restorative agriculture (Hutchinson *et al.*, 2007), and biomimicry (Benyus, 2002) are also steps in the right direction. These fields seek to address critical sink capacities like sequestering carbon, absorbing heat through eatable urban forests and urban agriculture, turning plastic waste into a resource without adding more energy, producing food with 10%nt of the freshwater of conventional agriculture, and generating energy through solar panels as thin as butterfly wings so food can grow underneath them. We can do all these things. Yet finding technical solutions for specific context conditions requires a vastly improved understanding of how nature works and what parameters prove most important in influencing sinks and their processing capacities (Kakovitch and O'Hara, 2021).

None of these things are easy. The self-assessment offered in the 2021 Sustainable DC progress report is telling. Nine years after Sustainable DC was launched, 73 of the admittedly ambitious 167 action items fall into the lowest-scoring categories of 'not started' and 'initiated' with less than moderate progress achieved. The examples list only one of the 73 underachieved actions for each of the 13 Sustainable DC topics. They illustrate how each action has a role to play in achieving at least one of the four principles of restorative economics including:

- Develop a citywide strategy for greening internal District Government operations to save money and improve environmental performance.
- Focus community engagement on communities that have been historically under-represented.
- Enhance programs to support businesses to open and operate in neighbourhood commercial corridors, focusing on vacant and underused spaces.
- Complete energy assessments of all District homes and buildings.
- Incorporate sustainability best practices into neighbourhood planning.
- Work with private partners to support new incubators and 'maker spaces' with a focus on communities most in need.
- Provide adequate support so that every student will have access to meaningful environmental experiences in elementary, middle, and high school.
- Remove all barriers to modernising electricity infrastructure to enable the deployment of neighbourhood-scale energy systems and distributed energy resources.
- Conduct a food waste assessment to identify the types and quantities of food that are thrown away in the District.
- Launch a comprehensive multi-level health literacy campaign across the District.

- Create a habitat connectivity plan to guide restoration of viable, native habitats throughout Washington DC and in coordination with surrounding jurisdictions.
- Develop design guidelines to ensure transit systems are resilient to climate change.
- Work with surrounding jurisdictions to develop and implement a regional approach to reducing plastic waste. Increase purchases of Environmentally Preferable Products and Services.
- Eliminate all lead service lines and plumbing from District Government-owned buildings.

These examples illustrate the challenges communities face in implementing the four principles of a restorative economics while the formal institutions and informal values of a growth and use-based economy remain firmly in place.

9.6 Conclusions

The concept of restorative economics addresses the deficits of the standard economic model that is typically used in accessing urban food systems and policies to improve them. Restorative economics adds several critical dimensions. The model brings commonly invisible social and environmental side effects of economic activity to the fore so that they can be reduced and ultimately turned into positive restorative capacities. The model starts with the familiar notion of production efficiency, broadens standard definitions of efficacy, and adds a focus on emissions and waste and the neglected sink capacities that are at work to process the emission and waste by-products of economic activity. The model thus addresses the conceptual, informational, and valuation deficits of standard economics, which acts as if it operates outside of the social and environmental context systems that sustain it by delivering resources and processing emissions and waste by-products.

Operationalising the restorative economics model will require considerable effort and a new urban policy agenda that includes improving the knowledge of decision-makers about the neglected sink dimensions of the economic process. The unique focus of the restorative economics model on sink capacities makes it possible to operationalise the restorative economy concept and focus on urban policies that address existing deficits and turn them into assets. The research needs of restorative economics are substantial and must include a focus on measures to operationalise a restorative economics and communication strategies to disseminate new insights among decision-makers, policy analysts, and the public at large.

Yet equally as important as formal policies is the adoption of mental maps that recognise the need for an ethical framework that affirms the value of thriving social networks and healthy ecosystems. Without the explicit commitment to restoring the environmental and social sink capacities even of our most densely populated cities we may not only face the rising costs of their decline, but the loss of economic options for future generations.

References

Benyus, J., 2002. Biomimicry: innovation inspired by nature. Harper Collins, New York, NY, USA.

Carson, R., 1962. Silent spring. Houghten Mifflin, Boston, MA, USA.

Case, A. and Deaton, A., 2020. Deaths of despair and the future of capitalism. Princeton University Press, Princeton, NJ, USA.

Costanza, R., d'Arge, R., De Groot, R., Farber, S., Grasso, M., Hannon, B., Limburg, K., Naeem, S., O'Neill, R. Paruelo, J., Raskin, R., Sutton, P. and Van den Belt, M., 1997. The value of the world's ecosystem services and natural capital. Nature 387: 253-260.

Daly, H., 1996. Beyond growth: the economics of sustainable development. Beacon Press, Boston, MA, USA.

Daly, H. and Cobb, J., 1989. For the common good: redirecting the economy toward community, the environment, and a sustainable future. Beacon Press, Boston, MA, USA.

DC Government, 2021. Sustainable DC – progress report. Available at: https://tinyurl.com/26p6vvzx.

DC Government, 2012. Sustainable DC. Available at: https://sustainable.dc.gov

Durant, A., LeQuere, C., Hope, C. and Friend, A., 2011. Economic value of improved quantification in global sources and sinks of carbon dioxide. Philosophical Transactions of the Royal Society A 369: 1967-1979. https://doi.org/10.1098/rsta.2011.0002

Georgescu-Roegen, N., 1984. Feasible recipes versus viable technologies. Atlantic Economic Journal 12: 21-31.

Gowdy, J. and O'Hara, S., 1995. Weak sustainability and viable technologies – Special Issue: Nicholas Georgescu-Roegen. Ecological Economics 22: 239-247.

Gowdy, J. and O'Hara, S., 1997. Weak sustainability and viable technologies – Special Issue: Nicholas Georgescu-Roegen. Ecological Economics 22: 239-247.

Hutchinson, J., Campbell, C. and Desjardins, R., 2007. Some perspectives on carbon sequestration in agriculture. Agricultural and Forest Meteorology 142(2-4): 288-302.

Intergovernmental Panel on Climate Change (IPCC), 2021. Climate change widespread, rapid, and intensifying. Available at: https://www.ipcc.ch/2021/08/09/ar6-wg1-20210809-pr/.

Kakovitch, T. and O'Hara, S., 2021. Water and the circular economy: learning from nature. Sustainability 13: x. https://doi.org/10.3390/su132212597

Lenton, T., Rockstroem, J., Gaffney, O., Rahmstorf, S., Richardon, K., Steffen, W. and Schellnhuber, H., 2019. Climate tipping points – too risky to bet against. The growing threat of abrupt and irreversible climate changes must compel political and economic action on emissions. Nature Nov. 27.

Mazzucato, M., 2020. The value of everything: making and taking in the global economy. Public Affairs. Perseus Books, New York, NY, USA.

Meadows, D., Meadows, D. Jorgen, R. and Behrens III, W., 1972. The limits to growth: a reports for the club of Rome's project on the predicament of mankind. Universe Books, New York, NY, USA.

Mellor, M., 1994. Ecofeminism and ecosocialism: dilemmas of essentialism and materialism. Capitalism, Nature, Socialism 3(3): 111-114. https://doi.org/10.1080/10455759209358511

O'Hara, S., 1984. External effects of nitrate fertilization – assessment problems and valuation of reduction measures under economic and ecological considerations (in German), Verlag Vauk, Kiel, Germany.

O'Hara, S., 1997. Toward a sustaining production theory. Ecological Economics 20(2): 141-154.

O'Hara, S., 2001. The challenges of valuation: ecological economics between matter and meaning. In: Cleveland, C., Costanza, R. and Stern, d. (eds.) The nature of economics and the economics of nature. Edward Elgar, Northampton, MA, USA, pp. 89-108.

O'Hara, S., 2010. Feminist ecological economics in theory and practice. In: Salleh, A. (ed.) Eco-sufficiency & global justice – women write political ecology. Pluto Press, London, UK.

O'Hara, S., 2014. Everything needs care: toward a relevant contextual view of the economy. In: Bjørnholt, M. and McKay, A. (eds.) Counting on Marilyn Waring: new advances in feminist economics. Demeter Press, Bradford, Canada, pp. 37-55.

O'Hara, S., 2016. Production in context: the concept of sustaining production. In: Farley, J. and Malghan, D. (eds.) Beyond uneconomic growth, Vol. 2: A festschrift in honour of Herman Daly. University of Vermont, Burlington, VT, USA, pp. 75-106.

O'Hara, S., 2017. The urban food hubs solution: building capacity in urban communities. Metropolitan Universities Journal 28(1). https://doi.org/10.18060/21477

O'Hara, S., 2018. The five pillars of economic development: a story of a sustainable future for ward 7 and 8 in Washington DC. College of Agriculture, Urban Sustainability and Environmental Sciences, University of the District of Columbia, Canada.

O'Hara, S., 2021. Restorative economics: considering pathways toward a Post-COVID economy. In: AI-Youbi, A., Zahed, A. and Ritzen, J. (eds.) The effects of the Corona pandemic and their impact on the education sector. Springer, London, UK.

Ostrom, E., 1990. Governing the commons: the evolution of institutions for collective action. Cambridge University Press, Cambridge, UK.

Perkins, P.E., 2007. Feminist ecological economics and sustainability. Journal of Bioeconomics 9(3): 227-244.

Pimentel, D., 2006. Soil erosion: a food and environmental threat. Environment, Development and Sustainability 8: 119-137. https://doi.org/10.1007/s10668-005-1262-8

Potsdam Institute fuer Klimaforschung (PIK), 2017. Tipping elements – the Achilles heels of the earth system. Available at: https://tinyurl.com/2een37nd.

Raworth, K., 2012. A safe and just space for humanity: can we live within the doughnut? Oxfam Discussion Papers. Oxfam International. Available at: https://www.oxfam.org/en/research/safe-and-just-space-humanity.

Rees, W., 2020. Ecological economics for humanity's plague phase. Ecological Economics 169: 106519. https://doi.org/10.1016/j.ecolecon.2019.106519

Smith, A., 1776. An inquiry into the nature and causes of the wealth of nations. Darnell, T. (ed.). Republished 2018.

Smith, A., 1759. The theory of moral sentiment. Hanley, R. (ed.) Republished 2010. Penguin Classics, London, UK.

Solow, R., *1956. A contribution to the theory of economic growth.* Quarterly Journal of Economics *70(1): 65-94.* https://doi.org/10.2307/1884513.

Stuiver, M. and O'Hara, S., 2021. Food connects Washington DC in 2050 – a vision for urban food systems as the centerpieces of a circular economy. Sustainability *13*(14): 7821. https://doi.org/10.3390/su13147821

Toussaint, E. and O'Hara, S., 2020. Food, fitness and fatalities. Human Rights Magazine 46(1). Available at: https://tinyurl.com/y93ty8w6.

United Nations Commission on Environment and Development, 1987. Our common future. Oxford University Press. Oxford, UK.

United Nations, 2015. Sustainable development goals. United Nations, New York, NY, USA. Available at: https://sdgs.un.org.

PART 4

CITIZENS IN THE SYMBIOTIC CITY

Chapter 10

Nature, health and well-being: evidence and examples

Sjerp de Vries

Abstract

One of the first times the effect of nature on stress relief was investigated, the results were so remarkable that the top journal Science devoted space to them. The study, published in 1984, showed that American patients who had undergone gall bladder surgery recovered faster when they could see a tree from their hospital window than when they could only see a blank wall. In 2003, in the Netherlands, one of the first epidemiological studies was conducted in which the amount of green space in the residential environment was objectively assessed and shown to be related to the health of residents. This chapter provides an overview of more recent research on the relation between nature in cities and the health and well-being of their inhabitants. Implications of the available knowledge for urban planning and design are discussed.

Keywords: health, well-being, nature, green space, research methodology, urban planning, landscape architecture

10.1 Introduction

This chapter provides a state-of-the-art overview of the relationship between nature in cities and the health and well-being of citizens. There is ample evidence of beneficial associations between the amount of nature in the (urban) living environment and the health and well-being of residents. However, in many cases the causality of the association has not been firmly established. Insight into how residential nature may enhance health and well-being helps to assess to what extent causal interpretations of observed associations are justified. Therefore, the question about the pathways through which residential nature is able to positively influence health and well-being is addressed. En route, a new framework for studying pathways is introduced. Subsequently, the related question about which type of greenery is particularly important for the well-being of city residents is dealt with. Related, because the answer is likely to depend greatly on which pathways provide the strongest links and generate the most beneficial health effects (the latter may depend on which health aspect one focuses on). The issue of the size of health impacts brought about by changes in the green infrastructure is briefly discussed, because of its relevance for urban planners. Benefits (and costs) of different ways to use limited space have to be weighed up, not only with regard to health, but also with regard to other functions, such as housing and climate resilience. Better insight into the health benefits of greenery makes it easier to take them

10.3920/978-90-8686-935-0_10, © Wageningen Academic Publishers 2022

explicitly into account when making trade-offs. The chapter ends with some suggestions for urban planners and landscape architects who wish to make use of the health benefits of nature in an urban residential context.

10.2 Nearby greenery and human health are positively associated

As shown by recent systematic reviews, human health and well-being is positively related to the amount of vegetation in the residential environment (Houlden *et al.*, 2018; Jabbar *et al.*, 2021; Twohig-Bennett and Jones, 2018; Van den Bosch and Ode Sang, 2017). The health aspects addressed in these studies range from the birth weight of neonates (Akaraci *et al.*, 2020; Hu *et al.* 2021; Zhan *et al.*, 2020) to the premature mortality of adults and longevity (Rojas-Rueda *et al.*, 2019; Yuan *et al.*, 2021). In between, the prevalence of many types of lifestyle-related and/or non-communicable diseases and disorders has been shown to be negatively associated with access to nature. For physical health, this includes being overweight (Luo *et al.*, 2020), type 2 diabetes mellitus (De la Fuente *et al.*, 2021), cardiovascular disease (Aerts *et al.*, 2020; Seo *et al.*, 2019). For mental health and cognitive functioning, beneficial associations have been established for depression and anxiety disorders (De Vries *et al.*, 2016; Hystad *et al.*, 2019). For ADHD among children and adolescents such associations have also been observed (Donovan *et al.*, 2019; Thygesen *et al.*, 2020), as well as for dementia among the elderly (Paul *et al.*, 2020; Slawsky *et al.*, 2022). To be clear, there are also studies that do not find a beneficial association. However, detrimental associations are rare.

10.3 But does increasing the amount of greenery lead to better health?

To summarise, the positive association between residential greenery and human health and well-being has been firmly established. But does this mean that increasing the amount of greenery within the residential environment will lead to improved health and well-being? To a large extent, this remains to be seen. Most of the studies conducted have a cross-sectional research design, with nature and health being measured at the same time. Based on such studies alone, it is not possible to firmly conclude that the higher amount of greenery in the residential environment is responsible for the better health and greater well-being: association does not necessarily imply causation. For example, wealthier people may have greener residential surroundings, and also be more healthy than poorer people, without the difference in greenery being responsible for their better health. Even though it is customary to control for the socio-economic status in cross-sectional studies, residual confounding may still be present.

Therefore, we will briefly examine research designs that give more insight into the plausibility that increasing the amount of greenery will actually enhance the health and well-being of residents. The classic example is the randomised controlled trial (RCT). The RCT design basically consists of an experimental condition and a control condition, with subjects being randomly assigned to one of the two conditions. The RCT is the gold standard in medical research, for example, for investigating the effect of a new medicine. People in the experimental condition get the new medicine and people in the control condition get a placebo. Differences in the outcome can only be attributed to the new medicine vs placebo difference. This type of research design has also been used with regard to the residential nature-health relationship. However, experimental studies usually take place in controlled (laboratory) settings, to rule out any kind of disturbance or interference during the experiment. As a consequence, they tend to focus on short-term effects, such as mood, stress and ability to concentrate. For outcomes of such experimental studies to be relevant for long-term effects in real-life residential context, the assumption has to be made that, over time, repeated contacts with associated short-term effects will lead to more durable improved health and well-being benefits. Another issue regarding experiments is what is generally called ecological validity: to what extent is the laboratory situation representative of real-life situations? (Note how the term 'ecological' is used here!) In experiments representations of natural settings are often used, with a strong focus on the visual aspect (pictures, videos). These representations are often not representative for residential greenery, but selected because of their naturalness, to create a stark contrast with built-up settings. This may lead to an overestimation of the effect, compared to that in residential settings. On the other hand, there are indications that the effects of (looking at) representations of nature are smaller than those of being exposed to real nature (McMahan and Estes, 2015; Stevenson *et al.*, 2018).

As a consequence, there is a call for research designs with a higher ecological validity, such as natural experiments and intervention studies. Both try to assess the impact of changes in the (residential) environment, usually by comparing pre- and post-change assessments of health and/or well-being. The usual question in both is whether increasing the amount of greenery (or its quality, however defined) leads to a subsequent improvement in health and well-being. However, looking at the impact of a decrease in the amount of greenery (or its quality) is also possible, and at least as informative. A potential weakness of these types of study design is that, besides the change in the greenery, other changes may take place and affect the outcomes. The more time passes between the intervention and the post measurement, the more likely it is that other changes and events have occurred (that may also affect health and well-being), making it difficult to assess long-term effects.

Another issue regarding the effect of environmental changes or interventions is that they are often rather small. From the perspective of an individual citizen, often only a small part of its residential environment has changed. Small changes are unlikely to lead to large effects. Vice versa, the service area of the changed site, in terms of the number of people that can reasonably be expected to visit the site or otherwise come into contact with it, may be small. Securing the participation of precisely those people in the study may prove to be difficult, leading to underpowered studies. A rare example of a large-scale natural experiment is provided by Jones (2021). He studied the effect of the first years of the Million Trees Programme in New York City (NYC). The large scale of the environmental change made it possible to make use of already available survey data, gathered for a sample of the population of NYC, as well as for similar populations outside NYC. Jones observed an increase in the quality of life of the inhabitants in NYC that did not occur in the other areas.

Also, longitudinal studies can shed more light on the causality of associations. Individual people or cohorts are followed through time. When their physical environment changes, their health and well-being is compared with that before the change. The idea behind it is that people (or cohorts) basically remain the same person(s) in a large number of respects. In that way, individual differences are controlled for to a substantial extent. A nice example of such a study is from Alcock *et al.* (2014). They investigated which people moved and whether this was to a greener or less green residential environment. They observed that the mental well-being of people moving to a greener residential environment increased, at least for three years after the move (after that, no data were available). However, the mental well-being of those that moved to a less green residential environment did not change.

One type of research design that offers strong evidence of a causal effect is twin studies. Monozygotic twins share their genetic make-up and usually also their upbringing: they have the same parents, same household and same residential environment. Therefore, individual differences are controlled for to an even larger extent. Later in life, the twins may relocate to different residential environments, which may also differ in the amount of vegetation. Cohen-Cline *et al.* (2015) conducted such a twin study in the United States. The twin member with a greener residential environment later in life scored lower self-reported depression than the member with a less green residential environment. The same was not observed for self-reported stress and anxiety.

10.4 Possible pathways linking residential greenery and human health

Another way to approach the issue of causality is by looking into the pathways through which greenery is expected to have a positive impact on human health and well-being. To the extent that a pathway is plausible, the causality of an observed association becomes more credible. However, this is not an all-or-nothing matter. The association can be partly causal and partly spurious, i.e. caused by a common third factor, affecting both. Pathways are usually empirically established by showing that an effect of A on C is mediated B: A affects B, B affects C, and in this indirect way A affects C. In other words, B mediates the association between A and C, and is therefore called a mediator. The word 'affects' is used here in a statistical sense, indicating predictive value rather than causality.

Several pathways have been identified or suggested in the literature. Markevych *et al.* (2017) have divided them into three categories: mitigation, restoration, and instoration. Examples of mitigating pathways are vegetation improving the air quality; or vegetation reducing heat stress by offering shade during the day, cooling the air by evaporation and/or absorbing less heat than most sealed surfaces. In some cases, contact with the greenery may influence how people respond to an environmental stressor. An example is the potential psycho-acoustic effect of greenery on noise annoyance: the same objective noise level may result in less noise annoyance if greenery is present (Dzhambov *et al.*, 2018; Mueller *et al.*, 2020).

Examples of restorative pathways are reducing stress, restoring one's mood (making it more positive/less negative) and/or ability to concentrate. In principle, in order to qualify as a restorative pathway, there needs to be something that is depleted or out of balance first, which implies that contact with nature is (only) effective if it takes place after the depletion has occurred. Otherwise, the pathway might be considered to belong to the third category, that of instorative pathways. Examples of instorative or capacity-building pathways are stimulating higher physical activity levels, facilitating social contacts and social cohesion within the neighbourhood, and improving the composition of the gut microbiome, and thereby the functioning of one's immune system.

We will briefly look at the empirical support for a number of these pathways. We start with the pathway of vegetation (A) improving the air quality, specifically by lowering fine dust concentrations (B), and in that way improving health (C). While it is true that vegetation can trap fine dust particles to some extent, with some types of vegetation doing so more than others, recent overviews suggest that the overall effect of urban vegetation on the concentration of fine dust is quite small. If the vegetation is wrongly positioned, blocking air flows through the city, its effect may even be detrimental (Nemitz *et al.*, 2020; Van Moorselaar and Van der Zee, 2020).

As for the pathway of (outdoor) vegetation improving health by reducing heat stress, it is important to make a distinction between daytime and night-time. During daytime, trees may offer cool spots because of the shade that they provide. During night-time, the effect of the vegetation on ambient air temperatures is more relevant, especially on that in the bedroom (assuming no air conditioner is present). This temperature affects one's sleep quality, and thereby one's well-being. Santamouris and Osmond (2020) posit that trees in the city may lower the maximum air temperature by at most 1.8 °C during daytime, and by 2.3 °C during night-time. These reductions are based on a tree coverage (projected tree crowns) of at least 60% of the surface, versus 0%. According to them, a more realistic 20%-point increase in tree cover would result in a reduction of 0.3 °C during daytime and 0.5 °C during night-time. Furthermore, they posit that a 0.1 °C decrease in the maximum daytime temperature will reduce the heat-related mortality by 3%. Pascal *et al.* (2021) concluded in a Parisian case study that in areas with a tree cover of 40%, the relative risk of heat-related mortality was 16% lower than in areas with a tree cover of 3%. Both studies suggest a similar magnitude of effect.

As for stress reduction, experimental studies tend to show beneficial (short-term) effects of looking at natural scenes when compared to looking at urban scenes, although the evidence for physiological stress measures seems less conclusive than that for self-reported stress levels (Bolouki, 2022; Corazon *et al.*, 2019; Mygind *et al.*; 2021). Kondo *et al.* (2018) arrived at similar conclusions based on field experiments in which participants were exposed to actual natural and built-up environments, rather than to representations. As for epidemiological research, recently researchers have started to look at the association between the amount of greenery in the residential environment and inhabitants' allostatic load levels. Allostatic load may be considered the physiological counterpart of chronic mental stress. Egorov *et al.* (2017) showed that on average people with a greener residential environment had a lower allostatic load level, i.e. had lower stress levels.

For mood improvement, there is experimental evidence that exposure to nature leads to (short-term) beneficial effects, although this evidence is not considered to be particularly strong (McMahan and Estes, 2015; Roberts *et al.*, 2019). Studies in daily-life settings also show that people's mood is more positive in natural than in built-up settings (De Vries *et al.*, 2021; MacKerron and Mourato, 2013). Also, the beneficial effect of exposure to greenery on restoring one's ability to concentrate (more quickly) is supported by experimental evidence (Mason *et al.*, 2021; Stenfors *et al.*, 2019; Stevenson *et al.*, 2018). There is also cross-sectional evidence showing beneficial associations between the greenness of the residential or school environment and the cognitive functioning/academic performance (Browning and Rigolon, 2019; Luque-Garciá *et al.*, 2021). The pathway from nature to ability to concentrate to health and well-being may seem less obvious than the mood pathway. However, it may be important for impulse control, leading to healthier choices (Berry *et al.*, 2020).

Another pathway is that nature offers attractive settings in which to conduct outdoor-recreational activities or influences the mode of transport, from passive (motorised) to active (e.g. by foot or bicycle), and in these ways stimulates people to be more physically active. There are no experimental studies looking into the effect of a natural setting on physical activity levels; it is hard to concoct such an experiment. According to Remme *et al.* (2021), in general reviews of studies looking into this pathway conclude there is a positive, yet weak, association between nature and physical activity, with the causality of the association being questionable. To give an example, Schipperijn *et al.* (2017) observed that when there are more parks within one kilometre from home, this is associated with 24 minutes more physical activity per week (objectively assessed) than when there are fewer parks nearby. A similar association was not observed for the distance to the nearest park. For children, nature in the residential area may be especially important as an activity setting, because of their limited action radius when it comes to autonomous play. Outdoor play can constitute an important part of their overall level of physical activity. A residential environment that offers safe and attractive (green) formal or informal play areas may help (Lambert *et al.*, 2019; Nordbø *et al.*, 2019).

To some extent, there may be countervailing forces in operation. Active transport is stimulated by having many destinations nearby, e.g. shops within an acceptable walking distance. On the other hand, walking for pleasure is likely to benefit from having attractive natural areas nearby. Destination density is likely to be higher in city centres, whereas natural areas are more likely to be present in suburbs. Shuvo *et al.* (2021) concluded that the amount of nearby greenery was even negatively associated with the score on a walkability index (an assessment of the extent to which an environment is conducive to walking for whatever purpose). Furthermore, with regard to health benefits, in terms of energy expenditure one's total amount of physical activity is what is most relevant, and not only that displayed in natural areas. People may prefer walking for pleasure in a more natural environment over walking in a predominantly built-up environment, but this does not necessarily mean that people without a nearby park will walk less. Triguero-Mas *et al.* (2017) observed that having a green area of at least a quarter hectare within 300 metres from home was associated with more physical activity in such areas, but not with more physical activity overall. At the same time, this outcome also shows that, although people may not be more active, the amount of time that they spend in a natural environment is likely to be higher when they have more greenery nearby. This may create opportunities for other pathways to enhance health and well-being, e.g. by way of stress reduction.

Nearby greenery may also facilitate social interactions and thereby increase social cohesion within the neighbourhood. An iconic natural experiment on this topic was conducted by Kuo *et al.* (1998). Residents were (quite randomly) assigned by the city hall to comparable apartment buildings. Originally, there was some grass and trees around all these buildings. However, over time this had disappeared around some of the buildings, but not around others. The study by Kuo *et al.* showed that residents living in a building that still had some grass and trees had more contact with fellow residents than residents of buildings where the grass was paved over. Wan *et al.* (2021) performed a systematic review of cross-sectional studies that looked at this specific pathway and concluded that most of these studies observed a positive relationship between the amount of residential greenery and the level of social cohesion within the neighbourhood. According to the authors, quality aspects, such as the level of maintenance, also play an important role.

A relatively new pathway focuses on the composition of the gut microbiome as mediator. As mentioned above, this composition is relevant for the functioning of the human immune system (Flandroy *et al.*, 2018). For example, both Roslund *et al.* (2020) and Sobko *et al.* (2020) showed a significant positive effect of nature-based interventions, respectively activities programmes, on the gut microbiome of children. However, there are also studies that do not observe similar associations. More in general, partly due to it being a relatively recent addition to the list of potential pathways, research on it is still limited (Brame *et al.*, 2021).

10.5 Relative importance of pathways linking residential nature and human health

In the previous section, several potential pathways with a diverse set of mediators linking residential nature to human health have been identified. The type of nature that is most effective in generating health benefits may differ by pathway. Therefore, it would help to know which pathways are the most important in producing health benefits. Up till now, there are only a few studies addressing this issue. De Vries *et al.* (2013) assessed the extent to which physical activity level, stress level and level of perceived social cohesion in the neighbourhood mediated the association between the amount of streetscape greenery in the residential environment and self-reported overall health. In this study, stress level and level of perceived social cohesion were shown to be more important mediators than physical activity level. Dadvand *et al.* (2016) looked at the extent to which the association between vegetation in the residential environment and self-reported overall health was mediated by physical activity level, level of perceived social support and mental health status. They concluded that the latter two were more important mediators than the physical activity level.

James *et al.* (2016) looked at the extent to which the beneficial association between the amount of greenery in the residential environment and mortality (rate) was mediated by either air quality (concentration of fine dust), physical activity, mental health (depression), and/or social engagement. Mental health and social engagement proved to be much stronger mediators of the greenery-mortality association than air quality and physical activity. All three studies have in common that the social pathway is observed to be stronger than the physical activity pathway when it comes to overall health and mortality. Note that the latter is mainly due to the first leg of that pathway, residential greenery (A) – overall physical activity level (B), being weak. Obviously, it does not mean that one's overall physical activity level (B) is not important for one's overall health (C).

There are more studies looking specifically at mediators with regard to mental health as endpoint. Zhang *et al.* (2021) performed a systematic review and concluded that there was empirical support for air quality, perceived stress and physical activity being mediators for mental well-being, whereas the support for social cohesion was not that strong. For mental illness, no or only weak evidence was observed for all mediators. It may be noted that Zhang *et al.* did not compare the extent of mediation between different mediators and that they also included studies looking at only one mediator. The authors themselves conclude that the empirical evidence regarding which pathways are important for mental illness and well-being in real-life situations, and to what extent, is still limited (Zhang *et al.*, 2021).

10.6 A new framework for studying pathways

To try and advance the research on pathways, we introduce a modified and extended version of the model proposed by Bratman *et al.* (2019) for effects of nature specifically on mental health. This new framework is meant to help structure thinking about which steps are important for a specific pathway, leading to a specific beneficial health impact. The first step, Existence, is about what is present in terms of nature, relative to one's location. Usually, the location chosen is the home of the individual. So, this step is about issues such as the amount of nature present within a certain distance, the size of and distance to specific green areas, the type(s) of nature and its characteristics. The second step, Exposure, concerns the amount of exposure to (parts) of this nature. Note that exposure is taken here to mean actual exposure: time spent in contact with nature, be it (only) by sight or (also) by other senses. In the literature, a proxy is often used for exposure, such as the amount of vegetation present within a certain distance, with this the proxy still being referred to as exposure. However, it is important to distinguish the two. For example, Schipperijn *et al.* (2010) showed that almost half of the people did not use the nearest urban green area the most, indicating that the distance to this nearest area is a rather poor proxy for actual exposure. The third step, Engagement, is about the type of contact with nature and/or the type of interaction with nature during the exposure. This may vary from looking at nature through one's living room window to taking a walk in a forest, to growing vegetables in an allotment garden, to give a few examples. The fourth step, Experience, is the experience resulting from the exposure and the engagement. It can be about how relaxing, stimulating or pleasurable the experience was, or about how meaningful it was to the individual. Finally, the resulting health effects may differ by health aspect. This new framework has been named the Natural Environment for Health Impact Assessment, or the Natural E4HIA Framework for short (Figure 10.1).

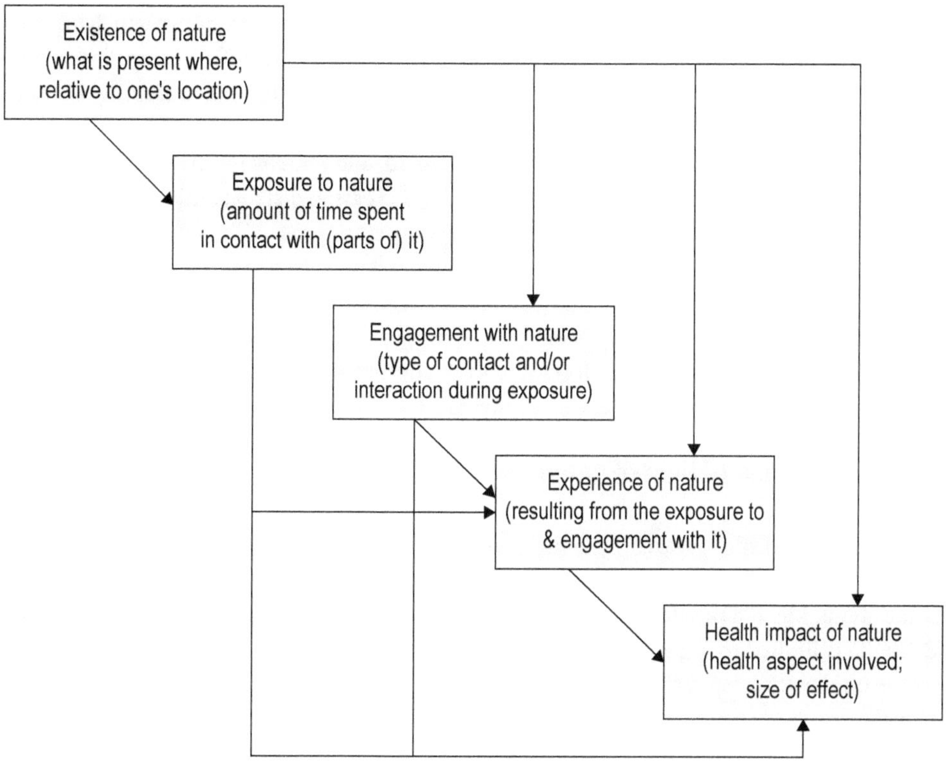

Figure 10.1. The Natural Environment for Health Impact Assessment Framework (Natural E4HIA Framework). The arrows in the framework indicate causal relationships (A affects B). There is no arrow between Exposure and Engagement because although Exposure is a precondition for Engagement, Engagement is not defined by it.

The proposed framework is meant for pathways towards physical health, as well as those towards mental health. At the same time, not all steps are equally important for each pathway in producing health benefits. This specifically pertains to Exposure, Engagement and Experience. As an example, consider the effect of residential greenery on heat stress. On the one hand, there is an effect of vegetation on the ambient air temperature. Night-time temperatures in surrounding homes (without air-conditioning) may be (somewhat) lower thanks to this, which is important for improving sleep quality. To profit from the effect, the individual does not need to be exposed to this residential greenery, and also the engagement with and experience of it are irrelevant. On the other hand, during daytime people may profit from the shade of trees, offering a relatively cool spot. To profit from this shade, the individual has to look up the trees: exposure is required. However, the activity employed during the visit (engagement) is less relevant for this specific effect. One could read a book or socialise with family and friends. On the other hand, when it comes to the pathway of increased physical activity, one's behaviour is essential. For the amount of energy expenditure, how the activity is experienced is less relevant. For the (mental) stress reduction pathway, the opposite is true: the experience is of paramount importance.

The second way in which the framework may be useful, is by helping to identify which aspects of the residential greenery (Existence) are likely to be relevant. This may differ by step. For example, the distance to a park is generally acknowledged as an important factor in determining how often one visits that park, and thereby the amount of exposure. However, it may be much less relevant for Engagement and Experience once the park is reached, and thereby for the health benefits per visit. Whereas Exposure is assumed to decrease with increasing distance, this does not hold for the quality of the experience, however defined. There is even a case to be made that the quality of the experience is likely to increase with increasing distance. In fact, the Travel Cost Method (TCM) for assessing the value of a visit to a park (in economic terms) assumes that if the travel costs increase, so must the minimal benefits of the visit, in order for the trip to still be worthwhile. Thus, there might even be a trade-off between the amount of exposure and the quality of that exposure, in terms of the produced health benefits.

10.7 The relevance of type of nature with regard to health impacts

There is relatively little research on which type of nature, with which characteristics, is most effective for what health aspect, and on the main pathway(s) through which the effect is expected to come about. Most of the aforementioned research on nature and human health and well-being in daily life is limited to accessibility and/or availability of nature on the one hand and a health outcome on the other hand (but see Lackey *et al.*, 2021, for some exceptions). Accessibility tends to focus on public greenery, for example the distance to the nearest park. Availability measures, on the other hand, usually include different types of greenery, in the sense of different land-use categories, or look at generic indices based on remote sensing images (aerial photographs or satellite images). A commonly used index is the Normalised Difference Vegetation Index (NDVI), based on the amount of photosynthesis taking place. The size of the buffer for which availability is assessed differs from study to study, but values are commonly between 300 metres and 1 kilometre from home.

The general conclusion is that better access and higher availability tend to be associated with better health and well-being. However, there are indications that availability is more consistently related to health than access (Ekkel and De Vries, 2017). A potential explanation is that higher availability is likely to go hand in hand with more variation in types of nature and the characteristics of different areas and elements, potentially resulting in more and/or better matches with what works well for different people, or even what works well for a single individual on different occasions. Needs and preferences differ between persons and can change over time within a person. Interestingly, there is relatively little research looking into the availability of nature on a per capita basis. The extent to which the same amount of nature has to be shared with other residents is likely to be relevant for pathways in which activity and/or experience are important steps. For example, a relaxing experience may be more difficult to achieve in a park that is overcrowded. Whether the total available amount of nature or the amount per capita is more important has considerable implications for planning purposes, such as what amount is considered enough.

Beute *et al.* (2020) conducted a systematic evidence mapping of studies looking into which characteristics of nature are important with regard to mental health. They concluded that almost all distinguished types of nature generally showed a beneficial association with or effect on mental health and well-being, with the possible exception of shrubland. Thus far, only few studies directly compared the effects or associations of different types of nature on/with mental health or well-being. Outcomes of (direct and indirect) comparisons were mixed. For example, in experimental studies forests tended to outperform parks, whereas in cross-sectional studies this tended to be the opposite. Nguyen *et al.* (2021) also conducted a systematic review and concluded that trees seem to be more effective than grass.

As for characteristics, the level of (perceived) biodiversity was the only characteristic that was studied frequently enough to draw a preliminary conclusion: a higher level appears to be more beneficial with regard to mental health (Beute *et al.*, 2020). Based on a systematic review, Reyes-Riveros (2021) also concluded that for public greenery higher levels of (perceived) biodiversity, but also naturalness, showed stronger beneficial associations with (mental) well-being. However, the term biodiversity is used in quite different ways in the literature. For example, Marselle *et al.* (2021) use it to denote the amount of living nature as well as its composition. Here, we use it as a characteristic of the available nature, regardless of its amount. Within empirical work biodiversity is often defined as the variation in the species that are present, in practice often limited to a subset of these species: the ones that are easy to observe and identify by lay people/ordinary citizens. Another frequently used definition is the variation perceived by research participants themselves. More ecologically relevant operationalisations, such as functional biodiversity, relevant for the health of the ecosystems, are rare in this type of research (De Vries and Snep, 2019). This raises the question about the extent to which the observed beneficial role of biodiversity with regard to health really aligns health promotion goals and policies with those for biodiversity. For example, will asking citizens to replace the exotic plants in their garden with native plants, which may make the garden less (bio)diverse in their perception, contribute to increased well-being or not?

10.8 Type of contact and interaction with nature (Engagement)

The way people come into contact with nature may differ substantially. It may even be by means of representations. Actually, representations of nature, such as photographs, videos and virtual reality, are quite commonly used in experimental studies. However, here we will focus on contact with real nature. Many of the studies that do look at actual exposure, look (only) at contacts in the form of visits to a public (urban) green areas, such as parks, usually for recreational purposes (see e.g. Lackey *et al.*, 2021). However, people also come into contact with nature while in transit. This includes not only the parks one travels through en route to one's destination, but also smaller natural elements such as street trees and vegetation in domestic front gardens. Such contacts are usually not intentional, but incidental. Research has shown that streetscape greenery is also beneficially associated with health and well-being, whether self-report measures of health are used (Helbich *et al.*, 2019; Van Dillen *et al.*, 2012), or the prescriptions of antidepressants (Marselle *et al.*, 2020).

Another type of contact is looking at the nature that is visible through one's living room window. This type of contact has also proven to be beneficially associated with one's mental health. For example, according to Honold *et al.* (2016) a window view of varied greenery is associated with lower levels of the stress hormone cortisol. Recent studies suggest that during the COVID-19 pandemic a window view of nature may have been even more beneficial, especially when a strict lock-down was in operation (Dzhambov *et al.*, 2021; Ribeiro *et al.*, 2021). Poortinga *et al.* (2021) observed a similar pattern for domestic gardens when the nearest public greenery is located further from home. In regions in which it was allowed, the number of visits to public greenery tended to increase considerably during the pandemic (see e.g. Venter *et al.*, 2020). A possible reason for the increased importance of contact with nature is that the COVID-19 pandemic and the associated measures taken by authorities are stressors, implying that there is more stress to reduce than before the pandemic. This may have increased the demand for relaxing environments, and thereby the visitation of green areas. However, the increased visitation may also partly be due to people feeling cooped up at home, with few alternative outdoor places to go to.

Thus, visual contact with nature already has a beneficial association with well-being. This is backed up by the many laboratory experiments in which the contact with nature was only visual. However, that is not to say that there are no differences in strengths of associations or effect sizes between different modes of contact. Browning *et al.* (2020) reviewed experimental studies that compared effects of virtual and real nature on mood. They found that an actual natural setting improved people's mood better than its virtual counterpart. The same may be expected to be true for looking at (real) nature through a window versus going out into nature. The latter experience is likely to be more immersive, involving more senses, to begin with. Also, the type of interaction is likely to differ between looking at nature through a window and being in a natural area. There is some evidence that active engagement is more beneficial than passive engagement (Holt *et al.*, 2019). Note that physical activity in itself is also beneficial for one's well-being, physically as well as mentally.

According to Bratman *et al.* (2019), around 150 different nature-human interaction patterns have been identified. Assuming that in the case of incidental contacts the interaction is limited to the sensory perception of nature, most patterns are likely to be related to purposeful visits of natural areas (or elements, e.g. climbing a tree). Perhaps interactions may be distinguished by the role that the natural area plays in the interaction. The behaviour displayed may be an activity in which the environment does not play a central role, other than offering an opportunity to perform the activity, such as playing a volleyball game. In that case the natural environment mainly functions as a backdrop, a decor in which the activity takes place. On the other hand, there may be activities in which one is actively and physically involved with the natural environment itself. Gardening and landscape maintenance activities come to mind. According to Van Lier *et al.* (2017), home gardening is associated with several health benefits (including healthier dietary habits). This appears to be the case for allotment gardening as well, with also benefits for social well-being (Genter *et al.*, 2015; Soga *et al.*, 2017). And then there is a category of conscious nature observation or study, which may not require physical contact, or at least not interfering with the natural environment. A first experimental study suggests that a higher awareness of the surrounding nature increases its restorative effect (Lin *et al.*, 2014b). But all in all, comparative research looking into the size of resulting health benefits of different types of interaction with nature in daily life is quite limited.

10.9 Differences between individuals and population segments

Up until now, research has paid little attention to individual differences and how these affect the influence of residential greenery on one's health. This is a question not so much about the effect of individual characteristics on health as such, but about how these characteristics interact with the four E's mentioned in the Natural E4HIA Framework in affecting health: Existence, Exposure, Engagement and Experience (Figure 10.1). We will only address some individual characteristics. The first is gender. Based on a systematic review, Bolte *et al.* (2019) concluded that there are no consistent gender differences in the impact of greenery on self-rated health. However, most of the reviewed studies only used Existence-based measures. Effects may differ by step. For example, when it comes to the visitation of urban green areas (Exposure), women tend to be more concerned with social safety (Experience) and visit such areas alone less often (Sreetheran and Konijnendijk-Van den Bosch, 2014).

Another characteristic is life stage. According to Li *et al.* (2021), there is no systematic evidence that exposure at any given life stage is more saliently associated with mental health outcomes than at others. As for physical health, based on their review Browning and Lee (2017) concluded that associations of the amount of residential greenery (Existence) with physical health outcomes were similar across the life span (but stronger for larger buffer sizes, up to two kilometres).

A third characteristic is socio-economic status (SES), usually defined in terms of level of education and economic wealth. SES is spatially relatively important, because spatial differences in the composition of the local population are larger on this dimension than, for example, for gender: there is a reason that we talk about low- and high-SES neighbourhoods, but not about female and male neighbourhoods. Based on their overview, Rigolon *et al.* (2021) conclude that low-SES people benefit more from residential greenery than high-SES people. One likely reason is that low-SES people tend to be confronted with more stressors in their daily life, for example in making ends meet financially. Therefore, they may have a higher need for stress reduction. Also, a lower mobility of low-SES people may play a role, making the residential environment more important. With regard to the latter, lower-SES people tend to have less nature of reasonable quality in their residential environment than their high-SES counterparts. Thus, SES is not only a moderator of the association between the amount of residential greenery and health and well-being, but also a determinant of the amount of neighbourhood greenery, as well as its attractiveness and its suitability for use. This is often referred to as a lack of (distributive) environmental justice with regard to access to nature (see e.g. De Vries *et al.*, 2020).

A final individual characteristic is how strongly a person feels connected to nature. The idea is that contact with nature may be more beneficial for one's well-being when one feels more connected to nature (also cognitively). According to Lin *et al.* (2014a), people that feel more strongly connected to nature visit parks more often. Martin *et al.* (2020) showed that park visits had a stronger association with self-reported general health among the people that felt more connected to nature. Especially with regard to this topic of nature connectedness, the possibility of reversed causality comes to mind. Is it the level of nature connectedness that drives the association with visitation and subsequent well-being effects, or does visiting nature more often leads to a higher level of nature connectedness? With regard to the latter, there are studies showing that people feel a higher level of nature connectedness after a nature visit, at the very least for a short time (Capaldi *et al.*, 2017), making it likely that the relationship is bi-directional (see also Wyles *et al.*, 2019). Note that this implies that nature connectedness might be enhanced by bringing people into contact with nature, resulting in larger well-being benefits of future visits.

The final moderator that we want to discuss is one at the neighbourhood level. It is the level of urbanity, commonly measured in terms of population density. Obviously, the level of urbanity is usually strongly negatively associated with the amount of nearby greenery. (This is even more so the case when looking at the amount per inhabitant.) The question is whether differences in the amount of residential greenery matter more at the lower end of the spectrum than at the upper end. Based on their review focusing on physical health, Browning *et al.* (2022) conclude that greenspace-health associations indeed tend to be stronger in more urban areas.

10.10 Size and valuation of health impacts

Urban planning involves making choices: how much space needs to be reserved for which type of land use or function? And where to locate this space? Since a lot of functions have to be accommodated in a limited amount of space, many trade-offs have to be made by urban planners. Therefore, it helps to be able to assess the advantages and disadvantages of certain choices for a specific function. This includes the quantification of such advantages and disadvantages: *how much* better or worse? A first step in this direction is to assess the size of the impact of a change in the urban green infrastructure on a specific health parameter. A subsequent step is the valuation of that impact for this specific health parameter. The previous sections have shown that, generally speaking, positive associations between residential greenery and human health and well-being have been firmly established, and that it is plausible that such associations are, at least partly, causal in nature. At the same time, they have also shown that the picture is quite nuanced.

Up until now, research has largely focused on the amount of vegetation and/or public green areas. Relatively little attention has been paid to the specific qualities of the vegetation and the green areas. The available evidence suggests that attractiveness matters, especially in the case of green areas, which are usually (purposefully) visited during leisure time. Furthermore, there are many different health aspects to consider, and impacts may not be the same for all aspects and all population segments. The same greenery may have a positive health impact on one aspect, but a negative impact on another. For example, a park that is highly suited to picnicking and barbecuing, and frequently used for this activity, may beneficially impact the social well-being of the visitors, but do little for their level of physical activity (the energy intake during a visit could even be larger than the energy expenditure). But this may be more the case for the adult visitors (sitting, eating and talking with each other) than for the accompanying children, engaging in physically active play.

For valuation purposes, rather than trying to value the impact per health aspect, it might be an option to look at measures of overall health or well-being, such as self-reported health, preventable mortality, quality of life, or quality-adjusted life years (QALYs). For example, Barboza *et al.* (2021) estimated the number of life years that could be gained by adhering to the WHO recommendation for access to greenery in about 1000 European cities, and arrived at a number of 245 years of life per 100,000 inhabitants per year. Jones (2021), who studied the impact of the New York City Million Trees Program, calculated that the increase in quality of life due to that programme could be valued at 506 US dollars extra household income per month.

10.11 Conclusions and recommendations for urban planners and landscape architects

Although the causality of beneficial associations between the amount of residential nature and health remains an issue, it is likely that the associations are at least partially causal, with the amount of nature driving the association. For some of the pathways linking contact with nature to health and well-being, such as stress reduction, this is supported by experimental evidence. For other pathways, it is supported by intimate knowledge of the process (fine dust concentration, heat stress). The amount of contact with nature is important, even if it is only visual contact. This is especially clear for mental health and well-being.

Relevant contacts with nature are not limited to visiting parks, forests, nature areas or the countryside, but include so-called superficial and unintentional contacts. Since most citizens spend a lot of time indoors, either at home or at work (or school), this makes having a (green) domestic garden and (window views on) street trees relatively important for their health and well-being. Also, most people spend little of their (leisure) time on outdoor recreational activities in natural settings. This implies that the effect of an individual visit needs to be substantial and long lasting, in order to positively impact long-term health and well-being.

Although there are differences between the different types of greenery and associated health and well-being benefits, thus far these differences appear to be relatively small. Contact with almost all types of nature tends to show a beneficial association, although trees may outperform grass in this respect. Therefore, it is highly likely that the amount of contact with residential nature of at least reasonable quality (safe, well-maintained) is more important for human health and well-being than the limited contact occurring during occasional visits to a high-quality nature area (e.g. protected area) located further away. In terms of the introduced Natural E4HIA Framework: differences in Exposure seem to matter more than differences in Engagement and Experience.

With regard to intentional visits to green areas, the attractiveness of the area is an important issue. What is considered attractive may differ from person to person. However, in general safety is important (but safety requirements may still differ, e.g. by age group). Second, attractiveness depends on the purpose for which people visit the area. Is it to find peace and quiet, to spend quality time with family members, to experience the outdoors/natural environment (fresh air), to study nature, or to be physically active (with some activities requiring specific amenities or facilities)? Given these different purposes, it is perhaps better to look into the suitability of the area for specific functions, rather than into its (generic) quality.

There are issues that have not been addressed in this chapter, for instance ways to increase contact with nature other than by increasing its amount or attractiveness. One such way is improving the access to existing natural areas. For example, it could be made easier for urbanites to reach the countryside. Another way is actively promoting the use of natural areas, for example by programming activities in these areas (Gianfredi *et al.*, 2021; WHO, 2017). Another such issue is the potential negative health and well-being effects of residential nature. Examples are nature harbouring pests (e.g. oak processionary moths) and vectors for zoonoses (ticks, mosquitoes), or its release of allergens and biological volatile organic compounds (BVOCs) that are detrimental to human health. Obviously, such negative side effects should be avoided as much as possible. A different type of negative side effect is eco-gentrification: poor residents being driven out of the neighbourhood and replaced by wealthier people, because of the urban greening resulting in higher house prices. According to Wolch *et al.* (2014), the solution might be making the neighbourhood just green enough, likely also in terms of its qualities. The prevention of eco-gentrification is especially important because, in terms of health, people with a low socio-economic status (SES) tend to profit more from residential nature than those with a high SES. According to Mitchell *et al.* (2015), greening low-SES neighbourhoods may therefore be an important means to reduce socio-economic health disparities, nowadays a common ambition in many national health policies.

Finally, residential nature may also contribute to other goals. One of these goals is to increase the level of biodiversity in the city as such. Another important goal is to increase the climate resilience of the city, with regard to water management and heat stress reduction. It is still largely an open question to what extent the same location, size and type of nature is the most optimal choice for all these functions, or that choices have to be made regarding which function to prioritise. In the event that trade-offs have to be made, the valuation of the health and well-being benefits of residential nature may help to take such benefits explicitly into account. However, this type of research is still very much under development. Since nature seems to affect many different aspects of health and well-being, we recommend using generic endpoints, such as QALYs, in valuation exercises, rather than the prevalence of a specific disease or disorder.

References

Aerts, R., Nemery, B., Bauwelinck, M., Trabelsi, S., Deboosere, P., Van Nieuwenhuyse, A., Nawrot, T.S. and Casas, L., 2020. Residential green space, air pollution, socioeconomic deprivation and cardiovascular medication sales in Belgium: A nationwide ecological study. Science of the Total Environment 712: 136426. https://doi.org/10.1016/j.scitotenv.2019.136426

Akaraci, S., Feng, X., Suesse, T., Jalaludin, B. and Astell-Burt, T., 2020. A systematic review and meta-analysis of associations between green and blue spaces and birth outcomes. International Journal of Environmental Research and Public Health 17(8): 2949.

Alcock, I., White, M.P., Wheeler, B.W., Fleming, L.E. and Depledge, M.H., 2014. Longitudinal effects on mental health of moving to greener and less green urban areas. Environmental Science & Technology 48(2): 1247-1255.

Barboza, E.P., Cirach, M., Khomenko, S., Iungman, T., Mueller, N., Barrera-Gómez, J., Rojas-Rueda, D., Kondo, M. and Nieuwenhuijsen, M., 2021. Green space and mortality in European cities: a health impact assessment study. The Lancet Planetary Health 5(10): e718-e730. https://doi.org/10.1016/S2542-5196(21)00229-1

Berry, M.S., Repke, M.A., Metcalf, A.L. and Jordan, K.E., 2020. Promoting healthy decision-making via natural environment exposure: initial evidence and future directions. Frontiers in Psychology 1682. https://doi.org/10.3389/fpsyg.2020.01682

Beute, F., Andreucci, M.B., Lammel, A., Davies, Z.G., Glanville, J., Keune, H., Marselle, M., O'Brien, L.A., Olszewska-Guizzo, A., Remmen, R., Russo, A. and De Vries, S., 2020. Types and characteristics of urban and peri-urban green spaces having an impact on human mental health and wellbeing: a systematic review. Technical report. EKLIPSE Expert Working Group. UK Centre for Ecology & Hydrology, Wallingford, UK.

Bolouki, A., 2022. Neurobiological effects of urban built and natural environment on mental health: systematic review. Reviews on Environmental Health. https://doi.org/10.1515/reveh-2021-0137

Bolte, G., Nanninga, S. and Dandolo, L., 2019. Sex/gender differences in the association between residential green space and self-rated health – a sex/gender-focused systematic review. International Journal of Environmental Research and Public Health 16(23): 4818.

Brame, J.E., Liddicoat, C., Abbott, C.A. and Breed, M.F., 2021. The potential of outdoor environments to supply beneficial butyrate-producing bacteria to humans. Science of the Total Environment 777: 146063.

Bratman, G.N., Anderson, C.B., Berman, M.G., Cochran, B., De Vries, S., Flanders, J., Folke, C., Frumkin, H., Gros. J.J., Hartig, T., Kahn Jr., P.H., Kuo, M., Lawler, J.J., Levin, P.S., Lindahl, T., Meyer-Lindenberg, A., Mitchell, R., Ouyang, Z., Roe, J., Scarlett, L., Smith, J.R., Van den Bosch, M., Wheeler, B.W., White, M.P., Zheng, H. and Daily, G.C., 2019. Nature and mental health: an ecosystem service perspective. Science Advances 5(7): eaax0903.

Browning, M. and Lee, K., 2017. Within what distance does 'greenness' best predict physical health? A systematic review of articles with GIS buffer analyses across the lifespan. International Journal of Environmental Research and Public Health 14(7): 675. https://doi.org/10.3390/ijerph14070675

Browning, M.H. and Rigolon, A., 2019. School green space and its impact on academic performance: a systematic literature review. International Journal of Environmental Research and Public Health 16(3): 429.

Browning, M.H., Rigolon, A. and McAnirlin, O., 2022. Where greenspace matters most: a systematic review of urbanicity, greenspace, and physical health. Landscape and Urban Planning 217: 104233.

Browning, M.H., Shipley, N., McAnirlin, O., Becker, D., Yu, C.P., Hartig, T. and Dzhambov, A.M., 2020. An actual natural setting improves mood better than its virtual counterpart: a meta-analysis of experimental data. Frontiers in Psychology 2200.

Capaldi, C.A., Passmore, H.A., Ishii, R., Chistopolskaya, K.A., Vowinckel, J., Nikolaev, E.L. and Semikin, G.I., 2017. Engaging with natural beauty may be related to well-being because it connects people to nature: evidence from three cultures. Ecopsychology 9(4): 199-211.

Cohen-Cline, H., Turkheimer, E. and Duncan, G.E., 2015. Access to green space, physical activity and mental health: a twin study. Journal of Epidemiology and Community Health 69(6): 523-529.

Corazon, S.S., Sidenius, U., Poulsen, D.V., Gramkow, M.C. and Stigsdotter, U.K., 2019. Psycho-physiological stress recovery in outdoor nature-based interventions: a systematic review of the past eight years of research. International Journal of Environmental Research and Public Health 16(10): 1711.

Dadvand, P., Bartoll, X., Basagaña, X., Dalmau-Bueno, A., Martinez, D., Ambros, A., Cirach, M., Triguero-Mas, M., Gascon, M., Borrell, C. and Nieuwenhuijsen, M.J., 2016. Green spaces and general health: roles of mental health status, social support, and physical activity. Environment International 91: 161-167.

De la Fuente, F., Saldías, M.A., Cubillos, C., Mery, G., Carvajal, D., Bowen, M. and Bertoglia, M.P., 2021. Green space exposure association with type 2 diabetes mellitus, physical activity, and obesity: a systematic review. International Journal of Environmental Research and Public Health 18(1): 97.

De Vries, S. and Snep, R., 2019. Biodiversity in the context of 'biodiversity-mental health' research. In: Marselle, M.R., Stadler, J., Korn, H., Irvine, K.N. and Bonn, A. (eds.) Biodiversity and health in the face of climate change. Springer, Cham, Switzerland, pp. 159-173.

De Vries, S., Buijs, A.E. and Snep, R.P., 2020. Environmental justice in The Netherlands: presence and quality of greenspace differ by socioeconomic status of neighbourhoods. Sustainability 12(15): 5889.

De Vries, S., Nieuwenhuizen, W., Farjon, H., Van Hinsberg, A. and Dirkx, J., 2021. In which natural environments are people happiest? Large-scale experience sampling in the Netherlands. Landscape and Urban Planning 205: 103972.

De Vries, S., Ten Have, M., Van Dorsselaer, S., Van Wezep, M., Hermans, T. and De Graaf, R., 2016. Local availability of green and blue space and prevalence of common mental disorders in the Netherlands. BJPsych Open 2(6): 366-372.

De Vries, S., Van Dillen, S.M., Groenewegen, P.P. and Spreeuwenberg, P., 2013. Streetscape greenery and health: stress, social cohesion and physical activity as mediators. Social Science & Medicine 94: 26-33.

Donovan, G.H., Michael, Y.L., Gatziolis, D., Mannetje, A.T. and Douwes, J., 2019. Association between exposure to the natural environment, rurality, and attention-deficit hyperactivity disorder in children in New Zealand: a linkage study. The Lancet Planetary Health 3(5): e226-e234.

Dzhambov, A.M., Lercher, P., Browning, M.H., Stoyanov, D., Petrova, N., Novakov, S. and Dimitrova, D.D., 2021. Does greenery experienced indoors and outdoors provide an escape and support mental health during the COVID-19 quarantine? Environmental Research 196: 110420.

Dzhambov, A.M., Markevych, I., Tilov, B., Arabadzhiev, Z., Stoyanov, D., Gatseva, P. and Dimitrova, D.D., 2018. Lower noise annoyance associated with GIS-derived greenspace: pathways through perceived greenspace and residential noise. International Journal of Environmental Research and Public Health 15(7): 1533.

Egorov, A.I., Griffin, S.M., Converse, R.R., Styles, J.N., Sams, E.A., Wilson, A., Jackson, L. and Wade, T.J., 2017. Vegetated land cover near residence is associated with reduced allostatic load and improved biomarkers of neuroendocrine, metabolic and immune functions. Environmental Research 158: 508-521.

Ekkel, E.D. and De Vries, S., 2017. Nearby green space and human health: evaluating accessibility metrics. Landscape and Urban Planning 157: 214-220.

Flandroy, L., Poutahidis, T., Berg, G., Clarke, G., Dao, M.C., Decaestecker, E., Furman, E., Haahtela, T., Massart, S., Plovier, H., Sanz, Y. and Rook, G., 2018. The impact of human activities and lifestyles on the interlinked microbiota and health of humans and of ecosystems. Science of the Total Environment 627: 1018-1038. https://doi.org/10.1016/j.scitotenv.2018.01.288

Genter, C., Roberts, A., Richardson, J. and Sheaff, M., 2015. The contribution of allotment gardening to health and wellbeing: a systematic review of the literature. British Journal of Occupational Therapy 78(10): 593-605.

Gianfredi, V., Buffoli, M., Rebecchi, A., Croci, R., Oradini-Alacreu, A., Stirparo, G., Marino, A., Odone, A., Capolongo, S and Signorelli, C. (2021). Association between urban greenspace and health: a systematic review of literature. International Journal of Environmental Research and Public Health 18(10): 5137. https://doi.org/10.3390/ijerph18105137

Helbich, M., Yao, Y., Liu, Y., Zhang, J., Liu, P. and Wang, R., 2019. Using deep learning to examine street view green and blue spaces and their associations with geriatric depression in Beijing, China. Environment International 126: 107-117.

Holt, E.W., Lombard, Q.K., Best, N., Smiley-Smith, S. and Quinn, J E., 2019. Active and passive use of green space, health, and well-being amongst university students. International Journal of Environmental Research and Public Health 16(3): 424.

Honold, J., Lakes, T., Beyer, R. and Van der Meer, E., 2016. Restoration in urban spaces: nature views from home, greenways, and public parks. Environment and Behavior 48(6): 796-825.

Houlden, V., Weich, S., Porto de Albuquerque, J., Jarvis, S. and Rees, K., 2018. The relationship between greenspace and the mental wellbeing of adults: a systematic review. PloS One 13(9): e0203000.

Hu, C.Y., Yang, X.J., Gui, S.Y., Ding, K., Huang, K., Fang, Y., Jiang, Z.X and Zhang, X.J., 2021. Residential greenness and birth outcomes: a systematic review and meta-analysis of observational studies. Environmental Research 193: 110599.

Hystad, P., Payette, Y., Noisel, N. and Boileau, C., 2019. Green space associations with mental health and cognitive function: results from the Quebec CARTaGENE cohort. Environmental Epidemiology 3(1).

Jabbar, M., Yusoff, M.M. and Shafie, A., 2021. Assessing the role of urban green spaces for human well-being: a systematic review. GeoJournal 20: 1-19. https://doi.org/10.1007/s10708-021-10474-7

James, P., Hart, J.E., Banay, R.F. and Laden, F., 2016. Exposure to greenness and mortality in a nationwide prospective cohort study of women. Environmental Health Perspectives 124(9): 1344-1352.

Jones, B.A., 2021. Planting urban trees to improve quality of life? The life satisfaction impacts of urban afforestation. Forest Policy and Economics 125: 102408.

Kondo, M.C., Jacoby, S.F. and South, E.C., 2018. Does spending time outdoors reduce stress? A review of real-time stress response to outdoor environments. Health & Place 51: 136-150.

Kuo, F.E., Sullivan, W.C., Coley, R.L. and Brunson, L., 1998. Fertile ground for community: inner-city neighborhood common spaces. American Journal of Community Psychology 26(6): 823-851.

Lackey, N.Q., Tysor, D.A., McNay, G.D., Joyner, L., Baker, K.H. and Hodge, C., 2021. Mental health benefits of nature-based recreation: a systematic review. Annals of Leisure Research 24(3): 379-393.

Lambert, A., Vlaar, J., Herrington, S. and Brussoni, M., 2019. What is the relationship between the neighbourhood built environment and time spent in outdoor play? A systematic review. International Journal of Environmental Research and Public Health 16(20): 3840.

Li, D., Menotti, T., Ding, Y. and Wells, N.M., 2021. Life course nature exposure and mental health outcomes: a systematic review and future directions. International Journal of Environmental Research and Public Health 18(10): 5146.

Lin, B.B., Fuller, R.A., Bush, R., Gaston, K.J. and Shanahan, D.F., 2014a. Opportunity or orientation? Who uses urban parks and why. PLoS One 9(1): e87422.

Lin, Y.H., Tsai, C.C., Sullivan, W.C., Chang, P.J. and Chang, C.Y., 2014b. Does awareness effect the restorative function and perception of street trees? Frontiers in Psychology 5: 906.

Luque-García, L., Corrales, A., Lertxundi, A., Díaz, S. and Ibarluzea, J., 2021. Does exposure to greenness improve children's neuropsychological development and mental health? A Navigation Guide systematic review of observational evidence for associations. Environmental Research 206: 112599. https://doi.org/10.1016/j.envres.2021.112599

Luo, Y.N., Huang, W.Z., Liu, X.X., Markevych, I., Bloom, M.S., Zhao, T., Heinrich, J., Yang, B.Y. and Dong, G.H., 2020. Greenspace with overweight and obesity: a systematic review and meta-analysis of epidemiological studies up to 2020. Obesity Reviews 21(11): e13078.

MacKerron, G. and Mourato, S., 2013. Happiness is greater in natural environments. Global Environmental Change 23(5): 992-1000.

Markevych, I., Schoierer, J., Hartig, T., Chudnovsky, A., Hystad, P., Dzhambov, A.M., De Vries, S., Triguero-Mas, M., Brauer, M., Nieuwenhuijsen, M.J., Lupp, G., Richardson, E.A., Astell-Burt, T., Dimitrova, D., Feng, X., Sadeh, M., Standl, M. Heinrich, J. and Fuertes, E., 2017. Exploring pathways linking greenspace to health: theoretical and methodological guidance. Environmental Research 158: 301-317.

Marselle, M.R., Bowler, D.E., Watzema, J., Eichenberg, D., Kirsten, T. and Bonn, A., 2020. Urban street tree biodiversity and antidepressant prescriptions. Scientific Reports 10(1): 1-11.

Marselle, M.R., Hartig, T., Cox, D.T., De Bell, S., Knapp, S., Lindley, S., Triguero-Mas, M., Böhning-Gaese, K., Brauback, M., Irvine, K.N., Kabisch, N., Kolek, F., Kraemer, R., Markevych, I., Martens, D., Müller, R., Nieuwenhuijsen, M., Potts, J.M., Stadler, J., Walton, S., Warber, S. and Bonn, A., 2021. Pathways linking biodiversity to human health: a conceptual framework. Environment International 150: 106420. https://doi.org/10.1016/j.envint.2021.106420

Martin, L., White, M.P., Hunt, A., Richardson, M., Pahl, S. and Burt, J., 2020. Nature contact, nature connectedness and associations with health, wellbeing and pro-environmental behaviours. Journal of Environmental Psychology 68: 101389.

Mason, L., Ronconi, A., Scrimin, S. and Pazzaglia, F., 2021. Short-term exposure to nature and benefits for students' cognitive performance: a review. Educational Psychology Review 1-39.

McMahan, E.A. and Estes, D., 2015. The effect of contact with natural environments on positive and negative affect: a meta-analysis. Journal of Positive Psychology 10(6): 507-519.

Mitchell, R.J., Richardson, E.A., Shortt, N.K. and Pearce, J.R., 2015. Neighborhood environments and socioeconomic inequalities in mental well-being. American Journal of Preventive Medicine 49(1): 80-84.

Mueller, W., Steinle, S., Pärkkä, J., Parmes, E., Liedes, H., Kuijpers, E., Pronk, A., Sarigjannis, D., Karakitsios, S., Chapizanis, D., Maggos, T., Stamatelopoulou, A., Wilkinson, P. Milner, J, Vardoulakis, S. and Loh, M., 2020. Urban greenspace and the indoor environment: pathways to health via indoor particulate matter, noise, and road noise annoyance. Environmental Research 180: 108850.

Mygind, L., Kjeldsted, E., Hartmeyer, R., Mygind, E., Stevenson, M.P., Quintana, D.S. and Bentsen, P., 2021. Effects of public green space on acute psychophysiological stress response: a systematic review and meta-analysis of the experimental and quasi-experimental evidence. Environment and Behavior 53(2): 184-226.

Nemitz, E., Vieno, M., Carnell, E., Fitch, A., Steadman, C., Cryle, P., Holland, M., Morton, R.D., Hall, J., Mills, G., Hayes, F., Dickie, I., Carruthers, D., Fowler, D., Reis, S. and Jones, L., 2020. Potential and limitation of air pollution mitigation by vegetation and uncertainties of deposition-based evaluations. Philosophical Transactions of the Royal Society A 378(2183): 20190320.

Nguyen, P.Y., Astell-Burt, T., Rahimi-Ardabili, H. and Feng, X., 2021. Green space quality and health: a systematic review. International Journal of Environmental Research and Public Health 18(21): 11028.

Nordbø, E.C.A., Raanaas, R.K., Nordh, H. and Aamodt, G., 2019. Neighborhood green spaces, facilities and population density as predictors of activity participation among 8-year-olds: a cross-sectional GIS study based on the Norwegian mother and child cohort study. BMC Public Health 19(1): 1-22.

Pascal, M., Goria, S., Wagner, V., Sabastia, M., Guillet, A., Cordeau, E., Mauclair, C. and Host, S., 2021. Greening is a promising but likely insufficient adaptation strategy to limit the health impacts of extreme heat. Environment International 151: 106441.

Paul, L.A., Hystad, P., Burnett, R.T., Kwong, J.C., Crouse, D.L., Van Donkelaar, A., Tu, K., Lavigne, E., Copes, R., Martin, R.V. and Chen, H., 2020. Urban green space and the risks of dementia and stroke. Environmental Research 186: 109520.

Poortinga, W., Bird, N., Hallingberg, B., Phillips, R. and Williams, D., 2021. The role of perceived public and private green space in subjective health and wellbeing during and after the first peak of the COVID-19 outbreak. Landscape and Urban Planning 211: 104092.

Remme, R.P., Frumkin, H., Guerry, A.D., King, A.C., Mandle, L., Sarabu, C., Bratman, G.N., Giles-Corti, B., Hamel, P., Han, B., Hicks, J.L., James, P., Lawler, J.J., Lindahl, T., Liu, H., Lu, Y., Oosterbroek, B., Paudel, B., Sallis, J.F., Schipperijn, J., Sosic, R., De Vries, S., Wheeler, B.W., Wood, S.A., Wu, T., and Daily, G.C., 2021. An ecosystem service perspective on urban nature, physical activity, and health. Proceedings of the National Academy of Sciences of the USA 118(22): e2018472118.

Reyes-Riveros, R., Altamirano, A., De la Barrera, F., Rozas-Vasquez, D., Vieli, L. and Meli, P., 2021. Linking public urban green spaces and human well-being: a systematic review. Urban Forestry & Urban Greening 61: 127105.

Ribeiro, A.I., Triguero-Mas, M., Santos, C.J., Gómez-Nieto, A., Cole, H., Anguelovski, I., Martins Silva, F. and Baró, F., 2021. Exposure to nature and mental health outcomes during COVID-19 lockdown. A comparison between Portugal and Spain. Environment international 154: 106664.

Rigolon, A., Browning, M.H., McAnirlin, O. and Yoon, H.V., 2021. Green space and health equity: a systematic review on the potential of green space to reduce health disparities. International Journal of Environmental Research and Public Health 18(5): 2563.

Roberts, H., Van Lissa, C., Hagedoorn, P., Kellar, I. and Helbich, M., 2019. The effect of short-term exposure to the natural environment on depressive mood: a systematic review and meta-analysis. Environmental Research 177: 108606.

Rojas-Rueda, D., Nieuwenhuijsen, M.J., Gascon, M., Perez-Leon, D. and Mudu, P., 2019. Green spaces and mortality: a systematic review and meta-analysis of cohort studies. The Lancet Planetary Health 3(11): e469-e477.

Roslund, M.I., Puhakka, R., Grönroos, M., Nurminen, N., Oikarinen, S., Gazali, A.M., Cinek, O., Kramna, L., Siter, N., Vari, H.K., Soininen, L., Parajuli, A., Rajaniemi, J., Kinnunen, T., Laitinen, O.H., Hyoty, H., Sinkkonen, A. and ADELE research group, 2020. Biodiversity intervention enhances immune regulation and health-associated commensal microbiota among daycare children. Science Advances, 6(42), eaba2578.

Santamouris, M. and Osmond, P., 2020. Increasing green infrastructure in cities: Impact on ambient temperature, air quality and heat-related mortality and morbidity. Buildings 10(12): 233.

Schipperijn, J., Cerin, E., Adams, M.A., Reis, R., Smith, G., Cain, K., Christiansen, L.B., Van Dyck, D., Gidlow, C., Frank, L.D., Mitas, J., Pratt, M., Salvo, D., Schofield, G. and Sallis, J.F., 2017. Access to parks and physical activity: an eight country comparison. Urban Forestry & Urban Greening 27: 253-263.

Schipperijn, J., Stigsdotter, U.K., Randrup, T.B. and Troelsen, J., 2010). Influences on the use of urban green space – a case study in Odense, Denmark. Urban Forestry & Urban Greening 9(1): 25-32.

Seo, S., Choi, S., Kim, K., Kim, S.M. and Park, S.M., 2019. Association between urban green space and the risk of cardiovascular disease: A longitudinal study in seven Korean metropolitan areas. Environment International 125: 51-57.

Shuvo, F.K., Mazumdar, S. and Labib, S.M., 2021. Walkability and greenness do not walk together: investigating associations between greenness and walkability in a large metropolitan city context. International Journal of Environmental Research and Public Health 18(9): 4429.

Slawsky, E.D., Hajat, A., Rhew, I.C., Russette, H., Semmens, E.O., Kaufman, J.D., Leary, C.S. and Fitzpatrick, A.L., 2022. Neighborhood greenspace exposure as a protective factor in dementia risk among US adults 75 years or older: a cohort study. Environmental Health 21(1): 1-10.

Sobko, T., Liang, S., Cheng, W.H. and Tun, H.M., 2020. Impact of outdoor nature-related activities on gut microbiota, fecal serotonin, and perceived stress in preschool children: the Play&Grow randomized controlled trial. Scientific Reports 10(1): 1-12.

Soga, M., Cox, D.T., Yamaura, Y., Gaston, K.J., Kurisu, K. and Hanaki, K., 2017. Health benefits of urban allotment gardening: improved physical and psychological well-being and social integration. International Journal of Environmental Research and Public Health 14(1): 71.

Sreetheran, M. and Konijnendijk-Van Den Bosch, C.C., 2014. A socio-ecological exploration of fear of crime in urban green spaces – a systematic review. Urban Forestry & Urban Greening 13(1): 1-18.

Stevenson, M.P., Schilhab, T. and Bentsen, P., 2018. Attention restoration theory II: a systematic review to clarify attention processes affected by exposure to natural environments. Journal of Toxicology and Environmental Health, Part B 21(4): 227-268.

Thygesen, M., Engemann, K., Holst, G.J., Hansen, B., Geels, C., Brandt, J., Pedersen, C.B. and Dalsgaard, S., 2020. The association between residential green space in childhood and development of attention deficit hyperactivity disorder: a population-based cohort study. Environmental Health Perspectives 128(12): 127011.

Triguero-Mas, M., Donaire-Gonzalez, D., Seto, E., Valentín, A., Smith, G., Martínez, D., Carrasco-Turigas, G., MAsterson, D., Van den Berg, M., Ambros, A., Martinez-Iniguez, T., Dedele, A., Hurst, G., Ellis, N., Grazulevicius, T., Voorsmit, M., Cirach, M., Cirac-Claveras, J., Swart, W., Clasquin, E., Maas, J., Wendel-Vos, W., Jerrett, M., Grazuleviciene, R., Kruize, H., Gidlow, C.J. and Nieuwenhuijsen, M.J., 2017. Living close to natural outdoor environments in four European cities: adults' contact with the environments and physical activity. International Journal of Environmental Research and Public Health 14(10): 1162.

Twohig-Bennett, C. and Jones, A., 2018. The health benefits of the great outdoors: a systematic review and meta-analysis of greenspace exposure and health outcomes. Environmental Research 166: 628-637.

Van den Bosch, M. and Ode Sang, Å., 2017. Urban natural environments as nature-based solutions for improved public health – a systematic review of reviews. Environmental Research 158: 373-384.

Van Dillen, S.M., De Vries, S., Groenewegen, P.P. and Spreeuwenberg, P., 2012. Greenspace in urban neighbourhoods and residents' health: adding quality to quantity. Journal of Epidemiology and Community Health 66(6): e8.

Van Lier, L.E., Utter, J., Denny, S., Lucassen, M., Dyson, B. and Clark, T., 2017. Home gardening and the health and well-being of adolescents. Health Promotion Practice 18(1): 34-43.

Van Moorselaar, I. and Van der Zee, S., 2020. Groen en luchtkwaliteit in een stedelijke omgeving: desktopstudie Academische werkplaats Milieu & Gezondheid. GGD Amsterdam, afdeling Milieu en Gezondheid, Amsterdam, the Netherlands.

Venter, Z.S., Barton, D.N., Gundersen, V., Figari, H. and Nowell, M., 2020. Urban nature in a time of crisis: recreational use of green space increases during the COVID-19 outbreak in Oslo, Norway. Environmental Research Letters 15(10): 104075.

Wan, C., Shen, G.Q. and Choi, S., 2021. Underlying relationships between public urban green spaces and social cohesion: a systematic literature review. City, Culture and Society, 100383.

WHO Europe, 2017. Urban green space interventions and health; a review of impacts and effectiveness. WHO regional office for Europe, Copenhagen, Denmark.

Wolch, J.R., Byrne, J. and Newell, J.P., 2014. Urban green space, public health, and environmental justice: the challenge of making cities 'just green enough'. Landscape and Urban Planning 125: 234-244.

Wyles, K.J., White, M.P., Hattam, C., Pahl, S., King, H. and Austen, M., 2019. Are some natural environments more psychologically beneficial than others? The importance of type and quality on connectedness to nature and psychological restoration. Environment and Behavior 51(2): 111-143.

Yuan, Y., Huang, F., Lin, F., Zhu, P. and Zhu, P., 2021. Green space exposure on mortality and cardiovascular outcomes in older adults: a systematic review and meta-analysis of observational studies. Aging Clinical and Experimental Research 33(7): 1783-1797.

Zhan, Y., Liu, J., Lu, Z., Yue, H., Zhang, J. and Jiang, Y., 2020. Influence of residential greenness on adverse pregnancy outcomes: a systematic review and dose-response meta-analysis. Science of the Total Environment 718: 137420.

Zhang, R., Zhang, C.Q. and Rhodes, R.E., 2021. The pathways linking objectively-measured greenspace exposure and mental health: a systematic review of observational studies. Environmental Research 198: 111233.

Chapter 11
Civic engagement as the corner stone of symbiotic cities

Menno Houtstra, Marian Stuiver, Joop Spijker and Clara McDonnell

Abstract

Local governments, scientists and urban planners need to create a sensitivity to greening initiatives of local citizens and their networks and learn to co-create locally relevant strategies with them, combining communal practices with broader policies and legislation and scientific efforts. We focus on three pathways to enhance this social capital needed for greening cities: communal learning; new types of ownership for citizens' initiatives; and making the impact of citizens' activities and practices more explicit. Three cases demonstrate each of these pathways. First, Steenbreek Foundation, an initiative of universities, municipalities and urban ecologists, exemplifies communal learning through local actions. Steenbreek Foundation aims to engage citizens and local municipalities in greening public and private gardens by demonstrating the added value of greening. In 2021 over 140 municipalities and local initiatives joined Steenbreek Foundation. Second, communal ownership is exemplified by the Kaskantine in Amsterdam, an urban farm café, where residents of the neighbourhood of Nieuw-West take over management of urban land and water to develop nature-based technologies for off-grid living in the city. Third, communal impact is exemplified by the Doughnut Coalition in Amsterdam, an organisation of residents that develop their own bottom-up measurement technique to measure the social impact of their neighbourhood activities. The three case studies show diverse ways of including communities in urban greening initiatives. Whether through encouraging community learning, creating strategies for communal ownership and management of resources, or through measuring and tracking the impact of community action, these initiatives have demonstrated the potential of civic engagement. The chapter concludes with three principles that local governments and urban planners need to embrace in order to involve communities in their greening strategies.

Keywords: neighbourhoods, civic engagement, social capital, impact, learning, distribution of urban resources, economies of scope

11.1 Introduction

In undertaking initiatives to green cities and care for nature and natural resources in urban environments, local governments and urban planners (but also scientists) should pay greater attention to the role citizens and their networks can play in such efforts. We believe that there is great potential for municipalities to embrace and endorse civic engagement in individual and

10.3920/978-90-8686-935-0_11, © Wageningen Academic Publishers 2022

collective actions to develop a symbiotic city. Such efforts could lead to the co-creation of locally relevant greening strategies with citizens and local social networks, combining the local civic practices with broader policies and legislation.

The role of citizens and their networks in maintaining the environments around them is frequently overlooked. Citizens can care for nature in their own urban environments, whether that is the street, a public square, a garden or a public park. In order to mobilise citizens for activities to green their neighbourhoods, social capital is crucial. We consider social capital to consist of the presence of social networks and new relationships among citizens, together with a shared understanding of norms and values that facilitate and support the co-operation among different people (OECD, 2001).

Building social capital is in itself an important response to the decline of personal ties and the alienation and isolation of people in urbanised cities. It is a form of resistance against cities that 'have become anonymous places where people sleep and work without much sense of community' (Putnam, 2000). A 2013 OECD project proposed four distinct interpretations of social capital that all play a role in civic engagement:

- *Visibility*: the activities and practices with which people engage in community life need to be mobilised and made more visible. These practices could include voluntary work, learning practices, or neighbourhood membership in the surroundings where they live.
- *Network development*: refers to building new relationships between people and the necessary social behaviours that contribute to establishing and maintaining these relationships.
- *Social network support*: refers to the resources – emotional, material, practical, financial, intellectual or professional – that are made available to each individual through the social networks. This support can motivate and mobilise people to get involved in initiatives or community efforts.
- *Trust*: refers to shared social norms and shared values that underpin the societal functioning of the citizens and that enable mutually beneficial cooperation between the different groups in the community. This also involves trust between the citizens and policy makers in acknowledging civic engagement as important for urban neighbourhoods.

These multiple understandings of social capital appear in various forms and combinations in the first author's own work with community initiatives. In this chapter, we focus on three case studies which use citizen initiatives and networks to green urban environments and improve the management of urban natural resources. The insights presented in each of these case studies are drawn from the authors' experiences within these initiatives, as founders or active participants. We consider each of these three cases to demonstrate a different route towards enhancing civic engagement in neighbourhoods: communal learning, new types of ownership for citizens' initiatives, and using strategies to illustrate and communicate concretely the impacts of citizen initiatives. First, we present Steenbreek Foundation, founded in 2014 as an initiative of universities, municipalities and urban ecologists, which exemplifies communal learning of citizens through local actions aimed at greening public and private gardens (Stobbelaar *et al.*, 2021). Second, we explore the possibilities of communal ownership, as exemplified by the Kaskantine in Amsterdam, where people in the neighbourhood organise their own greening initiatives, urban resources, and design all kinds of technologies for off-grid living in the city. Third, we demonstrate how communal impact can be measured and communicated, by presenting the case of the Doughnut Coalition in Amsterdam, which has developed its own bottom-up measurement technique to measure the social impact of their activities. The chapter will conclude with our vision for how local governments and urban planners can integrate and improve community engagement and initiatives into their greening strategies.

11.2 Communal learning: the case of Steenbreek Foundation

Learning occurs when citizens become involved in new networks and practices that they were not familiar with before. Steenbreek Foundation started in 2015 as an initiative to foster communal learning with the aim of greening cities. It aims to engage residents in new practices and networks through education and demonstrating the added value of greening for biodiversity, atmosphere, health and well-being, reducing heat and flooding risk. Each participating municipality/ community can choose their own approach, applied to their local context and needs.

Steenbreek Foundation originated from concerns about the negative ecological effects of garden 'paving' in the Netherlands: an increased use of paved surface areas or tiles in private gardens. As citizens are owners of these private gardens, this is typically a challenge in which citizens need to be engaged and informed. Private gardens have the potential to play an important role in greening cities, as in the Netherlands private property makes up approximately 30% of the ground surface in cities.

The concern for biodiversity loss based on these practices has been a motivating factor for Steenbreek Foundation since the beginning. After the Paris Agreement, the organisation considered a focus on climate adaptation to be its next logical step. The very name of the organisation, Steenbreek, 'Breaking stones', refers to replacing tiles with greenery so that the soil can more easily absorb rainwater, replenish the groundwater level, and avoid overloading sewers. Additionally, removal of stone surfaces prevents heat retention and thus overheating of urban areas.

Steenbreek Foundation's strategy is to guide local municipalities and communities with advice and to exchange knowledge and experiences amongst the network of municipalities on how to engage citizens in greening their environment.

At present almost 160 Municipalities are members of the Steenbreek Foundation and their approach differs: some let local NGOs take the lead, such as in The Hague, where Duurzaam Den Haag, the local Environmental Centre, coordinates their action. Other municipalities cooperate with NGOs, like Vogelbescherming (the national Bird Protection Association) or Groei en Bloei, an association of private gardeners in the Netherlands. Some municipalities choose to cooperate with local garden centres or even gardeners. Steenbreek Foundation has gained a lot of attention in the Dutch national and local news. With this attention, it has been able to grow the number of its members from just six municipalities in 2015 to more than 160 in 2021 (meaning that roughly half of all Dutch municipalities have joined). Additionally, various provinces and waterboards are also connected to the network of Steenbreek Foundation.

Together with local organisations, including municipalities, waterboards, provinces and local business and gardeners, the Steenbreek Foundation wants to make the living environment greener in order to increase biodiversity and other ecosystem services. Where possible, unnecessary surfacing in private and public space will be replaced with a variety of green spaces, for which the help of residents is desperately needed. Greening actions can include, for example, the creation of model gardens, the construction of insect hotels, the construction of facade gardens, or the greening of an entire street together with local residents.

Greening campaigns are regularly launched, which call for people to take action, for example, by replacing the tiles in their garden with greenery. These actions encourage awareness about greening, social contact and can also prevent loneliness. Moreover, because research has shown that green workplaces make working more pleasant and give employees the opportunity to get some fresh air outside in a pleasant way, Steenbreek Foundation has also started to dedicate more attention to increasing access to greenery around company sites, care locations and schoolyards.

In 2020, the cities of Rotterdam and Amsterdam called on their citizens to enter a competition to see which city could remove the most tiles and replace them with plants and trees. Rotterdam won the competition, removing a total of 47,942 tiles, in comparison with the 46,484 removed by Amsterdam. The competition served as a tool for public education and media attention, by inviting individual citizens to get involved with the green ambitions of the city.

In 2021, the Dutch competition *Tegelwippen* (removing tiles) was held as part of the campaign 'A greener Netherlands starts in your own garden'. It was supported by fifteen organisations including Steenbreek Foundation. During this national competition, a total of more than 1,500,000 tiles were removed across 90 municipalities and replaced with greenery. The Hague won the top prize of a 'Golden Spade' for removing a total of 200,000 tiles. Rucphen won the 'Golden Tile' for managing to remove the most tiles per inhabitant (TPI), removing a massive 989 TPI.

How do these efforts function as a form of communal learning? We understand communal learning as a variety of processes that people become engaged in through new relationships and activities. Through activities such as removing tiles, citizens engage in a learning process in which they develop knowledge relevant for greening their area. They do this in interaction with others and apply it to their specific context by removing tiles and greening their gardens.

There are different ways in which communal learning can be stimulated (Stobbelaar *et al.*, 2021). The first is to provide citizens with accurate information on the role that nature and natural resources perform for their health and well-being (see Chapter 10). There is a growing body of scientific literature demonstrating that a green environment promotes vitality, creativity, concentration, reduces stress, provides more varied play for children and reduces healthcare costs. Many studies have found, in various country contexts, a positive correlation between the amount of green in the living environment and the health of the local population (e.g. De Vries, 2016; De Vries *et al.*, 2003; Groenewegen *et al.*, 2018; Hartig *et al.*, 2014; Maas *et al.*, 2006, 2009; Mitchell and Popham, 2008).

The second way to stimulate communal learning is to encourage experiential learning, through increased contact with and use of green areas. Increased contact with natural environments can directly demonstrate certain benefits and foster learning through this experience. This can be especially important for more vulnerable groups of residents such as the elderly and children. The scientific literature indicates several ways through which (contact with) nature could influence the health and well-being of people. Hartig *et al.* (2014) and identify several possible mechanisms:

1. Stimulating physical activity, for example by making routes for active transport (walking/cycling) more attractive or by offering (attractive) space for recreational activities.
2. Increasing access to natural spaces to facilitate social cohesion of neighbourhoods, for example, by offering places where neighbours can meet or encounter each other in a pleasant environment.

The third way to stimulate communal learning is by encouraging political parties and urban planners to engage with citizens in greening initiatives, and to develop shared responsibility for taking care of nature in the city. Learning (how) to share responsibilities is a communal learning process in itself. Coalitions of or collaborations between very different parties can be, and are in practice, forerunners in putting greening initiatives into practice. This can include cities, architects and citizens or neighbourhoods, among others (Chapter 12). The next case study is a story about one such local community, in which citizens and residents take on the responsibility of greening their neighbourhood.

11.3 Developing communal ownership: the case of the Kaskantine in Amsterdam Nieuw-West

In 2014, a group of citizens established an 'urban farm café' in Nieuw-West, Amsterdam, and built a movable community centre made up of shipping containers, greenhouses and gardens that can be placed on temporarily available land. They experiment with a vast range of initiatives to contribute to the sustainability and well-being of the neighbourhood, including: the development of all kinds of technologies for off-grid appliances, the redistribution of unsold food from local shops, the creation of gardens, the production of food for the neighbourhood and the organisation of workshops and debates.

The Kaskantine is also actively petitioning the municipality of Amsterdam to form a new relationship with its citizens. The Kaskantine has worked to develop a 'Greening the neighbourhood plan', in which they propose that the Kaskantine can take over management of urban land and water in certain areas where these resources are currently underutilised and/or where the current management does not utilise the resources in ways that maximise the well-being of the neighbourhood. These can be temporary places, for example, empty lots, or other spaces to which natural functions (i.e. gardens or biodiverse spaces) could be added. The aim of this plan is to formalise these win-win situations into new innovative land- and water-use contracts ('commoning'). Therefore, the Kaskantine is a good case through which to understand the concept of communal ownership (Open Research, 2020).

The Kaskantine's first location was located in the ruins of an old sugar factory in a village halfway from Amsterdam to Haarlem, Halfweg, when two chefs decided to realise their dream of making affordable meals with local and self-produced ingredients. They acquired a second-hand greenhouse and an old shipping container and built a strawbale house. Their cooking attracted a lot of attention and the Kaskantine became a sort of 'post-apocalyptic pilgrim destination' for citizens, tourists and students from Amsterdam, interested in a self-sufficient low-impact lifestyle. The chefs set up regular activities, such as 'Friday night pizza nights', which were available 'on donation', meaning that visitors could leave a financial contribution ranging from whatever they could spare up to whatever they considered the meal was worth.

Since then, the Kaskantine has become a mobile community, setting up in the neighbourhoods on the Western side of Amsterdam, based on principles of self-sufficiency. Since its foundation, the Kaskantine has had four iterations, moving on the basis of the user agreements they arrange with landholders. In 2019, the Kaskantine moved to its present location, where a longer contract with the municipality is being negotiated. It is an old football field, which has been reserved for construction of a new neighbourhood in about seven years' time. The Kaskantine has grown exponentially with each new move, evolving from a set-up consisting of one shipping container with a lunchroom seating 20 people in 2016, into a community of 30 volunteers, 13 shipping containers and a restaurant-greenhouse with a capacity of 150 people in 2020. The Kaskantine is built with minimal investment, relying only on a small bank loan, a modest income generated by its donation-based activities, and some financial support from the landowners that support the Kaskantine in making the neighbourhood more attractive for the (future) renters.

The point of departure for the Kaskantine is the sustainable use of local natural resources, that are normally not integrated back into local ecological cycles and are to a large extent wasted. These include, for instance, food waste from the local grocery stores and garden surpluses but also solar energy and rainwater.

Based on their experience over the last eight years, the Kaskantine has developed five values which they consider to have been instrumental to their success in functioning as a local business and community organisation: Citizens share a strong purpose and motivation to protect and regenerate local values (motivational value); citizens positively value fluctuating ecological resource flows in their urban environment (redistributive value); citizens develop technologies that are able to fully integrate the different ecological resource flows of the production system (technological value); citizens get access to local resources (commoning value); citizens develop their social and economic activities by modular, organic, growth (growth value).

11.3.1 Motivational value

The activities engaged in by the Kaskantine are made possible through the commitments of volunteers and partner organisations. Without the prospect of financial benefit for their time and efforts, these volunteers nonetheless find fulfilment through these activities that align with their personal values. The personal motivations of volunteers frequently include seeking the possibility to contribute to a community project, the opportunity to gain knowledge and capabilities in self-sufficiency and/or sustainability, and the potential to build personal connections with a broader community.

The Kaskantine recruits long-term volunteers, asking prospective volunteers to commit to a minimum of six months, during which they volunteer approximately four hours per week. Even with these requirements, the Kaskantine has, throughout its eight years of operation, never struggled to recruit new volunteers. All types of citizens – from diverse ages, backgrounds, professions, etc. – become volunteers or visit the Kaskantine. The opportunity to build a social network seems especially appealing for those joining the project. Many volunteers have limited connections in the neighbourhood and join the project seeking to build those connections. To meet these needs, the Kaskantine has found that the provision of a local meeting space, especially one large enough to host larger groups of people, has been especially valuable.

11.3.2 Redistributive value

With access to nearly a hectare of organic urban land, the Kaskantine focuses on the redistributive capacity of nature and natural resources through their activities. Citizens engaged with the project take an active role in this approach to make adaptive decisions on the creation and management of the local natural systems at hand. Much of the communal knowledge has been gained through the process of designing their own food, energy and water production systems. Through this experience, as well as through the commitment to working within the restraints of the context in which they operate, they have created systems in which flows of energy and resources work in harmony with the human and natural sinks, instead of exploiting or depleting them. The Kaskantine has a smart water management plan, with water buffers, but also an energy system that is completely renewable. The Kaskantine invests in the topsoil of their land through their food processing and composting operations. By 'rescuing' food destined for the garbage from local stores and warehouses, redistributing it as much as possible through weekly donation-based food markets, and setting various composting systems (e.g. bokashi and vermiculture) to manage both food and garden waste, the Kaskantine aims to minimise waste in local food cycles, and return nutrients to its soil. Agricultural labour, including fertilisation, ploughing, reproduction, water conservation and pest control, is performed by carefully watching natural processes. The Kaskantine considers these activities to be a form of managed natural ecosystem services, performed with the ultimate aim of sustaining a nature-based production system from which communities can harvest and redistribute surplus. Through these activities, not only is the natural world cultivated, but social cohesion is also established, as those participating in these processes also contribute to caring for one another and the neighbourhood.

11.3.3 Technological value

Employing an ecosystem approach, which considers the whole system and all inputs and outputs, has allowed the Kaskantine to set up processes which co-exist symbiotically and serve to reduce the required inputs (Figure 11.1). By functionally integrating resource flows associated with different processes, surpluses of resources can be utilised to avoid waste and maximise the potential of the system as a whole. The Kaskantine experiments with functional integration at local and intermediate levels. For example, solar energy is used to power the building and core functions of the Kaskantine; however, when there is a surplus of solar energy, there are back-ups in place to allow for productive use of this surplus. Surplus energy can be channelled to the water pump, allowing it to work harder, and clean more of the water supply, resulting in greater access to clean water. Surplus food obtained from local shops, when it is too spoiled to eat, can be turned into compost which feeds the garden soil, and in turn produces fresh food for the neighbourhood. While relatively easy to implement at a local level, the Kaskantine considers that attention to and experience with integrating these resource flows can be an important stepping-stone to reforming resource flows at neighbourhood or even city levels (Rodríguez-Rodríguezad *et al.*, 2015). Employing an ecosystem approach has enabled the Kaskantine to understand how to integrate functions in a broader production system and experiment with technological innovations to enable this integration. This value could also be described as 'the micromanagement advantage of functional integration'. An in-depth manual created by the Kaskantine visually explains how the different functions support and relate to each other (Kaskantine, 2020).

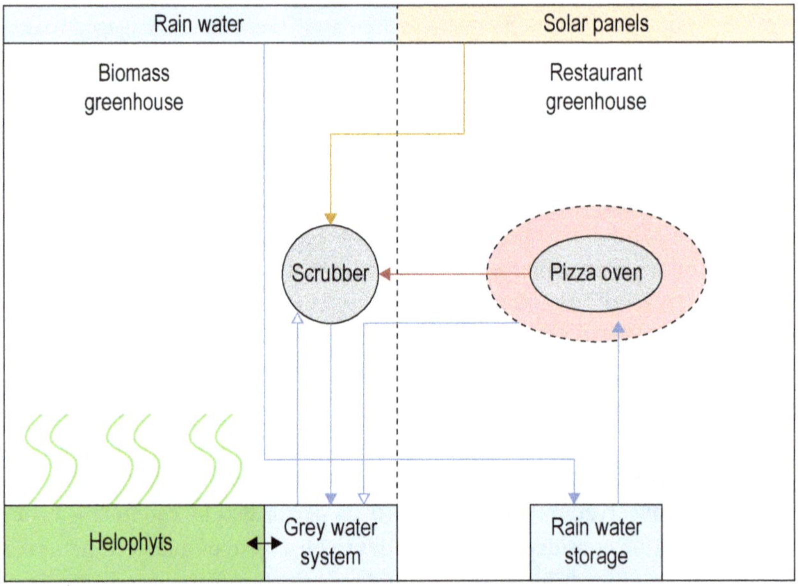

Figure 11.1. Closed loop system of CO_2 and water flow in the Kaskantine.

11.3.4 Commoning value

The Kaskantine claims that 'commoning' is an important value for the citizens to continue participating in this type of civic engagement activities. Commoning of natural resources guarantees that citizens who engage in neighbourhood activities to restore their local socio-ecological systems can have long-term access to considerable resources, including buildings, land, water, waste streams, wind and solar power (De Meent, 2020). One of the core goals of the Kaskantine is to use empty land or public land for the commoning of local resources. The municipality of Amsterdam has adopted a system of neighbourhood rights as part of their democratisation policy (Gemeente Amsterdam, 2021). In this system, citizens can apply for a trajectory to claim rights for land, to plan public space, and have the right to receive a budget to implement tasks normally undertaken by the municipality. Moreover, the requisite conditions for being able to take on a temporary lease of land are currently being researched by the municipality. In this way, one could consider that the municipality is now part of the commoning experiment, as this process could result in new conditions and types of neighbourhood contracts available for citizens to use and rent land. To date, some progress has been made with prioritising commons in tendering procedures for land property for housing projects.

11.3.5 Growth value

The Kaskantine aims for growth through economies of *scope*, in contrast to economies of *scale*. Economies of scope focus on the average total cost of production of a variety of goods, whereas economies of scale focus on the cost advantage that arises when there is a higher level of production of one good. An example of economies of scope can be seen even in the way the Kaskantine has been constructed. The urban café is built from repurposed shipping containers, which have the advantage of repurposing materials, being multi-functional in their design (the containers are equipped with rainwater collection and solar panels), as well as being portable. By leveraging access to second-hand materials, designing the building from a whole-system perspective, and leaving significant flexibility in the design, the Kaskantine is able to take advantage of land available at low rent, or no cost, even if this access is only temporary. In this way, the costs of land rental, access to utilities, and new building materials have been largely eliminated from the operation.

Rather than focusing on maximising the efficiency of production of a good, the Kaskantine promotes economies of scope by serving as a site for, or even incubator of, a diverse range of economic activities, like a bike repair shop, and the '*Voedselkringloop*' which redistributes food waste.

The Kaskantine also adopts an economies-of-scope perspective by considering the production of social and ecological value by the project, in addition to monetary value. The Kaskantine therefore developed their own impact assessment index based on the MAEX index. The MAEX index gives an indication of the value of the social impact of currently more than 1,200 organisations in the Netherlands. Most values of activities are estimated based on the market costs of those activities, while others are estimated on the basis of avoided environmental or health care costs (MAEX, 2021).

According to the MAEX index, the financial value of the total impact of the Kaskantine's activities in 2019 was 640,000 euro. This calculation is based on the value or costs of the activities that contribute to a specific interpretation of the SDGs. In the case of the Kaskantine, the most value was produced by upcycling food waste (±1 euro/kg), establishing a free public meeting space (±100 euro/m^2), management of green areas (±2 euro/m^2/year) and organising debates and workshops on climate adaptation and local solutions (±5 euro/participant). Considering the annual operational costs of all the Kaskantine's activities were a total of 40,000 euro, this means the social return on investment was more than 1,500%.

11.4 Community impact: the case of the Doughnut coalition

In 2021, eight Community organisations, based in four neighbourhoods in Amsterdam Nieuw-West, decided to create 'Neighbourhood Doughnuts'. Their approach is inspired by the Doughnut Economy approach conceptualised by Kate Raworth (Raworth, 2018). Kate Raworth has been directly involved as an advisor in Amsterdam's circular economy strategy and the development of an 'Amsterdam city doughnut', which aims to define an ecological and social zone within which the economy should develop.

The use of the neighbourhood doughnut has some key differences from Kate Raworth's methodology, which is a framework for change as well as a scientifically based monitoring instrument. The neighbourhood doughnut serves primarily as an instrument to mobilise/monitor civic engagement in taking care of social and natural resources in the neighbourhood. The MAEX index is a way to show the impact of the activities and identify which resources have the highest 'social return on investment'.

First, the goal of the communities to develop a neighbourhood doughnut is not a scientific exercise but is meant to be emancipatory. The neighbourhood doughnut makes visible the social capital in the neighbourhoods as well as natural resources. In many cases, natural resources have been in the hands of the government for such a long time that citizens need to relearn to recognise the natural resources, such as public land, roofs, food waste, parks and gardens. Access to these resources, and social capital can follow the old adage 'you don't know what you've got till it's gone'. A focus on (re)discovering the social capital in the neighbourhood can bring pride, a sense of belonging and unity and an urgency to act together – all of which we consider valuable in themselves.

Second, the neighbourhood doughnut is an investment in growing social capital, as it helps to bind people together in trust. With its focus on communal values, it enables mutually beneficial cooperation between groups that would normally not cooperate so easily. In this way it helps to overcome polarisation in the public debate. It is becoming increasingly obvious that the burden of urban measures is falling on the less wealthy. For example, they may have to continue to live in uninsulated houses, while being subject to increasing costs of living. In community meetings leading up to the neighbourhood doughnut, quite a lot of hostility was expressed towards taking initiative or making sacrifices for abstract or distant policy goals such as national CO_2 or Nitrate goals (personal communication). However, shifting the focus towards strengthening communal values seems to keep these neighbours on board. The neighbourhood doughnut also focuses on building new relationships between people and the necessary social behaviours that would contribute to establishing and maintaining these relationships. Ultimately, the hope is that citizens could start to cooperate together to optimise the local resources in the neighbourhood for their own, as well as for the communities' sake.

Third, measuring the communal values could lead to more appreciation, acknowledgement and even rewards for citizens that are already active in that field but have so far not received recognition. Social network support in the shape of emotional, material, practical, financial, intellectual, or professional resources are made available to each individual through the social networks of the neighbourhood doughnut. Some values can be translated to financial value, as the MAEX index demonstrates. This would present opportunities for the municipality to make 'doughnut deals'. Instead of open subsidy calls, organisations that make an impact in policy areas of the municipality could be automatically supported by being charged less for certain services by the municipality. The Kaskantine proposed investigating one such deal to the City Council for their land lease and was approved almost unanimously by the council at the end of 2019.

11.4.1 The Neighbourhood Doughnut of the Sierplein Buurt

Table 11.1 shows the Neighbourhood Doughnut of the Sierpleinbuurt, Slotervaart-Zuid, developed through several group sessions held in 2021.

The following types of civic engagement are specifically prioritised within this neighbourhood doughnut.

11.4.2 Management of natural resources and greenery

In this particular neighbourhood, there is for example the 'Zeis Kollektief', a group of ecologically-minded people, who advocate for managing green spaces in their own way, through using scythes (a traditional type of hand mower) and intermittent mowing, late in the season, to allow for more biodiversity in the greenery, as opposed to current standard mowing processes. The group agrees with the municipality to maintain a certain minimum quality of the green space. If this quality is not maintained, the municipality then has the right to take back control.

A second issue the neighbourhood doughnut aims to address is the problem of loss of green areas and the lack of connections between them, the so-called ecological corridors. Currently, the government approach is to guarantee a minimum of green surface of 6% in urban renewal projects; however, they do not specify any requirements for experience or biodiversity. On closer examination, these green areas are part of the original open structure of the so-called 'Garden City' (Hellinga, 2005). This city plan, originating in the early 1900s, is now considered extremely valuable and effective for addressing issues of climate change and biodiversity loss. Not only having green spaces, but also ensuring that they are interconnected, is crucial for allowing the migration of many species of flora and fauna. Of course, green spaces also provide important value

Table 11.1. Identified problems, values and indicators in Amsterdam-Sierplein.

Identified problems[1]	Neighbourhood values	Local action impact indicators
High unemployment (10%)	Strong solidarity in social networks	Number of hours of participating in programme for increasing opportunities on the labour market
High percentage has minimum income (20%), 11% of households below poverty line		Number of hours of consultancy in debt management and self-sufficiency
40% of adults and 22% of children are obese	Easy access to affordable supply of healthy food	m^2 of vegetable gardens available for the neighbourhood
Unsafe and dirty public space (litter, unsafe crossings, dark streets, etc.)	Hospitable streets	m^2 public space management programme making it more hospitable
Social cohesion is decreasing	Possibilities to meet each other	m^2 of physical location managed where people can meet each other
Cultural participation is very low	Rich cultural traditions	Number of people reached with deprived development opportunities participating in neighbourhood activities
Lack of exchange between social networks		Number of participants in neighbourhood events organised
Hardly any exchange between cultures		Number of participants in cultural activities and debates organised
30% of the citizens feel unsafe on the street and experience aggressive behaviour from drivers		Hours spent on neighbourhood watch and neighbourhood conflict resolution
More than 40% of the population has low literacy		Number of hours of study by participants of training programmes for basic and creative skills
10% of population is considered to be vulnerable/lonely		Number of neighbourhood services done for people in need
Inhabitants are not rewarded for positive action, and actually discouraged by the municipality from doing something	Mobilisation of citizens	m^2 underused spaced brought back into use for the neighbourhood
Green spaces are not connected, and undifferentiated mowing, rodent problem	Nature-friendly 'Garden City'	m^2 of habitat managed
More high-rise buildings causes de-greening and shade	Care for key species of flora and fauna	Number of species (re)introduced with management programme
CO_2/CO pollution	Clean air to breathe	Number of participants in climate debates/programmes
Soil is drying up quickly, water is drained too quickly to sewers	Clean water availability	Number of litres water storage

[1] Dark green areas are related to the natural environment.

for humans, as spaces for gathering, sport, and recreation. Local residents and nature lovers are often the ones that see, appreciate and act upon those values. In Sierplein Buurt, two municipality green maintenance officials now work shoulder to shoulder with the community organisations. They feel that as officials alone, without the citizens' groups, they could not manage the green 'in the proper way' (own communication).

11.4.3 Attractiveness and hospitality in the neighbourhood

Gan *et al.* (2017) studied 'green' environmental justice in two neighbourhoods in Amsterdam: the Bijlmermeer and Nieuw-West. The study showed that the green space within Amsterdam Nieuw-West is 27%, while the average green space in Amsterdam is 17%. In the seventies, local housing policies encouraged middle-class inhabitants in these districts to move to new towns around Amsterdam, while the new inhabitants of these neighbourhoods come from largely immigrant backgrounds. Gan *et al.* (2017) studied the role that this green space plays in peoples' lives. They concluded that the diversity of cultural groups in the neighbourhoods' value and use urban parks differently (see also Peters *et al.*, 2010 and Ozguner, 2011). The respondents of the study of Gan *et al.* (2017) recognised the positive effects greenery could have on their health, but they saw no contribution of greenery to social cohesion in the neighbourhood. These assessments were based on the quality of the park and its facilities and not on the amount of greenery. For instance, if a park or green area is perceived as unsafe, it excludes women from participating in the public space. The general maintenance level of urban green spaces also played a role in this, but in a different way. Not for the exclusion of a specific group, but for the exclusion in the neighbourhood as a whole: badly maintained parks contribute to a negative image of the whole neighbourhood.

Because of the degrading conditions of public spaces in Amsterdam Nieuw-West, the government took steps to implement more policing and cleaning, from the values of safety and cleanliness. However, a group of citizens experienced that the actions of the government did not have much effect on the overall neighbourhood. In response, a group of citizens started an initiative called the '*Vuilnisoproer*' (waste revolution)[3]. They organised people to make the public spaces more attractive and enjoyable and developed points of action for a clean and safe Nieuw-West. Their values focus on attractiveness and hospitality, rather than cleanliness and safety, which were the values of the government. These efforts demonstrated that citizens feel capable of addressing a local problem, but that the values they base their action on, and which they feel empowered to take responsibility for, are different from the values of the government.

The citizens bring a vastly different perspective to identified problems, values and solutions than that of the government. Residents are seeing problems through a lens of integrated values, while the government sees problems through the lens of sectoral technical solvability. The neighbourhood doughnut is a response to this gap between citizens and municipalities, as their integrated approach can produce local solutions that cannot be met by the more sectoral approach

[3] https://tinyurl.com/4trtuyw3.

of the government. The neighbourhood doughnut does not replace existing municipal rules and regulations but adds extra measures to embed citizens in their neighbourhoods and to become more self-organised so they can take care of their own social and natural capital. The Doughnut coalition positions communities alongside the market and the state as a new institution that has an impact in greening cities.

11.5 Conclusions

The three case studies presented above have explored three ways of including citizens in urban greening initiatives. Whether through encouraging community learning, creating strategies for communal ownership and management of resources, or through measuring and tracking the impact of community action, these initiatives have demonstrated the potential of civic engagement for developing an embedded symbiotic city. Based on the three case studies and our experiences with them, we have identified three principles which we consider crucial for developing different ways of involvement of citizens in greening the cities. While these are idealistic, and certainly would not be without their challenges, we consider that adopting aspects or combinations of these principles would greatly strengthen the efforts of planners, governments or scientists in implementing greening efforts (see also Stuiver, 2021).

11.5.1 First principle: keep on learning and dare to be radical

This principle focuses on learning as an essential reflexive step of all urban actors in order to introduce new forms of civic engagement that allow for the mobilisation of residents, their knowledge and their skills. It is essential that people start to see that they contribute their own expertise to a neighbourhood or a project. Local knowledge about the greenery and the neighbourhood is essential for making the cities green. All types of citizens have valuable knowledge to contribute: from local residents, to scientists asked to contribute ideas when designing a new neighbourhood or technology, to students and young researchers with innovative ideas.

Furthermore, strategic knowledge is needed on how to organise civic engagement in such a way that citizens can truly contribute and benefit from greening the city. Therefore, dialogue between actors in the urban domain, such as citizens, governments and scientists increases the opportunities to experiment with and learn about new organisational structures. Local governments frequently use rhetoric that calls for participation and working 'together with the neighbourhood', setting expectations that local citizens will play an active role in solutions. However, what these approaches usually boil down to is: 'we want more participation within our frameworks.' For example, in the last decade in the Netherlands, citizens have been invited to participate in council meetings. However, they are only asked to give their opinion about only two or three varieties of an urban plan, rather than being involved in the process of its development. The outcomes of such participation processes often lead to disappointment and a lack of support for the plans of municipalities. Therefore, institutions and other power structures need to be adapted towards radical civic engagement, because, in their current form, certain groups will always be excluded. The Kaskantine in Amsterdam and the Doughnut Coalition are interesting experiments in the transition to civic engagement and ways for governments to guarantee inclusion. This first principle raises questions on how to institutionalise these networks and the processes which favour inclusion, and how to ensure that the voices of residents are safeguarded in the long term. We suggest two more principles that can build the much-needed institutional networks in which citizens and their knowledge have a firm position.

11.6.2 Second principle: facilitate new types of communal ownership

This principle is based on the understanding that citizens take responsibility and ownership of the community, neighbourhood or a project when they have access to resources. This is one way of institutionalising the networks on a local level, as suggested by principle 2. Both local government and private companies can support citizens by compensating communities for their own ability to close their own ecological loops, for example by MAEX or a local 'doughnut fund', from which local civic organisations and networks can be structurally supported. A growing urban local economy could in such a way help to buffer seasonal and local fluctuations in resource flows that at present many companies are ill equipped to deal with.

A structural transfer of resources back to the local systems could be considered – what commoning theorist Michael Bauwens calls 'transvestment' (Bauwens, 2020). This type of transfer of resources (and responsibilities) could be a solution to the difficulties of the private and public sector in effectively (and rapidly) greening the urban economy. As traditional businesses are increasingly exposed to the consequences of their operations and deal with their so-called externalities, they are increasingly urged to manage these, for example through implementing CSR or sustainability policies. In the framework of such policies, it could be low-hanging fruit for businesses to support community initiatives, such as the Kaskantine, which already have experience in successfully dealing with compensating these consequences.

11.6.3 Third principle: endorse the impact of communities

In neighbourhoods citizens live their social lives. Neighbourhoods are the right level from which to accelerate the symbiotic city, as neighbourhoods provide the ingredients for creating new relationships through their practices, such as harvesting water and energy, the production of food, the ability to move freely and affordably without a car, and careful stewardship and reprocessing of materials (Hajer *et al.*, 2020).

Citizens' and communities' impact on the neighbourhoods they live in should be recognised and supported, while ensuring none are left behind. How do you ensure that green outdoor spaces remain available to citizens? The privatisation of public spaces often ignores the impact on citizens. When the municipality no longer determines what happens to that space, it becomes more difficult for citizens to become involved in its use. An approach that values the impact of citizens would let end-users propose what to do with that land in consultation with experts and civil servants.

Of course, this is not always easy. People with certain interests may assume their own views are universal. It is worth questioning what values and perspectives are being prioritised in debates on how to use a public space. This principle focuses on inclusion far beyond participation. You can have participation and the formulation of common goals in order, but that does not guarantee that these common goals and their actions are not hurting citizens that are not or wish not to be member of these communities.

While leaving space for disagreement, it is crucial that citizens and government maintain a dialogue with each other at the same level. We argue that governments should be open to citizens' initiatives and show adaptability to locally formulated values and impact, such as the MAEX index. On their side, citizen groups can invest in understanding the political-social context they operate in. Both groups must therefore work on the dialogue between politicians and citizens to create productive coalitions in the symbiotic city.

In co-creative processes that affect the broader neighbourhood, citizens that do not wish to take part should still be able to express themselves at clear moments. The solutions should not go against the decisions and policies of the representative democracy, but should be of added value or reinforce those policies. Additionally, when there is a large and clear resistance towards community-initiated projects, these projects should be put on hold and attentions should be geared towards the conflict. This is considered an insurance policy and a communal learning process that needs to be embraced instead of ignored, as is now so often the case.

References

Bauwens, M. and Pazaitas, A., 2020. P2P accounting for planetary survival. P2P Foundation, Guerrilla Foundation and Schoepflin Foundation, Berlin, Germany, 122 pp.

De Meent, 2020. Commons transitie plan. Amsterdam, the Netherlands. Available at: https://demeent.org/nieuws/47-commons-transitie-plan-amsterdam.

De Vries, S., 2016. Van groen naar gezond: mechanismen achter de relatie groen-welbevinden; stand van zaken en kennisagenda. Alterra-rapport 2714. Alterra Wageningen UR, Wageningen, the Netherlands, 46 pp.

De Vries, S., Verheij, R.A., Groenewegen, P.P. and Spreeuwenberg, P., 2003. Natural environments – healthy environments? An exploratory analysis of the relationship between greenspace and health. Environment and Planning A: Economy and Space 35(10): 1717-1731. https://doi.org/10.1068/a35111

Gan, Q., Jongen, L., Riharsya, A., Van Noorden, T. and Vaspintra, P. 2017. A study about the distribution of urban green space vs people's perception and exploitation in two neighborhoods of Amsterdam: Bijlmermeer and Amsterdam Nieuw-West. Wageningen UR, Wageningen, the Netherlands.

Gemeente Amsterdam, 2021. Beleidskader Buurtrechten. Amsterdam, the Netherlands. Available at: https://tinyurl.com/mr3sr5u4.

Groenewegen, P.P., Zock, J.-P., Spreeuwenberg, P., Helbich, M., Hoek, G., Ruijsbroek, A., Strak, M., Verheij, R., Volker, B., Waverijn, G. and Dijst, M., 2018. Neighbourhood social and physical environment and general practitioner assessed morbidity. Health & Place 49: 68-84. https://doi.org/10.1016/j.healthplace.2017.11.006

Hajer, M., Pelzer, P., Van den Hurk, M., Ten Dam, C. and Buitelaar, E., 2020. Neighbourhoods for the future, a plea for a social and ecological urbanism. Transity Valiz, Utrecht, the Netherlands.

Hartig, T., Mitchell, R., De Vries, S. and Frumkin, H., 2014. Nature and health. Annual Review of Public Health 35: 207-228.

Hellinga, H., 2005. Onrust in park en stad: stedelijke vernieuwing in de Amsterdamse westelijke tuinsteden, Het Spinhuis, Amsterdam, Nederland.

Kaskantine, 2020. Amsterdam, the Netherlands. Experimental Network Autonomy (ENA). Available at: https://ena.amsterdam

Maas, J., Verheij, R.A., Groenewegen, P.P., De Vries, S. and Spreeuwenberg, P., 2006. Green space, urbanity, and health: how strong is the relation? Journal of Epidemiology and Community Health 60(7): 587-592.

Maas, J., Verheij, R.A., De Vries, S., Spreeuwenberg, P., Schellevis, F.G. and Groenewegen, P.P., 2009. Morbidity is related to a green living environment. Journal of Epidemiology and Community Health 63(12): 967-973.

MAEX, 2021. MAEX Social Handprint. Rotterdam, the Netherlands. Available at: https://tinyurl.com/2s455h8c.

Mitchell, R. and Popham, F., 2008. Effect of exposure to natural environment on health inequalities: an observational population study. The Lancet 372(9650): 1655-1660.

Open Research, 2020. Commons en de omgevingsvisie. Amsterdam, the Netherlands. Available at: https://tinyurl.com/2p8r5tzr.

OECD, 2001. The wellbeing of nations: the role of human and social capital, education and skills. OECD Centre for Educational Research and Innovation, Paris, France.

Ozgüner, H., 2011. Cultural differences in attitudes towards urban parks and green spaces. Landscape Research 36(5): 599-620.

Peters, K., Elands, B. and Buijs, A., 2010. Social interactions in urban parks: stimulating social cohesion? Urban Forestry & Urban Greening 9(2): 93-100.

Putnam, R.D., 2000. Bowling alone: the collapse and revival of American community. Simon & Schuster, New York, NY, USA, 544 pp.

Raworth, K., 2018, Doughnut economics, seven ways to think like a 21st century economist. Chelsea Green Publishing, White River Junction, VT, USA, 320 pp.

Rodríguez-Rodríguezad, D., Kaina, J.H., Haaseb, D., Baróc, F. and Kaczorowska, A., 2015. Urban self-sufficiency through optimised ecosystem service demand. A utopian perspective from European cities. Futures 70: 13-23.

Stobbelaar, D.J., Van der Knaap, W. and Spijker, J., 2021. Greening the city: how to get rid of garden pavement! The 'Steenbreek' program as a Dutch example. Sustainability 13(6): 3117.

Stuiver, M., 2021, A green city for everyone. Wageningen UR, Wageningen, the Netherlands. Available at: https://www.wur.nl/en/show-longread/A-green-city-for-everyone.htm.

Zuijderwijk, L.M.A., 2019. Public space: always under construction. University of Rotterdam, Rotterdam, the Netherlands, 196 pp.

Chapter 12
Urban architecture for well-being: a design canvas for inclusive green cities

Anna Franz, Marian Stuiver, Eric Harris and Shahryar Sarabi

Abstract

Landscape and urban planners, designers, as well as architects, can take up inclusive planning and design processes that acknowledge human needs and dignity, and foster participation in shared decision-making. Underpinning these processes are values of empathy, placation, accessibility, and identity essential to imagining and creating inclusive green cities. This chapter provides a deeper understanding of the different dimensions of an inclusive green city and how social equity is an integral part of any design effort. We begin by exploring our vision and the values necessary for green urban design. A framework for three aspects of equity – recognitional equity, procedural equity, and distributional equity – is presented with selected case studies that serve as evolving good practices for equitable green urban design. From this framework and resulting values, we built on a value-inclusive design canvas and present design principles that planners, designers and architects can adopt as their own.

Keywords: inclusion, equity, greening cities, architecture, urban planning, equity, accessibility, placation, identity, empathy

12.1 Introduction: the urgency of inclusive green cities

Citizens need to be part of the design process to co-create their neighbourhoods, wards, and districts. In the coming years, it is essential for landscape and urban planners, designers and architects to orchestrate inclusive planning and design processes that acknowledge human needs and dignity, explore ideas, foster participation in shared decision-making, and result in healthy environments that enhance well-being and access for all individuals. The city of the future, therefore, develops wildlife and biodiversity, as well as social capital (Bourdieu, 1972; Putnam, 2000).

This matter is becoming more pressing as the current world population is approximately 7.9 billion (United Nations, 2021) and by 2057, the population is estimated to reach 10 billion, with growth mostly occurring in urban areas, causing economic, ecological, and environmental challenges for citizens to live in these cities. The proportion of people living in cities is expected to increase from 54% in 2018 to 68% by 2050 globally as a result of urbanisation. Migration issues, population growth, and socio-economic challenges lead to shelter, food, education, job insecurity, and ultimately, social degradation and exclusion (UNHCR, 2018). The design of cities and urban

10.3920/978-90-8686-935-0_12, © Wageningen Academic Publishers 2022

spaces either divides or brings people together based on socially equitable design decisions. We must recognise the significance of the opportunity to transform our local communities into green cities through co-creation and 'value-inclusive' design approaches with an empathic lens and perspective for well-being.[4]

According to Travis Price (2006) in *The Archaeology of Tomorrow*, archaeology is the architecture of today. He concludes that how time treats the cities we build this century is our legacy and what endures can create awe or leave nothing. He describes three distinct characteristics of significant architecture: (1) Stillness – the eternal mythical time; (2) Movement – the transient, changing aspects of time, change; and (3) Nature – time's perpetual consistency. These design characteristics, continually at play, need to become a balanced element of what we call, inclusive design to capture the legacy of architecture hovering between the dual infinities of time past and time future.[5]

This chapter aims to support landscape and urban planners, designers, as well as architects to develop a deeper understanding of the different dimensions of an inclusive green city and how social equity is an integral part of any design effort. We begin by exploring our vision and the values necessary for green urban design. A framework for three aspects of equity – recognitional equity, procedural equity, and distributional equity – is presented with selected case studies that serve as evolving good practices for equitable green urban design. From this framework and resulting values, we built a value-inclusive design canvas and present design principles that planners, designers and architects can adopt as their own.

12.2 Point of departure: green urban design

Urban design is the design of public spaces, buildings and infrastructure for people. From the beginning of time, architecture has explored various means for developing form, addressing function, shaping space and experience, and using materials in new ways – all to create places and space for humans to meet their needs. In recent years, more and more voices have been advocating green urbanism, to provide healthy environments that are safe for all and promotes wellness to its occupants. Architects, planners, designers, scientists, and other stakeholders need to promote green urbanism and create sustainable and resilient spaces for the quality of life. Green urbanism means that cities are eco-friendly, cut waste and emissions, use sustainable construction materials, and promote electrified mobility. Green urbanism aims to minimise the use of energy, water, and materials at each stage of the city's life cycle.

The theory of ecological design was conceived in the early 1970s by Dr Kenneth Yeang. The theory commenced the industry's vigorous focus on sustainability, shifting the paradigm for standard practice, changing the future course of design, and establishing sustainability as a new standard for success (Franz, 2013; Franz *et al*, 2013). His partitioned matrix, a systems model depicting

---

[4] E. Harris, personal communications.

[5] https://travispricearchitects.com/philosophy.

internal interdependencies, total inputs, total outputs, and external interdependencies was later enhanced by 'red', 'grey', 'blue', and 'green' eco-infrastructures (Yeang, 1980, 2010). This approach, highlighting the importance of energy-efficient and environmentally sound architecture and the built environments' connection of people, society, and culture to nature and water, has grown to become a standard of sustainable practice for the industry (Franz, 2013; Franz *et al*, 2013).

Steffen Lehmann and Gaëll Mainguy (2010) define green urbanism as 'a conceptual model for zero-emission and zero-waste'. They formulated a series of principles based on different schools of thought and practices that started in the United States in the 1990s, promoting compact energy-efficient urban development. Their 15 Principles address all the key aspects needed to establish green urbanism and form a sustainability matrix that can empower the urban designer to implement measures on a neighbourhood and district scale (Lehman, 2010).

In *Cities and Health*, Helen Pineo proposes an urban design and planning framework, outlining how theory translates to praxis. The framework presents three core principles for healthy urbanism describing it as inclusive, equitable, and sustainable (Pineo, 2020). Details of this high-level framework for inclusion are further outlined in *The International Journal of Sustainable Development & World Ecology* where key features and dimensions of an inclusive city are presented in a systematic bibliometric analysis and literature study (Liang *et al.*, 2021). Five dimensions of the model include economic, social, political, spatial, and environmental inclusion. Combined, these dimensions create equitable and inclusive green spaces which enhance wellness and quality of life.

Galen Cranz, a social architect once said: 'If architecture is the integration of social purpose, materiality, and aesthetics, then social factors are fundamental to design. Social purpose, behaviour, and feelings are not a constraint on artistic freedom, but rather a basis for inspiring new forms or refining traditional ones. Any design that fails to integrate social, technical, and sensory dimensions is immature' (Sinha, 2018). Architecture is unavoidably social. And as for science (rather than as a practice or art which architecture also can be), an architectural design is a type of hypothesis about how people might respond to a building or a public space and the conditions it creates.

Designs are grounded in theories about how people behave and what organisations, families, and individuals need spatially. These theories are taught in architectural schools and developed through historical studies of building types and their layouts as well as empirical research into how people use buildings and public space. And architects develop their theories too, from their experience of the world and how people live in it (Wood, 2015). Many of these theories have emerged from approaches such as user-centred design, design thinking, and co-creation.

Robert Gutman, an architect and sociologist once stated, 'there has never been an architect who was not, in some sense, a student and critic of society.' (Gutman, 1972, 2010). The idea of architecture as a social science might seem odd but there isn't much that more powerfully places, joins, separates, and patterns people, and relations between people than the built spaces we live, work and learn in (Wood, 2015). Empathy, therefore, is a crucial component of design thinking that allows architects to solve rather intricate, sensitive, and complex problems.

In general, though, the social sciences are reticent about architecture itself, its buildings, and particularly its interiors in terms of what they (attempt) to do to people (or allow people do to themselves, depending on your take on things) and how people respond to those stipulations/invitations. Similarly, although architecture frequently deals with social questions in the design of public buildings and institutions such as housing, hospitals, and schools, where the social here is commonly a synonym either for 'lots of people and/or for 'government-funded', it nowadays tends to avoid serious issues of internal space, relationality and the issues discussed here (Wood, 2015). However, addressing, acknowledging, and beginning the process of solving these intricate, sensitive, and complex social problems of human needs and dignity is essential to design inclusive urban environments and promote healthy living.

12.3 Three building blocks for inclusive design

Our framework for design is based on the three building blocks of social equity presented in the Tripartite Framework for Social Equity (Meerow *et al.*, 2019). This framework was initially used to study the extent to which cities focus on equity and found that the dominant conceptions of equity are generally tied to a distributional orientation, with less focus on the recognitional and procedural dimensions.

The difference between equity and equality is relevant (Cook and Hegtvedt, 1983; Reeskens and Van Oorschot, 2013). In short: equality refers to equal measures, despite the outcome; equity refers to a fair outcome that does not have to be equal. For instance, for health reasons, an equity approach would promote measures such as green space, healthcare, and social work in deprived neighbourhoods, while an equality approach would provide each neighbourhood with the same number of measures (De Haas *et al.*, 2021). Nonetheless, inclusive design requires both equitable and equal approaches.

In this chapter different types of equity are explored. Case studies are presented to highlight the types of equity (while recognising that elements of each of the cases may fit into another category).

12.3.1 Recognitional equity

The first block is Recognitional Equity. This aspect of equity values human differences and comprehensively considers the needs of all races, cultures, and populations in design. It considers the importance and representation of history and culture in place-making, and place-finding. It eliminates structural racism and promotes equal access and use. Success is measured through changed attitudes among people.

For example, Rotterdam in the Netherlands is recognised as a diverse or a 'hyper-diverse' city where the diversity of the population manifests not only in socio-economic, social, and ethnic aspects, but also in terms of lifestyle, attitudes, and activities (Tersteeg *et al.*, 2017), with Feijenoord as one of the most diverse neighbourhoods in this city. During the past years, there have been several local and municipal initiatives to improve the liveability of the public spaces in this neighbourhood. One of these initiatives was the redevelopment of the Afrikaanderplein, an area previously notorious for street crime. The redevelopment, which was carried with close communication between authorities, designers, constructors, and local stakeholders, turned this area into a flexible multi-use space that people from different age groups, ethnicity, and gender can use.

12.3.2 Procedural equity

The second block is Procedural Equity. Procedural equity engages different members of the public and stakeholders to create ideas, imagine solutions, and decide and plan initiatives and development. Success is measured through community participation and co-creation of improvements in the built environments and services.

For example, Rotsoord is a former industrial area on the bank of the river Vaartsche Rijn just outside the centre of Utrecht, the Netherlands. This area has been converted into a popular hotspot and a breeding ground for the creative sector and local initiatives in recent years (Duineveld *et al.*, 2021). Rotsoord is an example of procedural equity, where the municipality of Utrecht adopted a participatory approach to include citizens in the planning and decision-making process along with businesses and societal organisations.

This project was implemented in the context of Motion 99, which was the policy adopted by the Utrecht city council in 2016, focusing on the inclusion of social values in area development. This motion was started by a group of citizen-initiated discussions among civil society, governments, and developers and encouraged collaboration in the redevelopment of neighbourhoods (Duineveld *et al.*, 2021).

One other example is Steenbreek Foundation in the Netherlands, which also is addressed in Chapter 11: this is a foundation that organises awareness-raising campaigns and helps municipalities and citizens with greening private gardens.[6] This initiative is helping municipalities to provide support for access to green across the urban areas, so citizens have almost equal opportunities to receive the benefits of living near green areas. In this initiative, citizens are playing an active role in the development actions, which can be recognised as a form of procedural equity.

12.3.3 Distributional equity

The third block is Distributional Equity. Distributional equity demonstrates strategies to fairly distribute high quality, sufficient and accessible resources for food, affordable housing, education, and health care across a city and within communities. It provides planning and development for infrastructure, ensuring ample public transportation to essential facilities and services, and equal access to land for food production and distribution. Economic opportunities are offered in policy

---

[6] https://steenbreek.nl.

and programmes, providing jobs through education and training along with strategies to fund and enhance business opportunities. Success is measured through resources, infrastructure, and economic improvements.

The 15-minute city is an initiative in Paris, France that has received much attention, especially after the COVID-19 pandemic. This initiative aims to provide car-free access to necessities and decentralise city life and services. It means reducing unnecessary travel across the city, increasing the liveability of streets and public areas, improving health and well-being, and many more benefits. This concept developed by Carlos Moreno in Paris has been adopted increasingly by cities, especially across Europe, for a better recovery from the COVID-19 pandemic and an equitable sustainable transition (Moreno *et al.*, 2021).

In Paris, the 15-minute city was a key pillar of the elected Mayor's election campaign in 2020. As the flagship of this initiative, Paris aims to achieve carbon neutrality by 2050 with the help of this approach (Moreno *et al.*, 2021). One of the main goals of this approach in Paris is to reduce the share of urban spaces occupied by cars and dedicate them to other uses. The construction of pedestrian and cycling paths has been significantly scaled up. As another action, school playgrounds have been made accessible to the public outside school hours to improve accessibility to green areas.

The green initiative 'Oost Indisch Groen' in Amsterdam, the Netherlands (East Indian Green) is an example in which all three types of equity are addressed. This is a citizen initiative for transition to a more sustainable and liveable neighbourhood through active citizenship. The initiative started with the creation of a community garden, followed by the creation of a community group.[7] The municipality of Amsterdam provided funding for the creation of the community garden and recognised the power of citizens to implement and maintain a public space. This community garden provides an environment in which people with different ethnicity, gender, and socio-economic backgrounds can collaborate and co-create.

In the United States, Miami's Inclusive Underline Project, as shown below in Figure 12.1, is also an example of all three types of equity. The Underline's northernmost portion, the Brickell Backyard, is a procession of unique rooms around the park's dedicated bike and pedestrian paths. Each room is designed to respond to the specific needs of local community members (Oberliesen *et al.*, 2021). The Underline's community priorities, for example, include natural elements, like a pollinator park; gathering places, including a 50-foot dining table and smaller tables with dominoes; and playground and recreational areas. It also offers free amenities for entertainment and a variety of venues for artistic and cultural expressions. Free, intentional programming, such as Miami Ballet performances and yoga classes, offer options regardless of income or ability. A gym, track, and flex sports court serve those needs too. And because visitors can get to the park without driving, actively engaging with the community's youth and senior populations has been a priority. The

[7] https://oostindischgroen.nl.

Underline has taken this into consideration in various ways, including plans for a mix of affordable housing for seniors directly across from tall condos (Oberliesen *et al.*, 2021). And, they also have plans for youth outreach and training.

Meg Daly, founder of the project says that high school students will soon test their tour guide skills and plant species knowledge as part of a youth engagement programme funded by a grant from the Nahmad Family Foundation (Oberliesen *et al.*, 2021). 'It's an experiment to train juniors and seniors to be ambassadors, while also teaching them about 'the power of smart plants that help with drainage,' Daly says. Another key part of the Underline, flexibility, is a valuable tactic for creating a welcoming atmosphere. Flexibility also can help 'future proof' design, and one of the things planners and urban designers have learned from the COVID-19 pandemic is that things can shift dramatically, and often, quickly. Designing inclusivity from the start can help public spaces and communities adapt and thrive in a rapidly changing world (Oberliesen *et al.*, 2021).

12.4 Value driven inclusive green design

A model with four values underpinning the inclusive design processes is shown below in Figure 12.2. The model intersects the three aspects of social equity and related dynamics of Meerow *et al.* (2019) with four values that emerge as essential to achieving value-inclusive green design: identity, placation, accessibility, and empathy.

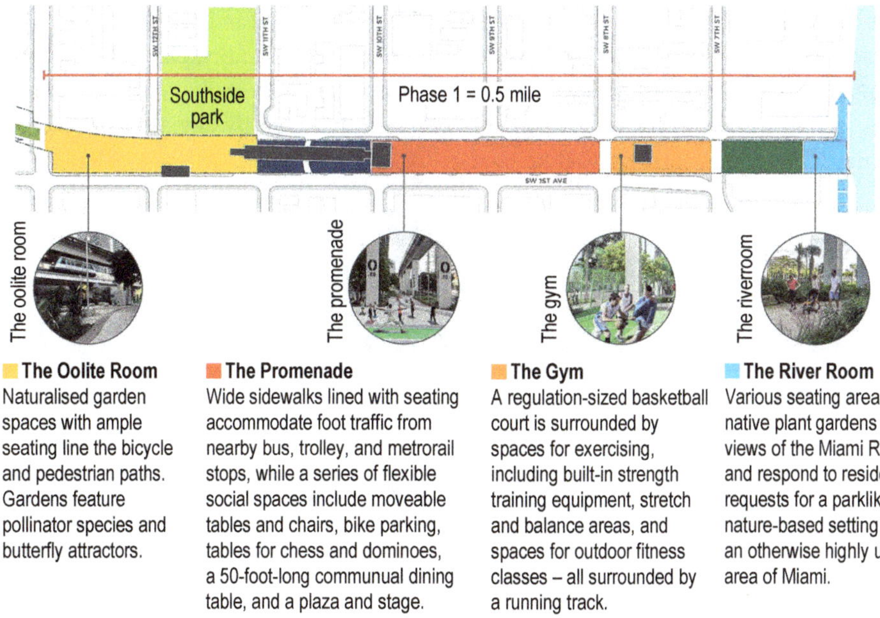

Figure 12.1. Miami's inclusive underline project (Oberliesen *et al.*, 2021).

- Identity equates to the values of human needs and dignity and the celebration of different lifestyles and identities. This is at the 'scale of the individual' and captures the understanding of the diversity in human histories, skills, and needs as related to physical, social, mental, and spiritual health. Identity means that people can express themselves without any hesitation.
- Placation is the value of making everybody visible and overcoming anonymity in buildings, technologies, private and public areas as the basis of healthy living. Placation looks at the 'scale of the location' to make everyone visible and equally be represented. It acknowledges 'co-existence' for all creatures (human and non-human) in the city to truly co-create public environments.
- Accessibility is closely connected to both humans and non-humans having access to the city and is at the 'scale of urban resources' Accessibility allows equal distribution of resources in architecture, landscape, and interior and circular design.
- Empathy is the central value in the figure as the basis for urban architecture for well-being. Every inclusive design begins with empathy and recognises both the needs of human and non-human actors (Latour, 2017) in the city. It looks at the 'scale of relationships' and explores social capital solutions by applying building blocks for improved quality of life for all creatures that inhabit the city.

The values of empathy, placation, accessibility and identity are foundational to any design process. Inclusive design depends on the extent that planners, designers, architects, and the community at large acknowledge, embrace and apply these values in designing the urban fabric. We propose that the four values are a necessary first step in commencing inclusive design. To translate the four values to inclusive design, we present a value-inclusive design canvas mapping: (1) physical, social, mental and spiritual health; (2) interactive settings for quality of life; and (3) green inclusive cities. Then, we examine representative images for the four values of identity, placation, accessibility, and empathy (while recognising that the images shown may include multiple values). From these images, design principles for each value are presented that identify key design objectives for achieving urban architecture for well-being.

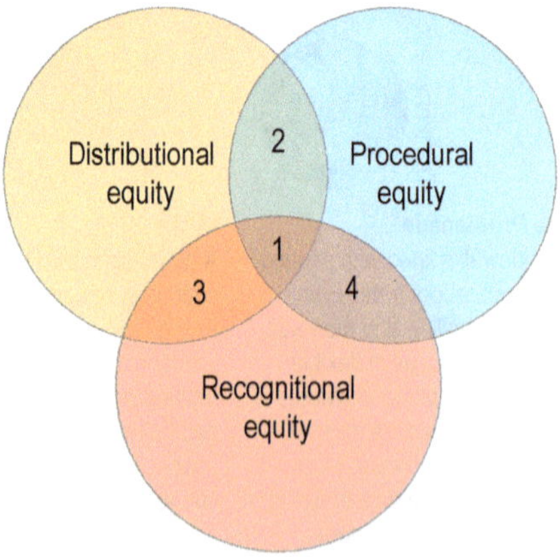

Figure 12.2. Four values that intersect with Meerow's model for social equity (2019) 1 = empathy; 2 = placation; 3 = accessibility; 4 = identity. This figure was developed by M. Stuiver, S. Sarabi, M. Takken, L. Rondard, R. Valkenburg and T. Yuksel for https://ewuu.nl.

One could argue that the four values associated with social equity are the drivers of urban design or one could argue that urban designs drive these four values in society. We propose that the four values and accompanying design principles and key design objectives are integrated and need to become part of an iterative process to create an urban architecture for well-being. Catalytic change and transformation of the urban fabric as discussed in *Architects for Community* by Torti *et al.* (2017) are denoted below in Figure 12.3, imagining inclusive design.

12.4.1 A value-inclusive design canvas

To achieve a user-centred and inclusive design, we present a value-inclusive design canvas in Figure 12.4. Planners, designers, architects and community participants must first co-create and align their goals, visions and theories on design processes and outcomes (Franz, 2013; Franz *et al.*, 2013). Then, their theories need to be translated into practice through design thinking, an iterative approach between theory and practice that evaluates multiple design considerations. This comprehensive process is displayed as a value-inclusive design canvas (Figure 12.4), a visual map connecting knowledge areas associated with recognitional, procedural, and distributional equity (horizontal direction) to theory and practice (vertical direction). Stakeholders can make choices and learn about the values of identity, placation, accessibility, and empathy with the help of this canvas to achieve inclusive green cities expressing an urban architecture for well-being.

The value-inclusive design canvas is meant as a dynamic and iterative tool for design. Imagining inclusive design involves creating sustainable and aesthetically pleasing places that bring new meaning and vitality to urban communities. Using smart technologies and envisioning fresh approaches in the built environment also are essential to drive change. Examples are spaces with indoor and outdoor environmental quality, spaces for energy, nature and water conservation,

Figure 12.3. Imagining inclusive design (illustrations courtesy of Torti Gallas + Partners, 2021).

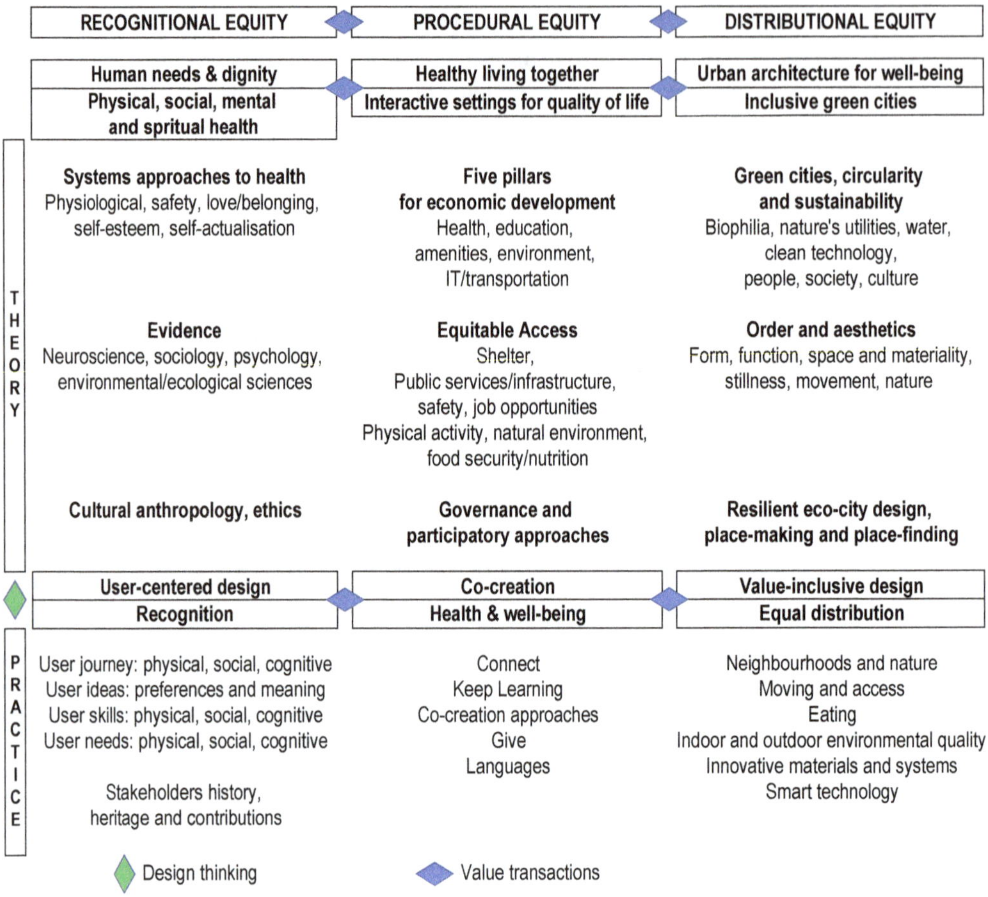

Figure 12.4. Design canvas for inclusion.

and spaces that make use of innovative materials and systems. Ultimately, connecting social capital with green cities in urban design promotes physical, social, mental, and spiritual health for individuals, society, and communities.

12.4.2 Recognitional equity: recognising human needs and dignity to promote physical, social, mental, and spiritual health

Research has uncovered an extensive body of knowledge for promoting physical, social, mental, and spiritual health which contributes to human health and well-being. Psychology, neuroscience, sociology, and the environmental and ecological sciences have all revealed parameters that enhance human health, perception, and experience of architecture and landscapes. Depicted in the design canvas is a collection of evidence for design and well-being as summarised in the five ways to well-being (Aked *et al.*, 2008) and research on systems approaches to health, urban design and well-being (Anderson, 2014). Identity is an important value of recognitional equity

(Figure 12.2). Cultural anthropology, the study of human culture, reveals that while humans are very different, they also share the same values for recognition and dignity. Research on ethics shows the importance of social equity in the advancement of human civilization. Recognitional equity develops an understanding of human needs and dignity through user-centred design. Translating this into practice requires a thoughtful user-centred assessment of user needs, ideas, and views. Ideas and views are shaped by culture, heritage, and history. Design is best when it responds to human and community culture and promotes inclusion for all participants and users. This occurs by understanding the historical, cultural and communal backgrounds of different groups, the identities they have and the problems that they face in the accessibility (Figure 12.2) of the design to their needs and wishes.

However, this all begins through an empathetic framework. Empathy is described as a phased process that gives insight into a designer's experience and its impact on design when there is empathy with the user (Kouprie and Sleeswijk Visser, 2009). Understanding human needs and dignity through user-centred design fosters collaboration to explore user ideas and preferences, which initiates a storytelling process for describing the user's journey, skills, and needs from a physical, social, and cognitive perspective. And as a result, it acknowledges personal biases and constraints by listening to others. Empathic understanding goes beyond knowledge: when empathising you do not judge, you 'relate to [the user] and understand the situations and why certain experiences are meaningful to these people' (Battarbee, 2004), a relationship that involves an emotional connection (Battarbee and Koskinen, 2005). Empathy informs and inspires designers to create products that fit user needs (Kouprie and Sleeswijk Visser, 2009).

12.4.3 Procedural equity – creating healthy living and interactive settings for the quality of life

Research also has uncovered an extensive body of knowledge for enhancing healthy living through developing interactive settings (or places; see Figure 12.2) for the quality of life. The Five Pillars for Economic Development (O'Hara, 2020; O'Hara and Vazquez, 2006) recognises the human need for and importance of health, education, amenities, environment, and technology/transportation for promoting quality life. This translates into a body of knowledge relating to equitable access for shelter, public services and infrastructure, safety, job opportunities and includes spaces that contribute to experiencing nature and the natural environment, and landscapes that can be used to promote food security and nutrition. Research from health and well-being reveals how human behaviour and feelings of well-being are influenced by context. Translating this into active participation and interaction of people that together explore design thinking is essential as it leads to solutions that can connect people and their activities, enhance continuous learning, and promote collaboration. It also offers opportunities to give back to the communities that were involved, and through the lens of different human languages and different views, new ideas can emerge for living and interaction.

Promoting healthy living through design thinking is a systematic process via an empathic lens. Empathic techniques cover direct contact, communication, and stimulating ideation by enhancing imagination (Kouprie and Sleeswijk Visser, 2009). By drawing on user experiences as well as historical and cultural context, it is a way to bridge the impact gaps of communities that will connect them with local municipalities, stakeholders, community leaders, and designers. This includes collaborating with these users to analyse and develop options to satisfy their needs. Acknowledging different types of knowledge present in the stakeholder community and learning from it focuses attention on pro-social solutions to foster a positive impact on happiness and well-being within these communities.

12.4.4 Distributional equity – designing the urban architecture for well-being and inclusive green cities

Designing user-centred spaces cultivates empathic experiences that honour and dignify historical and cultural context. These elements provide a sense of 'place' to all inhabitants of these spaces by integrating their human activities and their needs for access to resources within the built environment. When observing the user in the user's environment, the designer stays beside the user. One reason is that the designer is aware of his intervention in the user's context and has a researcher's role to play (Kouprie and Sleeswijk Visser, 2009).

Being empathic includes a range of activities where designers must imagine what it would be like to be in the position of the user of the end-design. What does the design provide for all people in terms of distributive power and access to resources in that location? It provides a fundamental understanding of the mental process of achieving empathy and using that understanding in designing (Kouprie and Sleeswijk Visser, 2009). For example, in the schematic phase of designing, architects meet with various stakeholders and present conceptual solutions through vignettes, which consist of the research and data collected during a survey. These vignettes engage the community of stakeholders to ensure that their voices are heard and that their human needs are met. Data represented in stories, storyboards, or personas serve very well in allowing designers to step into and walk around in the end-user's world, and connect with the end-user (Pruitt and Adlin, 2006).

Roman architect Vitruvius Pollio identified three elements necessary for a well-designed building: *firmitas*, *utilitas*, and *venustas*. Firmness or physical strength secured the building's structural integrity (University of Chicago Library News, 2011). It should stand up robustly and remain in good condition (Shackell, 2018). The utility provided an efficient arrangement of spaces and mechanical systems to meet the functional needs of its occupants. It should be useful and function well for the people using it. And *venustas*, the aesthetic quality associated with goddess Venus, imparted style, proportion, and visual beauty. It should delight people and raise their spirits. Rendered memorably into English by Henry Wotton, a seventeenth-century translator, 'firmness, commodity, and delight' remain essential components of all successful architectural design (University of Chicago Library News, 2011). This creates sustainable, resilient, and inclusive architecture, addressing a multitude of issues and supporting a multitude of user groups (Shackell, 2018; University of Chicago News, 2011).

Figure 12.5. Torti Gallas + partners design charrette – co-design and value-inclusive design (illustrations courtesy of Torti Gallas + Partners, 2021).

12.5 Design principles for identity, placation, accessibility and empathy

Recognitional, Procedural, and Distributional equity are integral to the holistic approach for the inclusive green design of cities. The next sections discuss the design principles for the four values as depicted in the model of Figure 12.2. The design principles for identity, placation, accessibility and empathy are explained and translated into possible design objectives.

12.5.1 Design principles for identity

Identity is the place where recognitional and procedural equity meet (Figure 12.2). Identity in the context of the design canvas means recognising the diversity of different stakeholders, hearing their voices, and including them in the design and development process. Figure 12.5 shows Torti Gallas + Partners conducting a design charrette – everyone comes together, everyone shares ideas, and everyone is heard in the decision-making process. Design solutions represent the collective community where all are welcome. Identity equates to human needs and dignity and the celebration of different lifestyles and identities. Identity is connected to the *scale of the individual* and captures the understanding of the diversity in human histories, skills, and needs as related to physical, social, mental, and spiritual health, and how these diverse identities can be expressed without any hesitation.

Design principles for identity include designing for safety, providing activity settings, and creating complete streets accommodating all users. Citizens should see their identities and preferences represented in the design process so they will be motivated to play an active role in shaping the urban environment and benefiting from it. Figure 12.6 provides an illustration of design principles for identity.

Design principles for identity

Providing safety, fostering activity and taking notice, and enhancing complete streets

Figure 12.6. Inclusive green cities, identity (illustrations courtesy of Torti Gallas + Partners, 2021).

Safety is the first design principle for identity. It does not matter where you are in the city, but there is a sense of community and safety, people can have their ways of doing things, they are safe day and night, there is an absence of violence. Different perceptions of safety need to be recognised in the design process. Safety can have different meanings for different groups, and a mission of the inclusive design process is to find the synergy among these different perceptions to provide a safe environment where people can express themselves in their social activities.

Social activity is a design element in green cities that contributes to wellness and enhances the quality of life of the people that inhabit them. Urban residents have different needs and preferences regarding the form of the space they want to be active in. These different preferences should be considered in the design process to provide urban environments where residents of all ages and backgrounds share their time, passion knowledge, and efforts. People's skills and abilities matter in how the city looks and fostering co-design and co-creation of urban spaces is necessary for equitable design for communal activities. Social activities integrated into the urban fabric contribute to the development of a concept called complete streets.

A complete street is a street designed to provide a safe and more efficient transportation network for all users. A complete street is designed considering the needs, preferences, and vulnerabilities of different users in mind. In addition to reducing sedentary lifestyles and calming traffic, complete streets provide significant opportunities for the creation of social capital by placing more eyes on the street (particularly pedestrians and cyclists), therefore creating a 'lively' street (Jones, 2012). According to the US Department of Transportation (2015), some of the health benefits of complete streets include:

- reduced chronic disease (e.g. asthma, diabetes, heart disease);
- increased physical activity and improved safety;
- reduced human exposure to transportation-related emissions;
- reduced motor vehicle-related injuries and fatalities; and,
- reduced contribution of transportation to air pollution.

Approaches to design vary based on community context and address a wide range of provisions for elements such as sidewalks, bicycle lanes, bus lanes, public transportation stops, crossing opportunities, median islands, accessible pedestrian signals, curb extensions, modified vehicle travel lanes, streetscape, and landscape treatments (US Department of Transportation, 2015). Jones (2012) found that mid-block crosswalks increased safety and Reynolds *et al.* (2009) noted that complete streets reduce motor vehicle-related crashes and pedestrian risk, as well as cyclist risk when well-designed bicycle-specific infrastructure is included. Careful design of the urban fabric's transportation infrastructure promotes walking and cycling by providing safer places in which to achieve physical activity through transportation. One study found that 43% of people who have a place to walk reported that they were significantly more likely to meet current recommendations for regular physical activity than were those reporting no place to walk (Powell *et al.*, 2003).

12.5.2 Design principles for placation

Placation is where procedural and distributional equity meet (Figure 12.2). It is defined as the art of overcoming anonymity and make a visible space and place for all people in the urban design. Placation is connected to the *scale of the location* to provide spaces in the urban fabric that facilitate full representation of all people and nature. Placation deals with space for interactions between nature and people as the creator and/or beneficiaries of that space.

Design principles for placation include harmonisation of the urban and natural environment, providing universal access, and creating communal spaces. Placation implies that citizens should be provided with equitable access to the green city and contribute to shaping it. Figure 12.7 provides an illustration of design principles for placation.

Design principles for placation

Connecting the urban environment with nature, providing universal access and creating communal spaces

Figure 12.7. Inclusive green cities, placation (illustrations courtesy of Torti Gallas + Partners, 2021).

Streets are very essential elements of green cities. They are multifunctional spaces used as the means to commute, as the place for activity and leisure, but also where different challenges exist. The benefit that people get from nature usually depends on the proximity of people with green elements, and green streets can provide the benefits of greenery to a large population considering the daily presence of citizens in the streets of a city. In addition, natural elements can greatly contribute to addressing different biophysical challenges present in urban streets like heat islands effect, air pollution, noise pollution, and social challenges like violence. Considering the impact that green streets can have on the daily life of citizens, it is also more motivating for different groups to actively engage in the design process.

Communal spaces that provide a welcoming place for gathering and social interaction have a significant impact on quality of life. It goes beyond barrier-free design. Communal spaces tie us together and enhance the common experience in the built environment. The design fosters relationships and conversations, creating a sense of place and purpose. Reimagining the urban landscape and the communal spaces using co-creation establishes a partnership between designers and the community. This is envisioned within the natural urban environment by creating accessible and communal spaces for all human and non-human inhabitants.

12.5.3 Design principles for accessibility

Accessibility is the place where distributional and recognitional equity meet (Figure 12.2). It is closely connected to the *scale of urban resources,* such as locations, knowledge and technologies, allowing equitable distribution of resources in architecture, landscape, and interior design. Accessibility is concerned with providing equitable opportunities for all the people from different backgrounds to benefits from the green city. Design principles for accessibility include accessible food and farming, accessibility and accessible design, and access to economic opportunities. Figure 12.8 provides an illustration of design principles for accessibility.

Design principles for accessibility
Using available land for food production and gardening,
connecting movement and urban functions and providing economic oppurtunity

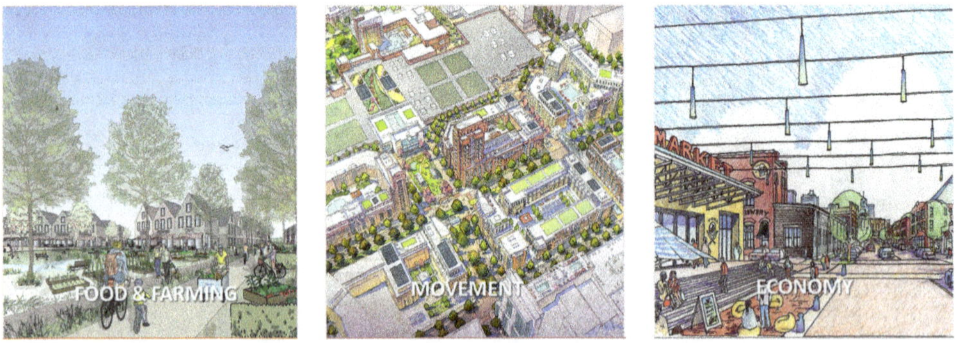

Figure 12.8. Inclusive green cities, accessibility (illustrations courtesy of Torti Gallas + Partners, 2021).

Food production and urban farming are some of the most tangible benefits that can be provided by green cities. Cities can become places where people have access to food and can have farming activities closer to their own homes. The city provides places where people can get income, education, and opportunities for food production and farming irrespective of their background. Demand for locally grown food and seasonal produce becomes a significant revenue source and creates spaces where direct marketing and consumption of the urban food is visible in the streets.

Accessibility and accessible design are both important aspects of inclusive design and have the same aim: to accommodate a diverse range of people (Levanier, 2021). Accessibility, however, involves the word 'access' and this provides an extra important dimension to inclusive design. Accessibility describes whether there are any literal barriers that prevent someone from experiencing something and having access to something. Accessibility focuses not only on social, political, spatial, and environmental dimensions but also on economic accessibility. This includes food production, the use of green spaces, co-located activities, job creation, new businesses, and sustainable transportation, to name just a few. Therefore, accessibility can add to the success of the urban economy by recognising the problems of the community and devising solutions that will enhance the quality of life of all citizens and in this way add to the resources of the city.

12.5.4 Design principles for empathy

Empathy is the basis for all three types of equity (Figure 12.2) and is connected to the *scale of relationships*. It explores how people can meet, get to know each other and their environments in such a way that there is space for both humans and non-humans in the urban fabric. Figure 12.9 depicts design principles for empathy – providing views, designing points of interest, and walkable public space.

Design principles for empathy
Providing views, designing points of interest and walkable public space

Figure 12.9. Inclusive green cities, empathy (illustrations courtesy of Torti Gallas + Partners, 2021).

Buildings can bridge the exterior nature with the interior environments of humans through providing views, such as places with windows and balconies. These can create a point of interest that gives meaning and purpose to a building structure by displaying views of the natural and building environment interchangeably. This includes entry, circulation, programming, form, as well as function which all contribute to the urban vernacular of inclusivity in building design. Defining special moments in the landscape, interior, and exterior of the urban environment promotes activity, play, and cultural exchanges. Integrating sustainable and resilient design solutions into the community has long-term positive effects. And it provides a welcoming destination for all to inhabit and experience.

Public space like parks, plazas, and trails are no different (Oberliesen *et al.*, 2021). Beyond safety and basic usability, the look, feel, and function of these spaces can significantly affect whether a person wants or is even able to spend time and engage there. Planners, designers and architects need to understand the inherent challenges of creating this kind of space that wins the hearts of their communities. COVID-19 restrictions and social distancing requirements have increased the need for public space and re-emphasised the importance of inclusive, multi-generational design. There is no specific design checklist to follow because each community is vastly different. As urban designers continue to assess how public spaces serve communities and how to make them inclusive for everyone, it's important to understand how designs can influence who feels supported and welcome, and who doesn't (Oberliesen *et al.*, 2021).

Interest starts when spaces begin creating a 'welcome feeling' for all users (Oberliesen *et al.*, 2021). There's more to it than investing in lighting, security, and for instance the Americans with Disabilities Act requirements, though those are all vital. Creating a welcoming feeling is about giving people a space where they feel comfortable and where their needs for engagement and expression are met. So, empathy is very closely connected to placation. A key purpose of public space is to create room for people to take part in civic, physical, and social activities so that they can start to communicate and understand each other. Designs should include as many members of the community as possible (Oberliesen *et al.*, 2021).

Below are a few examples of inclusive design and how it impacts a community:
- design for differing abilities;
- consider gender dynamics;
- provide amenities for cultural activities;
- recognise public spaces as storytellers; and,
- make room for retail and restaurants.

Figure 12.10 also provides additional design principles for empathy – integrating urban fabric, finding urban play space, taking time to share with people. Spaces that integrate activities, provide for play, and enhance human activity, embrace the human culture, foster heritage and promote physical, social, mental, and spiritual health. It is important to note that the design of the urban fabric enhances social interaction and human activity when spaces for activities, such as play are integrated into the design. Designing urban spaces that are receptive to human interactions and

Design principles for empathy
Integrating urban fabric, finding urban play space and taking time to share

Figure 12.10. Inclusive green cities, empathy (illustrations courtesy of Torti Gallas + Partners, 2021).

enhance human experience and spaces for art, music, festivals, and farmer's markets contribute to the richness of urban living and enjoyment. Play is a critical instrument for the physical, social, and emotional development of children. Designing an inclusive and safe play space for kids with different backgrounds, abilities, and skills is essential for enhancing equity in society especially considering the higher vulnerability of kids and their limited access to resources.

12.6 Conclusions

Inclusive design is the ongoing process of designing solutions to accommodate the viewpoints, experiences, and situations of people that were not accommodated before (Levanier, 2021). In the cities of today, inclusion shares an intense relationship with its opposite term: exclusion (De Haas *et al.*, 2021). Therefore, anyone working on inclusive green design will encounter obstacles of old designs that were based on exclusion from dated philosophies and ideologies. Inclusive design is largely about eliminating these points of exclusion. These can be physical or emotional, permanent, temporary, situational, or non-situational. Factors that are physical, permanent, and non-situational might be along the lines of race, gender, mobility, and age. Inclusive design, therefore, also can and needs to adjust the present urban built environment beyond its exclusion mechanisms.

The value-inclusive design canvas presented in this chapter is very much an appeal to people's empathic skills, assumptions and viewpoints. It challenges the designer's untested assumptions about the 'ideal' and 'universal' person s/he is designing for. Designing inclusively doesn't mean you're making one thing for all people. You're designing a diversity of ways for everyone to participate in an experience with a sense of belonging (Levanier, 2021) as depicted in Figure 12.11. Developing a deep understanding of equity can be done through learning about the distinctive four values of identity, placation, accessibility, and empathy.

Figure 12.11. Urban architecture for well-being (illustrations courtesy of Torti Gallas + Partners, 2021).

Finally, we want to stress that developing cities through a value-inclusive design is urgent and necessary for creating green urban architecture for the well-being of humans as well as non-humans. In the Foreword to Revisiting 'Social Factors': Advancing Research into People and Place (Lindsay and Morhayim, 2015: xv), the author asks, 'How do we measure the relationship between humans and their environment? How do we measure the hyphen in person-environment relations?' (Sinha, 2018). Deeply looking into human needs and the importance of empathetic relationships between humans as well as nature can shape healthy urban environments for the people that inhabit them and set a new vision for the future development of vibrant symbiotic cities.

References

Aked, J. Thompson, S., Marks, N. and Cordon, C., 2008. Five ways to well-being. The evidence. New Economics Foundation, London, UK.

Anderson, J., 2014. Urban design and well-being. Doctoral thesis. University of Cambridge, Cambridge, UK.

Battarbee, K., 2004. Co-experience: understanding user experience in social interaction. Doctoral dissertation. Series ILMARI A51. University of Art and Design, Helsinki, Finland.

Battarbee, K. and Koskinen, I., 2005. Co-experience: user experience as interaction. CoDesign 1(1): 5-18.

Bourdieu, P., 1972. Outline of a theory of practice. Cambridge University Press, Cambridge, UK.

Cook, K.S. and Hegtvedt, K.A., 1983. Distributive justice, equity, and equality. Annual Review of Sociology 9: 217-241.

De Haas, W., Hassink, J. and Stuiver, M., 2021. The role of urban green space in promoting inclusion: experiences from the Netherlands. Frontiers in Environmental Science 9: 618198. https://doi.org/10.3389/fenvs.2021.618198

Duineveld, M., During, R., Aue, S., Fulgenzi, A., Wang, B., Simanjuntak, D.F., Litjens, E., Uosukainen, M., Van Schriek, M. and Hall, P., 2021. Rotsoord: mag het misschien wat groener en inclusiever? – een analyse en 6 aanbevelingen. (Rapport Wageningen University & Research, Wetenschapswinkel; No. 358). Wageningen University & Research, Wetenschapswinkel, Wageningen, the Netherlands. https://doi.org/10.18174/515655

Franz, A., Sarkani, S. and Mazzuchi, T., 2013. Whole system design and evolutionary 21st-century American buildings + infrastructure. International Journal of Architectonic, Spatial and Environmental Design 9: 48.

Franz, A.Y., 2013. Whole system design and whole system design and evolutionary 21st-century American buildings + infrastructure. ProQuest Dissertations Publishing, Morrisville, NC, USA.

Gutman, R. (ed.), 1972. People and buildings. Basic Books, London, UK.

Gutman, R., 2010. The questions architects ask. In: Cuff, D. and Wriedt, J. (eds.) Architecture from the outside in selected essays by Robert Gutman. Princeton Architectural Press, New York, NY, USA, pp. 152-185.

Jones, D., 2012. The behavioral impacts of urban street modifications. A case study of East Blvd in Charlotte, NC. Ph.D. in design dissertation, North Carolina State University, Chapel Hill, NC, USA.

Kouprie, M. and Sleeswijk Visser, F., 2009. A framework for empathy in design: Stepping into and out of the user's life. Journal of Engineering Design 20: 437-448. https://doi.org/10.1080/09544820902875033

Latour, B., 2017. On actor-network theory. a few clarifications, plus more than a few complications. Philosophical Literary Journal Logos 27(1): 173-197. https://doi.org/10.22394/0869-5377-2017-1-173-197

Lehmann, S. and Mainguy, G., 2010. Green urbanism: formulating a series of holistic principles. Sapiens 3(2). Available at: http://journals.openedition.org/sapiens/1057.

Levanier, J., 2021. The principles of inclusive design. Available at: https://99designs.com/blog/tips/inclusive-design/.

Liang, D., De Jong, M., Schraven, D. and Wang, L., 2021. Mapping key features and dimensions of the inclusive city: a systematic bibliometric analysis and literature study. International Journal of Sustainable Development and World Ecology 29(1): 60-79. https://doi.org/10.1080/13504509.2021.1911873

Lindsay, G. and Morhayim, L. 2015. Revisiting 'social factors'. Advancing research into people and place. Cambridge Scholars Publishing. Cambridge, UK.

Meerow, S., Pajouhesh, P. and Miller, T., 2019. Social equity in urban resilience planning. Local Environment 24(9): 793-808. https://doi.org/10.1080/13549839.2019.1645103

Moreno, C., Allam, Z., Chabaud, D., Gall, C. and Pratlong, F., 2021. Introducing the '15-Minute City': sustainability, resilience and place identity in future post-pandemic cities. Smart Cities 4: 93-111. https://doi.org/10.3390/smartcities4010006

Oberliesen, E., Peinhardt, K., and Storring, N., 2021. Well-designed public spaces are inclusive ones. American Planning Association, Chicago, IL, USA. Available at: https://tinyurl.com/bdfdknct.

O'Hara, S., 2020. The five pillars of economic development. A study of a sustainable future-forward 7 and 8 in Washington, D.C. Available at: https://www.fivepillarsdc.org/.

O'Hara, S. and Vazquez, J., 2006. The five pillars of economic development: a study of best practices for the Roanoke Valley. Roanoke College, Salem, VA, USA.

Pineo, H., 2020. Towards healthy urbanism: inclusive, equitable, and sustainable (THRIVES) – an urban design and planning framework from theory to praxis. Cities & Health. https://doi.org/10.1080/23748834.2020.1769527

Powell, K.E., Martin L. and Chowdhury, P.P., 2003. Places to walk convenience and regular physical activity. American Journal of Public Health 93: 1519-1521.

Price, T., 2006. The archeology of tomorrow. Earth Aware Editions.

Pruitt, J. and Adlin, T., 2006. The persona lifecycle. Keeping people in mind throughout product design. Morgan Kaufmann, San Francisco, CA, USA.

Putnam, R.D., 2000. Bowling alone: America's declining social capital. Journal of Democracy 6(1): 65-78. https://doi.org/10.1353/jod.1995.0002

Reeskens, T. and Van Oorschot, W., 2013. Equity, equality, or need? A study of popular preferences for welfare redistribution principles across 24 European countries. Journal of European Public Policy 20(8): 1174-1195. https://doi.org/10.1080/13501763.2012.752064

Reynolds, C.C., Harris, M.A., Teschke, K., Cripton, P.A. and Winters, M., 2009. The impact of transportation infrastructure on bicycling injuries and crashes: a review of the literature. Environmental Health 8: 47.

Shackell, G., 2018. Three principles of good architecture. Available at: https://tinyurl.com/2jx2eu8u.

Sinha, A., 2018. Social dimension of design and design education. In conversation with Galen Cranz. Tekton 5: 88-95. Available at: https://tekton.mes.ac.in/issues/volume-5-issue-1/dialogue/.

Tersteeg, A., Bolt, G. and Van Kempen, R., 2017. Divercities: dealing with urban diversity – the case of Rotterdam. Available at: https://doi.org/10.5281/zenodo.437536.

Torti, J., Gallas, T. and O'Neill, C., 2017. Torti Gallas + Partners, Architects of community. Vendome Press, New York, NY, USA.

United Nations, 2021. World Population Prospects 2021, Online Edition. Available at: https://population.un.org/wpp/.

UNHCR, 2018. Global trends forced displacement in 2018. Available at: https://www.unhcr.org/statistics/unhcrstats/5d08d7ee7/unhcr-global-trends-2018.html

US Department of Transportation (2015). Complete streets. Available at: https://www.transportation.gov/mission/health/complete-streets.

University of Chicago Library News, 2011. Firmness, commodity, and delight: architecture in special collections. Exhibition on view from May 9, 2011 – July 29, 2011. Available at: https://www.lib.uchicago.edu/collex/exhibits/firmness-commodity-and-delight/

Wood, A., 2015. Architecture as a Social Science? Architecture and Education. Available at: https://architectureandeducation.org/2015/10/27/architecture-as-a-social-science/.

Yeang, K.K., 1980. Theoretical Framework for incorporating ecological considerations in the design and planning of the built environment. Ph.D. Thesis. University of Cambridge, Department of Architecture, Cambridge, UK.

Yeang, K.K., 2010. Ecological design and planning. The University of Cambridge, Martin Centre Research Seminar Series. Available at: https://tinyurl.com/4hsssytp.

Chapter 13
Being a voice of nature in urban transformations

Marian Stuiver

Abstract

The transformation process needed to achieve the symbiotic city requires a whole subset of transformative changes in cities. These range from re-embedding cities in the biotic and abiotic foundations of water, soil and organisms to building new institutional, technical, circular and socio-ecological networks in the city. Examples of these transformative practices have been described in the previous chapters with case studies in which people express different voices of nature. First, they point to the importance of the agency of urban actors, i.e. their ability to change existing urban practices in symbiotic ones. Second, they stress the importance of new networks, including governance arrangements that favour nature-based thinking. Third, economics needs to be re-embedded into society and ecology, resulting in new economic narratives that embrace nature. Finally, the inclusion of all people and their expressions of the values of nature as well as their heterogeneous knowledge is central to enforcing the transformations. Grouped in six categories of citizens, businesses, NGOs, corporates, municipalities and knowledge organisations, actors can work together on the transformation to make the symbiotic city happen. In this chapter we introduce these actors and we address the challenges that every type of actor can work on starting from tomorrow, from strategic, to tactical, to operational and reflexive activities that can be performed to become a voice of nature in urban transformations.

Keywords: transformation, urban eco-innovation network, citizens, businesses, NGOs, municipalities, knowledge institutes, regime change, knowledge management

13.1 Introduction

The transformation to symbiotic cities is essential for developing a future for next generations to come (UN, 2015; UN-Habitat III, 2017; WCED, 1987). This requires a whole set of changes, from re-rooting our cities in biotic and abiotic foundations of water, soil and organisms to building new institutional networks, technical networks, circular networks and social-ecological networks in the city. All actors in the urban eco-innovation network can work together on this transformation to make the symbiotic city happen (Gunderson and Holling, 2002; Hölscher *et al.*, 2021; Termeer *et al.*, 2017). They can be inspired from the critical practices that have been described in the previous chapters: (1) the integration of cities with the abiotic and biotic foundations of nature (Chapters 4, 5 and 6); (2) new urban governance arrangements (Chapters 7, 8 and 9); and (3) inclusion of all urban actors (Chapters 10, 11 and 12).

10.3920/978-90-8686-935-0_13, © Wageningen Academic Publishers 2022

In Section 13.2 the six categories of actors in the urban eco-innovation network will be described, as well as their role in the transformation towards symbiotic cities: citizens and neighbourhood communities, businesses, NGOs, large corporates, government and knowledge organisations. In Section 13.3 different questions are answered that can guide actors to work on the transformation practices. These practices can vary from strategic, to tactical, to operational and reflexive activities. In Section 13.4 specific attention is paid to the role of new governance arrangements and heterogenous knowledge production to make the needed urban transformations happen.

13.2 Together we make the symbiotic city

The Quadruple double helix in Figure 13.1 exemplifies the six categories that can take action to build this symbiotic city: citizens and neighbourhood communities, businesses, NGOs, corporates, governments and knowledge institutes.

13.2.1 Citizens and neighbourhood communities

We all live in a certain place, street, a neighbourhood, a village or a town. We as citizens make the city, and either by ourselves or in neighbourhood communities we hold the local knowledge and power that is needed to make change happen. In Chapter 11 we have seen that citizens can form alliances to do so and in Chapter 12, we have read why it is so important to include all citizens and

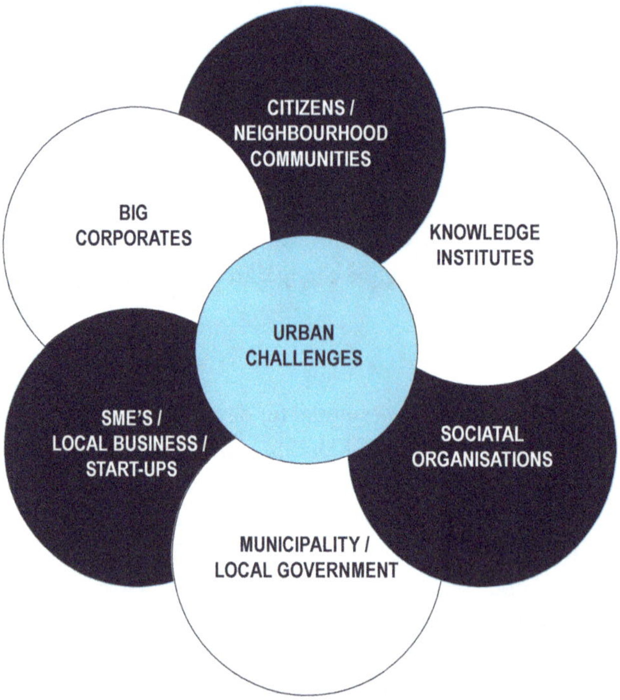

Figure 13.1. The double triple helix (courtesy of Pakhuis de Zwijger).

give them distributive power, recognitional equity and means of participation in the symbiotic city. Residents are best situated to enact change in their neighbourhoods and engage others to do the same (Anguelovski, 2013; Gruezmacher and van Assche, 2022; Knapp *et al.*, 2013). They can implement all types of measures in their homes and gardens. Planting succulents on flat surface roofs, climbing plants against fences or planting hedges instead, and less concrete surface area in the garden are all examples of how citizens can improve their surroundings. And they can be a voice of nature in their unique way and contribute to creating new socio-ecological and circular networks as well as institutional and technical networks. This will be written in detail in Section 13.3.

13.2.2 Businesses

Businesses can be the local grocery store, the pub, business start-ups, but also farmers (Cascone *et al.*, 2018; Westerink *et al.*, 2021). They represent the commercial bodies of the city. Depending on their business they have their offices or buildings in the city centre or are placed at the outer borders of the city in business parks (Snep, 2014). Business parks often serve as a stepping stone for nature, due to their locations, often between the countryside and the city. The people working in these businesses are crucial for the transformation as they can provide new embedded economic narratives on how to connect the city to the natural environment. They can start to see their businesses as sites for ecosystem development, or they can decide to build nature-inclusive buildings, as we have seen with the examples in Chapter 4. There are three types of tactical reasons for businesses to invest in greening their properties and businesses. First, when rules and regulation dictate, because otherwise they will not receive a permit or they may get a fine. Second, where the green space created yields an economic revenue, for instance when it is paid for and used by residents. Think of accessible roof gardens in high-rise buildings away from the noise, quiet and not public, and often with a beautiful view. Third, when the investor is convinced that it is the right thing to do, for instance for the users of the services and goods, for biodiversity and for the city. In other words, when the investor has decided to become a voice of nature.

13.2.3 Non-governmental organisations

Non-governmental organisations (NGOs) operate independently of governments and are usually non-profit organisations. NGOs are often organised around a specific theme and are strategically active and lobbying in sectors such as environment, health or human rights. Some organisations are especially famous for their efforts to protect nature, such as the World Wildlife Fund, Greenpeace and Friends of the Earth (Burgers, 2021). They are institutionalised voices of nature, lobby for nature and aim to make citizens aware of how politics works. They are heterogeneous in their expressions in terms of the visions of nature they portray, or the groups they target. Friends of the Earth, for instance, targets large companies and nation states that do not achieve sustainability goals. Over the last decades NGOs have gained a significant role in connecting with organisations from government and businesses to develop plans that help to strengthen the voice of nature in cities. They also often participate in projects to connect citizens to nature. In Chapter 11 we had the example of the Steenbreek Foundation for greening of urban environments in collaboration with citizens and authorities (Steenbreek, 2021).

In the last few decades, the attribution of rights to entities of nature has been put on the global agenda. Take, for example, the rights of mountains, rivers or nature reserves, but also seas and trees. On the website of the UN there is a list of examples in 30 countries worldwide where nature has received a legal right of its own (Anonymous, 2021). For instance, as of March 2017 in New Zealand the river Whanganui is a living being and is given the same rights as a human being. Both the government as well as the Maori people can give voice to the needs of the river in order to protect it. More and more organisations and individuals argue that non-human actors should have the same legal rights as people. The idea behind it is that when a natural entity acquires a legal personality, this means that the entity becomes a legal subject instead of a legal object and in this way can voice its own rights itself in court (Burgers and Outer, 2021). Of course, in connection with that is the debate about who can represent the voice of these entities and non-human actors and who can meet the demands of being an eco-centric voice of nature.

13.2.4 Governments

Every city and town has a local government that has a major influence on the design of the city, with the local rules and regulations that exist in their political realms. They have their own offices and real estate and they also decide about other people's land or estates, for instance in local rules and regulations on the construction of houses. They have responsibilities over the urban grey, green and blue infrastructure in the management of the city or town.

Local governments can be among the first to address the consequences of nature in the city for their citizens (Atelier groenblauw). They are crucial as they are the central body for making very practical decisions about the urban planning of their municipality; for example, building a park, constructing a cycle path or building houses. In addition, the municipality also implements many national laws through co-rule and diverse governance arrangements (Loorbach *et al.*, 2015; Termeer *et al.*, 2017). The role of governments is to reflect on whether existing rules and regulations are helpful in the transformation towards this symbiotic city (Kemp *et al.*, 2001).

13.2.5 Corporates

Corporates and investors can play leading role in the implementation of strategic plans for biodiversity and the integration of nature-based solutions in other economic sectors such as regional food production (Dubbeling *et al.*, 2016). Through their high capital assets and land, they can invest and be an example to other economic actors. Heineken has set a high-profile example in the Netherlands and beyond in the field of Business & Biodiversity (Snep, 2014). Social housing corporations are especially important because they house many citizens in the areas (Aalbers *et al.*, 2015) and in this way own a lot of private parcels in the city. More greenery makes the neighbourhood more climate-proof (Cities Biodiversity Center, 2017). Residents from vulnerable groups mainly benefit from more greenery nearby: healthy air, less stress, an invitation to exercise (De Vries *et al.*, 2020). Three cases recently calculated by Wageningen University and Research show that urban greenery in Amsterdam is worth more than housing in some other locations (Dijkshoorn-Dekker *et al.*, 2018).

One challenge for investors is that nature-inclusive, sustainable, climate-adaptive or future-proof buildings might require a higher initial investment and generate a multitude of values that often do not convert into measurable cash flow back into the investor's bank account. In other words, there is an economic disconnect between the investment of the investor and the gains for a wide range of beneficiaries. For instance, if an investor invests in green roofs of nesting boxes, this is good for biodiversity and the liveability of many that have not necessarily contributed to the investment. Overcoming this economic narrative is only possible if one acknowledges that values of nature often cannot be expressed in classic economic narratives that focus on making a financial profit and that exclude other values. Corporates and investors can produce more socially embedded economic narratives that focus on crucial values such as the generation of fertile soils, the provision of clean air and water and the aesthetic and intrinsic value of nature.

13.2.6 Knowledge centres

Knowledge centres are organisations such as universities, but also consultancies, architects and designers. The people that work for these knowledge centres can support others with knowledge for building the symbiotic city (de Haan and Rotmans, 2018; Rip, 2002; Yudina, 2017) through their knowledge activities. Knowledge organisations can focus on the discovery of new scientific insights such as the relationship between nature and climate change, or the relationship between nature and health. They can also focus on the integration of knowledge into new ways of thinking (for instance, in terms of nature-based food systems). They can play a role in the application of knowledge in measures, tools, maps and their effects. Finally, they have a large role to play in the education of future generations (Boyer, 1996). When scientific knowledge has to be translated into practically applicable measures, they have to consider the various roles of science and heterogeneous modes of knowledge production (Cash *et al.*, 2003; de Haan and Rotmans, 2018; Dorst *et al.*, 2019; Hölscher *et al.*, 2021; Lagendijk and Cornford, 2000). See also Section 13.4.

13.3 Challenges to make the symbiotic city

This paragraph shows a set of challenges that every type of actor can work on, starting from tomorrow. These can vary from strategic, to tactical, to operational and reflexive challenges that actors can take on to develop the symbiotic city.

13.3.1 What values do I attach to nature?

It starts with discovering your own relationship to nature. What are your preferences, values and fears? What happens if I start working together with nature in the places where I can have an impact? It starts with understanding your own perception of the value of nature and how to give voice to the values of nature. This book has given eight expressions of the values of nature that can guide your discovery of your own preferences.

The value of nature for adaptation

The first value of nature is adaptation: in the city we can build connections between soil, water, biodiversity and technical networks that improve the adaptivity of cities toward future shocks, such as pandemics and climate challenges (Voskamp *et al.*, 2021). Included are the necessary social-ecological and institutional networks between rural and urban regions (Arnaiz-Schmitz *et al.*, 2018; Ulanowicz, 2021). This value is exemplified in Chapter 6 and 8 of this book.

Communal values of nature

The second value of nature is to express communal values of nature and promote the social inclusion of all people in the green urban domain. Social diversity is needed to stabilise communities in the same way that biodiversity stabilises ecosystems (Cook and Hegtvedt, 1983; Ewall, 2012; Hunter *et al.*, 2019; Jennings *et al.*, 2016; Kruize *et al.*, 2019). This value is exemplified in Chapter 11 and 12 of this book.

The value of nature for biodiverse neighbourhoods

The third value of nature focuses on the development of green neighbourhoods and districts incorporating nature-based designs of houses and streets to restore the foundations of soil and water and sustain the health of ecosystems (Cities Biodiversity Center, 2017; Roggema *et al.*, 2021). This value is exemplified in Chapter 4 and 11 of this book.

The intrinsic value of nature

The fourth value gives space to the intrinsic value of nature through practices of urban forestry, restoration and regeneration, creating biodiversity hotspots (Leclère *et al.*, 2020; Snep, 2014). These practices focus on rewilding in urban areas (Arts *et al.*, 2022) and include the reintroduction of native plant and animal species, building parks on empty lots, incorporating nature-based design or simply allowing nature to reclaim space. This value is exemplified in Chapter 4 of this book.

The value of water for water smart cities

The fifth value focuses on water and gives space to water smart cities, and the development of nature-based smart water management to restore the foundation of the city (Arden and Ma, 2018; Cascone *et al.*, 2018; Cirkel *et al.*, 2018; Ulanowicz, 2021). This value is exemplified in Chapter 5 of this book.

The value of nature-based solutions for urban development

The sixth value focuses on the functionalities of nature to produce high-tech nature-based technologies and circular innovations that enhance natural resources and strengthens sink capacities rather than depleting them (Duncan and Pascucci, 2016; Morseletto, 2020; Pascucci, 2020; Prieto-Sandoval *et al.*, 2018; Stuiver and O'Hara, 2021). This value is exemplified in Chapter 5, 8 and 9 of this book.

The value of nature in personal lifestyles and health

The seventh value focuses on individual values of nature, personal lifestyles and health, ensuring lifestyles that integrate nature in their experience (De Vries *et al.*, 2003, 2020; Hartig *et al.*, 2014; Maas *et al.*, 2006), This value is exemplified in Chapter 10 and 12 of this book.

The value of nature in restorative urban food production

The eight value focuses on the value of nature for food production systems in the urbanscape that is nature based, sustainable, and able to link urban, peri-urban, and rural communities. The food economies are embedded in local biodiverse landscapes, green technology networks and green infrastructure (Alrøe *et al.*, 2006; Gupta *et al.*, 2018; Kaufman and Bailkey, 2000; Wiskerke, 2015). This value is exemplified in Chapter 7 of this book.

13.3.2 What can I do to build a symbiotic city?

It is important to understand our agency, that is the understanding that we all are citizens and live in a certain place where we can reconnect ourselves better to our natural surroundings. We make the city, and either by ourselves or in neighbourhood communities we can do a wide range of things that are needed to make the symbiotic city a reality.

Take care of the foundation of water and soil

First of all, look at the water and soil situation in your own built environment. The soil is often characterised by impervious surfaces, so a large percentage of rainfall becomes water run-off. Reducing the amount of impervious surface (Figure 13.2) will reduce the amount of run-off (EC, 2012). This can be done by replacing pavement with semi-pavement or gravel. It is also important to prevent soil compaction. Building activities are nowadays mostly accompanied by heavy traffic.

This means that the soil becomes very compacted and unsuitable for plant growth. For that reason, the traffic should be restricted to areas outside those that are designed to become green. It is crucial to increase soil organic matter: higher organic matter content increases soil fertility and water-retention capacity, and by sequestering carbon it contributes to climate neutral cities (Pauleit and Duhme, 2000). Therefore, composting is an activity that should be reintroduced in cities, not only for agricultural reasons, but also for making living soils (Sonneveld, 2004). Another advantage is that organic waste produced by citizens gets a new life. Finally, choose your vegetation in accordance with the technical networks in the soil. Many soils in cities contain networks of drainage and cable systems. Above-ground vegetation should not destroy this underground infrastructure and, vice versa, the infrastructure should not interfere with plant growth and vegetation development (FAO & ITPS, 2022).

In order to manage the water, there are a couple of measures for retaining or storing water (Chung *et al.*, 2021). Increasing soil humus content by planting and management results in better water retention and infiltration. Infiltration crates can be used to store water. This is an underground rainwater storage system with delayed release to the subsoil. Under-road infiltration is achieved by road cuttings: an excavated section, often filled with sand. Drain rainwater to surface water, if present and if permitted. A wadi stores water in a low area either without plants or with plants: a semi-wet riverbank and pioneer vegetation, (water) insects, amphibians, birds. If the soil and elevations are suitable, construct a pond with space for water buffering but also to further promote aquatic and riparian vegetation, aquatic insects, amphibians and birds (Van Hattum *et al.*, 2016).

Figure 13.2. Semi-pavement for bicycles (courtesy of Nanda Sluijsmans).

Develop social-ecological networks

Where and when possible, you can create biodiverse grassland, place thickets and plant trees. The seedlings used should have a variety of at least five different, preferably native, species. Trees and thickets should be flowering and fruit-bearing as they are more likely to attract insects and small animals. Extensively manage grassland and use a local biodiversity mix: seasonal flowering plants, rich in nectar and pollen. Only implement weed control if necessary and use an ecological method. In order to be more hospitable to insects, either install insect hotels or create beehives or nesting sites in the soil especially for insects that nest there. Use the leftover pavement to build stone piles that insects can inhabit, and put in place sand heaps or nesting mounds. Also, leave leaves where possible or create shelters by heaping up branches. For building-dwelling birds and bats process nesting stones in or place nest boxes on the wall. Limit street lights for nocturnal animals. Create small nests and hiding places by placing, for example, hedgehog houses. Think about making fauna passages to create wildlife crossings or nature bridges for animals to pass an obstacle/barrier (Cirkel *et al.*, 2018).

Choose your targets well and focus on a type of nature that would fit both the space as well as the value that nature can have for the use of the urban area (Figure 13.3). Also take into consideration the perception of the place by others and communicate about it. Decide on the selection of target species and biotopes together with those that develop the site as well as those that will inhabit the site (see also Chapter 12 and 13 of this book). If you want to develop wildlife in cities, find arguments for habitat development. Link this habitat development not only to the value of nature for biodiversity goals but also to other values that people find important such as healthy lifestyles and adaptation.

Adapt technical networks

When you build a house, you can focus on the garden, the house and the roof (Figure 13.4) to introduce nature and rewild your living environment (Arts *et al.*, 2022). The easiest way to get a green roof (Figure 13.5) is to plant a vegetable layer of sedum (succulents). Through design and lay-out you can improve the vegetation layer to match the ecological environment or even leave the substrate surface to self-sow by wind and avian seed dispersal, to exactly mimic the wild environment and rewild the roof (Cascone *et al.*, 2018). Make sure that you perform soil-bound planting (rooted in the open ground) against the wall either directly or with support constructions (for example, cables, depending on the type of climbing plant). In new buildings a completely green façade can be created with the construction of a vertical garden or wall garden covered by various plants.

Create a small urban forest or a larger one. Invest in trees and shrubs. Fences can become greener by planting climbing plants up against them. Even better is to replace fences with hedges or, if space allows, thickets (shrubs). Beware of the shade effect of the buildings in relation to the vegetation to match the desired effects in the environment. Adapt the building and the planting of trees and shrubs for controlling the effects of wind action and the cooling of the environment. Additions of wood, stones and sand heaps can help to create a 'rougher' terrain. Remember that one tree can change an entire neighbourhood (de Smet and van Reusel, 2018).

Figure 13.3. Green in the city of Utrecht (courtesy of Kees Rutten).

Figure 13.4. Rooftop garden (courtesy of Sabine O'Hara).

Figure 13.5. The roof as a garden.

Invest in circular networks

First of all, take a look at the circular networks you already or potentially could engage in. What is your own relationship with waste? Do you know what waste you produce and how to decrease it, or reuse it? Start to get to know the organisms that work for you, like worms or moles. They keep the soil in your garden healthy (De Kimpe and Morel, 2000). Worms turning up soil is good for drainage. The mole also helps you to keep pests in check: a mole feasts on insects and they also eat baby mice (Figure 13.6).

If you like gardening, start composting and reuse your organic waste to improve the soil's conditions (Agudelo-Vera *et al.*, 2011). Build houses of biobased materials or wood. Store water in a fence with delayed drainage and retain water in rain barrels. Build a green-blue roof: a vegetated roof with extra water storage capacity such as retention crates. A good example is the polder roof where water can be reused, stored or drained. This can also have a cooling effect (Cirkel *et al.*, 2018).

Figure 13.6. A mole is a very good worker for nature.

Build new institutional networks

This type of activity is more strategic in nature. Suppose you work for an organisation or company that itself owns real estate or land. You want to speed the process up and build new linkages and institutional networks that support the symbiotic city. Real estate companies, project developers, housing corporations are controlled by the boardroom, i.e. by those in power who make decisions about the strategy and vision of the organisation. Put forward the argument that every company can decide to become a voice of nature and a forerunner in the market. For instance, developing nature-inclusive real estate benefits from good examples and if your company sets that example, this generates a new profile in the market, but also in the minds of your customers and residents (Stuiver and Koffijberg, 2022c).

Onno Dwars, the CEO from Ballast Nedam Development in the Netherlands is very pro-active in building new institutional networks for nature-inclusive housing (NL Greenlabel, 2021). He says: 'Only by pushing the standards for nature inclusive housing, streets and neighbourhoods ourselves, we can determine how we make the world a better place and push our boundaries. Therefore, we are taking the next step to develop nature-inclusive real estate. We call on partners to join this new normal and challenge each other to go further and further.'
Another example of a new institutional network is Green Circles, initiated by the province of South Holland in the Netherlands. It is an extensive network of innovative companies, knowledge institutions, governments and social organisations. The partners work together on

sustainability solutions, in which nature plays an important role. Their aim is to bring nature and the economy into better balance and actively contribute to a better living environment, a circular economy and a healthy future (Lemstra *et al.*, 2020).

13.3.3 How do you become a leading voice?

How do you take energetic steps in the direction you want to go? How can you become a leading voice of nature? Everyone has a role to play in the transformation. For instance in the Netherlands, during the elections of 2022 (Figure 13.7), a study revealed that all political parties find greening the city equally important (Rutten and Stuiver, 2022).

Develop an appealing and clear narrative

Don't leave the story of what your city looks like to other people. Create a hopeful story, a dream image that you want to achieve, and involve residents in making that dream. Ask what residents really benefit from, how they can live well, what is valuable for their lives now and in the immediate future. Inspiration can very well come from the question: how can we be good ancestors for the children and grandchildren who will live in the city after us (Krznaric, 2020)? Backward-planning can help creating a desired future scenario for the city or village in 2050, 2100, with a visualisation, co-created in a series of meetings with citizens, administrators, policy makers, public and private organisations (see Chapter 3 for inspiration).

Figure 13.7. Dutch election board with local parties in the city of Utrecht (courtesy of Kees Rutten).

Tell your story with great enthusiasm. What works very well is to have an inspirational icon of nature. Choose butterflies instead of wasps, focus on the beaver and not the wolf. Use species with a feelgood factor – herb-rich grass is good for bees and butterflies, but also for blackbirds and hedgehogs. Make sure that you are aware of other people's emotions regarding nature. Make sure that the measures are not presented too much as a hobby of nature lovers. Broaden the story. Link it to other things that people enjoy. Children can play in a more exciting way (Figure 13.8), you can find blackberries for jam, and you can pick flowers and leaves for making teas (Stuiver and Koffijberg, 2022d).

Invite other parties to participate with you

More and more parties (project developers, green contractors, knowledge institutes) are intrinsically driven to take part in greening and are rapidly developing knowledge and experience. Do you use it? They are very eager to participate and to put greening into practice. Ask these parties to actively participate, invite them. Municipalities hold an important key to constructing new buildings that are nature-inclusive. They can state this in the programme of requirements for the buildings to be constructed. This challenges architects and contractors to build nature-inclusive properties. Architects and contractors who include nature in their plans based on intrinsic motivation are therefore rewarded for their green approach (Stuiver and Koffijberg, 2022f). Opt for these motivated parties in concrete projects; that makes implementation a nicer experience as you cooperate towards the same goal. Reward the forerunners and encourage them to take an extra step. Think over all steps in planning, design, development and maintenance of the urban location: who should do what when? Address potential opportunities and bottlenecks in advance, and involve actors in the use and maintenance phase early in the project (Stuiver and Koffijberg, 2022d).

Figure 13.8. Children's playground (courtesy of Nanda Sluijsmans).

Realise that it is possible: still adjust designs

There are many examples where even on the construction site it has been decided to apply nature-inclusive measures. A well-known example of this is Amersfoort station in the Netherlands, where dozens of stones in which swifts can nest have been bricked in, and these were quickly occupied too. So, be sure to start a conversation about what is still possible: nest boxes, facades, the entrances of the buildings, the fences, soil management, roof greenery, bird-resistant glass or open surfaces at the entrance of houses (Figure 13.9). They seem like small measures, but for some species they can make a big difference (Stuiver and Koffijberg, 2022f).

13.3.4 How do you include citizens from the start?

It is crucial to involve citizens from the start in plans for more ecological management of greenery. They will only adopt measures if they have had an impact. They know where all kinds of practical bottlenecks are. And they have a lot of knowledge.

Take a firm stance on inclusion

Everyone should have access to quality greenery regardless of their social position, economic standing, class, race, gender, or creed (FAO, 2017). Such a rights-based perspective may require a more active role of the public sector to ensure that all are represented, especially in times of crises. Leaving nature development in the city to market forces alone can result in increased risks, especially in highly centralised and global markets. The most vulnerable will be the first to suffer from gentrification when a street or neighbourhood is made greener. The social inclusion of all stakeholders and actors is therefore crucial, because nature is a right and not a privilege.

Figure 13.9. Station Amersfoort, the Netherlands (courtesy of Nanda Sluijsmans).

How to develop inclusive leadership? First, it starts with the acknowledgement that not putting an effort into social equity will potentially lead to more exclusion of certain groups and communities. Second, be bold and dare to address the topic of inclusion in your own life and personality. Develop an understanding of the terms and language that is used. Do you know what your blind spots are? What are the barriers that prevent the participation, power and recognition of certain groups in your own life and projects? What does your network look like? Do you use inclusive or exclusive language? Third, dare to learn about your own work and get inspiration from inclusive leaders that have already worked on this topic. Learn from their experiences. Look at your day-to-day or work practices; how inclusive are you in your own work and can you improve it? Fourth, build an inclusive world. Build your own database of inclusive case studies and best practice stories. Finally, do not be afraid to be different. The dominance of social norms in existing institutions is one of the causes of exclusion. Develop your skills to contribute to an inclusive world.

Invest in people that do the actual greening

Invest in the people that must do the actual greening in streets, public parks, the bus stops (Figure 13.10) and so on. Exchange knowledge about ecological management among those responsible. Have locals, experienced gardeners and ecologists accompany the people who manage the greenery in your city, so that they can explain on the spot what a different working method yields for the biodiversity at this specific location. And let them tell you why that is important. Or organise a good, inspiring in-company course. And start with a pilot: a test piece somewhere in a part of a neighbourhood or along the side of a road when it comes to trying things out together and learning from it. What helps here is that the district managers themselves can indicate the promising locations for ecological management in their district (Stuiver and Koffijberg, 2022c).

Encourage people that are working on greening the city and support them. In this time of transformation, give people the confidence and space, and reward them with space to do things differently. Beautiful integral designs are often made, but they then must be executed by people with limited time and budget. These limitations often kill off a lot of creativity and effort. You can also reward someone who wants to do it differently, because this person must stand firm. Take them along in the new designs and the narrative of the symbiotic city (Stuiver and Koffijberg, 2022a).

Encourage tenants to keep/make their garden green

Housing associations can encourage tenants by making rules about garden maintenance, for example how departing tenants must leave the garden when they move out of the house or rules about the height of trees. More greening can be achieved by supporting tenants that want to green the gardens. An example is Havensteder in Vrijlandt in Rotterdam Lombardijen. Here, principles of the garden village, based on the works of the garden city (Howard, 1898), are used. The fences are composed of different types of hedges and the new tenants will find a garden that is green and neat. During the contract phase, the tenants receive information about mutual expectations of

Figure 13.10. New public green: bus stop in Eindhoven, the Netherlands (courtesy of Nanda Sluijsmans).

how to make and maintain the garden. They encourage tenants to keep their garden as green as possible. For example, by putting them in touch with the green manager or other partners from the Havensteder network for advice (Stuiver and Koffijberg, 2022e).

Chapter 12 of this book argues that it is crucial to integrate the core values of empathy, placation, accessibility and identity in any inclusive urban design process (Figure 13.11). Inclusive design depends on the extent that planners, designers, architects, and the community at large can

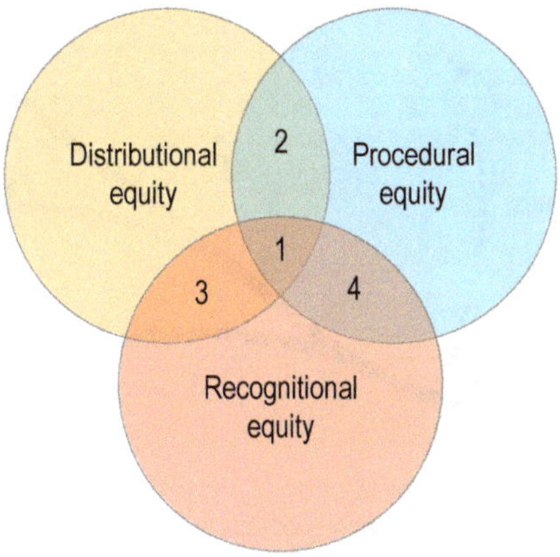

Figure 13.11. Four values that intersect with Meerow's model for social equity (2019) 1 = empathy; 2 = placation; 3 = accessibility; 4 = identity. This figure was developed by M. Stuiver, S. Sarabi, M. Takken, L. Rondard, R. Valkenburg and T. Yuksel for https://ewuu.nl.

acknowledge, learn to understand and apply these values in designing the urban fabric. Identity means that we need to learn how to acknowledge, appreciate and celebrate different lifestyles and identities. This is at the 'scale of the individual' and captures the importance for all of us to understand the diversity in human histories, skills, and needs as related to physical, social, mental, and spiritual health. Placation acknowledges the value of 'co-existence' for all creatures (human and non-human) in the city to truly co-create public environments. Accessibility is the value to work on the equal distribution of resources in architecture, landscape, and interior and circular design. Empathy is the central value in the figure. Every inclusive design begins with empathy that recognises both the needs of human and non-human actors in the city and truly aims to understand and connect with others in the city.

13.4 Become part of the transformation

We can all be actors who change the rules of the game and work together to change the urban landscape into a nature-inclusive landscape for all people. The end goal of this collective transformation process embraces the symbiotic city with new governance arrangements that reflect the different voices of nature (Kemp *et al.*, 2001; Roosendaal *et al.*, 2020; Rotmans, 2014).

The institutional approach to regime change distinguishes three levels of change as part of the transformation process (Figure 13.12). First, there is the level of niches where actors test novel approaches and alternatives in a protected setting. This book has given a multitude of examples where actors developed nature-based urban pathways. Second, there is the level of regimes (i.e. the set of rules and regulations that are produced and reproduced in coalitions, projects, research and government bodies, that is to say, everywhere where actors mobilise and form an alignment, and can change new rules and regulations). A regime is also expressed by new technologies and new

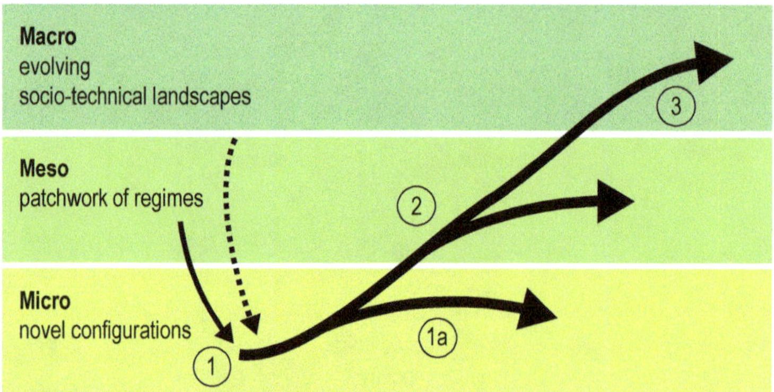

1. Novelty shaped by existing regime
1a. Novelty does not develop, or is only succesful in a local context
2. Novelty evolves, is taken up, may modify regime
3. Landscape is transformed

Figure 13.12. Stages of regime change (Rip and Kemp, 1998).

design options (Deuten, 2003). Third, there is the level of the landscape where different regimes (for instance of food production, housing, energy, transport and mobility) shape the landscapes on the macro level (Rip and Kemp, 1996).

Regime change is institutional change, i.e. the rules on how to act or on how not to act and it is also epistemological change, e.g. the ideas on 'what to know' and 'how to know' (Deuten and Rip, 2000; Rip, 2000, 2002).

The first aspect of institutional change concerns the actual diagnosis of the regime. How do you perceive existing rules and regulations on how nature gets a place in the city? Fundamental to the diagnosis are the norms and values of nature that you express. What is your opinion about the actual changes (in politics, markets, technologies and science) that need to take place? Are your proposals based on incremental or fundamental changes?

The second aspect of institutional change is the niches that you engage in to change the existing regime. These include the innovative practices with new social-ecological networks and technical networks as described in the former paragraphs, as well as the development of innovative institutional arrangements (Buizer *et al.*, 2011) and discourse coalitions (Hajer, 1993) to increase urban nature (Resilience Alliance, 2010). See which relevant partners of your company or organisation and in your network also opt for nature-inclusive construction. Make coalitions with them. Examples of such coalitions that are presented in this book are 'Green Healthy City', an initiative of Green Circles (Chapter 2), the Bloei & Groei initiative (Chapter 2) and the Doughnut Coalition in Amsterdam (Chapter 11) as well as the Climate Adaptive Building programme with nature (KAN) that focuses on new nature-inclusive construction of houses (Chapter 2).

The third aspect of institutional change is the knowledge you think is needed to change the regime. What type of knowledge is needed and in what way? Principles of adaptive governance (Dietz *et al.*, 2003) and knowledge governance (Gerritsen *et al.*, 2013) can be used to emphasise the agency (Westley *et al.*, 2013) of humans to learn and adapt to changing relationships between society and ecosystems (FAO, 2021; Resilience Alliance 2010). Key characteristics are to increase the agency, reflexivity, learning and cooperation among stakeholders (Chambers *et al.*, 2021; De Haan and Rotmans, 2018; Horlings *et al.*, 2020), for instance about the worries and expectations of the citizens that are affected by the regime change.

Knowledge of residents about greening the city can be very practical, for instance in relation to their objections to certain types of greening in the area. They may, for example, object to overhanging blackberries and nettles along cycle paths and intersections. Their objections can also be more strategic as we have seen in Chapter 11 with the Doughnut Coalition, where neighbourhood associations want to be part of the neighbourhood team with the green contractors to manage the greenery in the city and identify possible hotspots for biodiversity.

If you want to incorporate the knowledge of citizens, it is important to learn from other municipalities that have been working on ecological management while involving citizens (Pahl-Wostl and Hare, 2004; Stuiver and Koffijberg, 2022b).

In developing the symbiotic city, we focus on a transformation that is accessible for all groups of actors and makes use of their knowledge. Transformation goals need to address heterogeneous forms of knowledge explicitly. An important question related to this is how we ensure that the transformation is not having adverse and undesirable effects, such as on climate injustice for certain groups of people in the world (IPCC, 2022), and the exclusion of large parts of people (De Groot *et al.*, 2010). Therefore, local knowledge, viewpoints, stories and interests of all citizens come increasingly to the foreground in design activities (Arnaiz-Schmitz *et al.*, 2018; De Smet and van Reusel, 2018; Haaland and van den Bosch, 2015). Heterogeneous knowledge management is another useful angle to address that there are several modes of knowledge needed for transformation (Kagono, 1985; Nonaka, 1994; Nonaka and Takeuchi, 1995).

Figure 13.13 exemplifies that these different modes of knowledge are exchange circulation, natural history and controlled circumstances. The latter mode is well known; it is knowledge as the outcome of (experimental) findings yielded under controlled circumstances, much like laboratories. The biggest advantage of this mode of knowledge production is that variability, which might otherwise confuse or hinder aggregation efforts, can now be managed in these laboratory settings with clear boundaries to the outside world. However, this mode of knowledge is not sufficient to grasp the insights of all people, especially not those outside academia. Exchange circulation is a mode of knowledge production that helps to understand that knowledge is a by-product of actions and interactions between local practices. Trans-locality is the result of (years of) experiences and exchanges with others. Often, there is no clear conceptual grasp of how the performance is realised. Typically, there is no theory involved; it is experiential knowledge (cf. learning-by-doing; learning-by-interacting) (Stuiver, 2008).

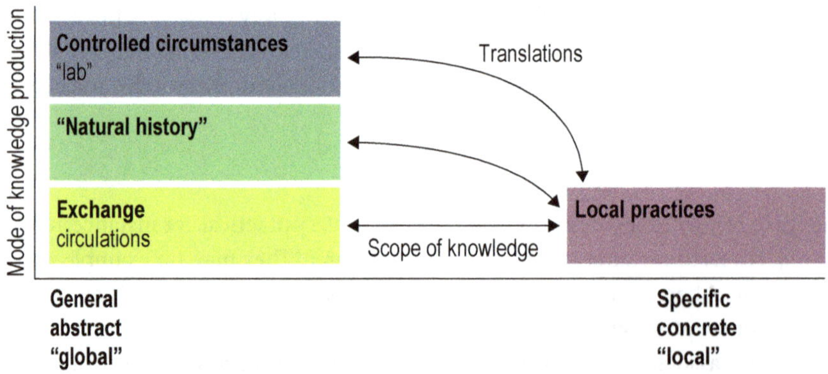

Figure 13.13. Different modes of knowledge (Nonaka and Takeuchi, 1995).

Citizen science is an example of active involvement of the public and specific stakeholders in knowledge development. Citizen science can be an important way of collecting information for biodiversity conservation goals. Through citizen science, citizens are engaged to increase biodiversity and reduce health risks and they can be consulted about their own knowledge and perceptions of nature. The Netherlands has one of the most developed citizen science networks for natural systems. In most of these networks the volunteers are involved in crowd sourcing. Citizen science projects include monitoring birds in urban environments, hopping cameras for surveying mammals in private backyards, the annual garden count and reporting dead animals.

The last mode of knowledge production is called natural history, where knowledge is the outcome of intentional collection and accumulation of experiences and findings across space and time. For example, the knowledge of plants and trees in arboreta or natural history museums, where the emphasis is on collecting, systematising and classification. It can also involve mapping new visions on the city of the future (Timmermans *et al.*, 2022) and documenting medicinal practices and knowledge of endogenous peoples (Batista *et al.*, 2021) (Van 't Hooft, 2014). Furthermore, it can combine history and art with landscape architecture, such as in Museum Voorlinden in the Netherlands with a garden of landscape architect Piet Oudolf (Figure 13.14). All these expressions of natural history are slowly emerging as important modes of knowledge production to give nature a voice, and they go hand in hand with forms of exchange circulation of all urban actors.

Figure 13.14. Museum Voorlinden in Wassenaar, the Netherlands (courtesy of Nanda Sluijsmans).

One specific way to give expression to the different voices of nature is by making so-called immutable mobiles, either texts, books, graphs or maps, or 3D presentations of elements of a city, such as lands, trees and birds, that are made immutable and can then travel and in this way become mobile (Latour, 1987). Figure 13.15 is one such strong representation (Leclère *et al.*, 2020) that was presented in the scientific journal Nature to portray what needs to be done in order to stop biodiversity decline. This visual reminds us why it is so urgent and compelling to be the voice of nature for urban transformations towards the symbiotic city.

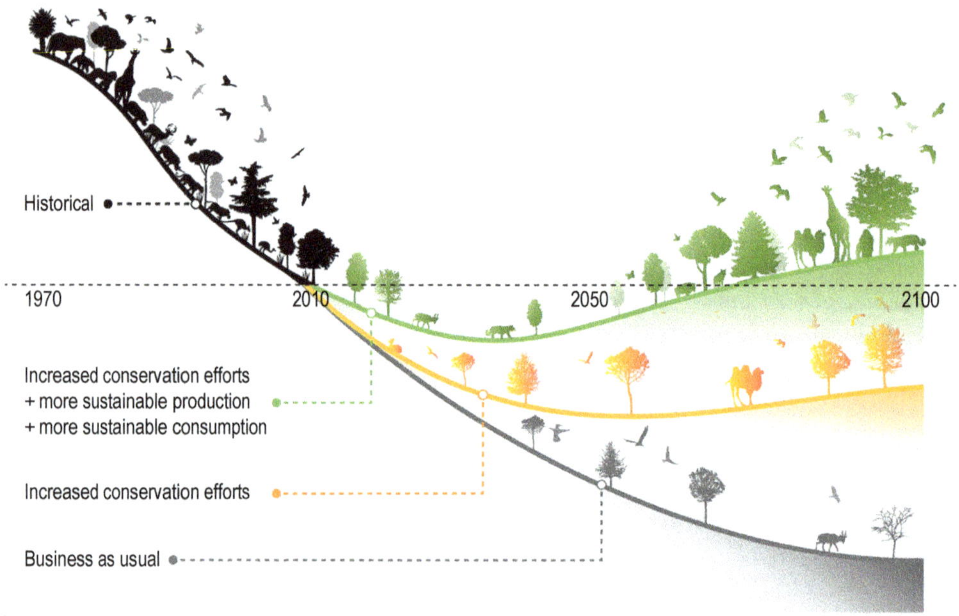

Figure 13.15. The urgency of increased efforts for biodiversity (Leclère *et al.*, 2020). This artwork illustrates the main findings of the article, but does not intend to accurately represent its results (https://doi.org/10.1038/s41586-020-2705-y).

References

Agudelo-Vera, C.M., Mels, A.R., Keesman, K.J. and Rijnaarts, H.H.M., 2011. Resource management as a key factor for sustainable urban planning. Journal of environmental management 92: 2295-2303. https://doi.org/10.1016/j.jenvman.2011.05.016

Alrøe, H.F., Byrne, J. and Glover, L., 2006. Organic agriculture and ecological justice: ethics and practice. Global development of organic agriculture: challenges and prospects. CABI, Wallingford, UK. https://doi.org/10.1079/9781845930783.0075: 75-112

Anguelovski, I., 2013. New directions in urban environmental justice: Rebuilding community, addressing trauma, and remaking place. Journal of Planning Education and Research 33: 160-175. https://doi.org/10.1177/0739456X13478019

Anonymous, 2021. Harmony with nature. Available at: http://www.harmonywithnatureun.org/rightsOfNature/.

Arden, S. and Ma, X., 2018. Constructed wetlands for greywater recycle and reuse: a review. Science of the Total Environment 630: 587-599.

Arnaiz-Schmitz, C., Schmitz, M.F., Herrero-Jáuregui, C., Gutiérrez-Angonese, J., Pineda, F.D. and Montes, C., 2018. Identifying socio-ecological networks in rural-urban gradients: diagnosis of a changing cultural landscape. Science of the Total Environment 612: 625-635. https://doi.org/10.1016/j.scitotenv.2017.08.215

Arts, K., Bakker, L. and Buys, A., 2022. Rewilding in Nederland. KNNV Uitgeverij, Zeist, the Netherlands.

Atelier groenblauw, Sociaal-maatschappelijke en economische waarde van groen en blauw. Available at: https://nl.urbangreenbluegrids.com/social/.

Batista, M.L.P., Macêdo, E.M., Silva, A.J.d. and Barros, R.F.M.d., 2021. Endogenous knowledge, culture, and nature in the approach to sustainable local development in a rural community in Northeastern Brazil. Research, Society and Development 10: e55310616237. https://doi.org/10.33448/rsd-v10i6.16237

Boyer, E.L., 1996. From scholarship reconsidered to scholarship assessed. Quest 48: 129-139. https://doi.org/10.1080/00336297.1996.10484184

Buizer, M., Arts, B. and Kok, K., 2011. Governance, scale and the environment: the importance of recognizing knowledge claims in transdisciplinary arenas. Ecology and Society 16. https://doi.org/10.5751/ES-03908-160121

Burgers, L., 2021. Friends of the Earth Netherlands versus Royal Dutch Shell: all companies must act against climate change, Transformative Private Law Bog. ACT, Amsterdam, the Netherlands.

Burgers, L. and Outer, J., 2021. Rights of nature – case-studies from six continents: compendium, Ambassade van de Noordzee.

Cascone, S., Catania, F., Gagliano, A. and Sciuto, G., 2018. A comprehensive study on green roof performance for retrofitting existing buildings. Building and Environment 136: 227-239. https://doi.org/10.1016/j.buildenv.2018.03.052

Cash, D.W., Clark, W.C., Alcock, F., Dickson, N.M., Eckley, N., Guston, D.H., Jäger, J. and Mitchell, R.B., 2003. Knowledge systems for sustainable development. Proceedings of the National Academy of Sciences 100: 8086-8091. https://doi.org/10.1073/pnas.1231332100

Chambers, J.M., Wyborn, C., Ryan, M.E., Reid, R.S., Riechers, M., Serban, A., Bennett, N.J., Cvitanovic, C., Fernández-Giménez, M.E., Galvin, K.A., Goldstein, B.E., Klenk, N.L., Tengö, M., Brennan, R., Cockburn, J.J., Hill, R., Munera, C., Nel, J.L., Österblom, H., Bednarek, A.T., Bennett, E.M., Brandeis, A., Charli-Joseph, L., Chatterton, P., Curran, K., Dumrongrojwatthana, P., Durán, A.P., Fada, S.J., Gerber, J.-D., Green, J.M.H., Guerrero, A.M., Haller, T., Horcea-Milcu, A.-I., Leimona, B., Montana, J., Rondeau, R.,

Spierenburg, M., Steyaert, P., Zaehringer, J.G., Gruby, R., Hutton, J. and Pickering, T., 2021. Six modes of co-production for sustainability. Nature Sustainability 4: 983-996. https://doi.org/10.1038/s41893-021-00755-x

Chung, M.G., Frank, K.A., Pokhrel, Y., Dietz, T. and Liu, J., 2021. Natural infrastructure in sustaining global urban freshwater ecosystem services. Nature Sustainability 4: 1068-1075. https://doi.org/10.1038/s41893-021-00786-4

Cirkel, D.G., Voortman, B.R., Van Veen, T. and Bartholomeus, R.P., 2018. Evaporation from (Blue-) green roofs: assessing the benefits of a storage and capillary irrigation system based on measurements and modeling. Water 10: 1253. https://doi.org/10.3390/w10091253

Cities Biodiversity Center, 2017. How cities are using nature-based solutions for sustainable urban development. Available at: https://cbc.iclei.org/cities-using-nature-based-solutions-sustainable-urban-development/.

Cook, K.S. and Hegtvedt, K.A., 1983. Distributive justice, equity, and equality. Annual review of sociology 9: 217-241.

De Groot, R.S., Alkemade, R., Braat, L., Hein, L. and Willemen, L., 2010. Challenges in integrating the concept of ecosystem services and values in landscape planning, management and decision making. Ecological Complexity 7: 260-272. https://doi.org/10.1016/j.ecocom.2009.10.006

De Haan, F.J. and Rotmans, J., 2018. A proposed theoretical framework for actors in transformative change. Technological Forecasting and Social Change 128: 275-286. https://doi.org/10.1016/j.techfore.2017.12.017

De Kimpe, C.R. and Morel, J.-L., 2000. Urban soil management: a growing concern. Soil Science 165: 31-40. https://doi.org/10.1097/00010694-200001000-00005

De Smet, A. and van Reusel, H., 2018. How one tree can change the future of a neighbourhood: the process behind the creation of the Boerenhof Park as an example for tactical urban planning. Urban Forestry & Urban Greening 30: 286-294. https://doi.org/10.1016/j.ufug.2017.09.001

De Vries, S., Buijs, A.E. and Snep, R.P., 2020. Environmental justice in The Netherlands: presence and quality of greenspace differ by socioeconomic status of neighbourhoods. Sustainability 12: 5889.

De Vries, S., Verheij, R.A., Groenewegen, P.P. and Spreeuwenberg, P., 2003. Natural environments – healthy environments? An exploratory analysis of the relationship between greenspace and health. Environment and Planning A: Economy and Space 35: 1717-1731. https://doi.org/10.1068/a35111

Deuten, J., 2003. Cosmopolitanising technologies. A study of four emerging technological regimes. Twente University Press, Enschede, the Netherlands.

Deuten, J.J. and Rip, A., 2000. Narrative infrastructure in product creation processes. Organization 7: 69-93. https://doi.org/10.1177/135050840071005

Dietz, T., Ostrom, E. and Stern, P.C., 2003. The struggle to govern the commons. Science 302: 1907-1912. https://doi.org/10.1126/science.1091015

Dorst, H., van der Jagt, A., Raven, R. and Runhaar, H., 2019. Urban greening through nature-based solutions – Key characteristics of an emerging concept. Sustainable Cities and Society 49: 101620. https://doi.org/10.1016/j.scs.2019.101620

Dubbeling, M., Carey, J. and Hochberg, K., 2016. The role of private sector in city region food systems, RUAF Foundation, Leusden, the Netherlands.

Duncan, J. and Pascucci, S., 2016. Circular solutions for linear problems: principles for sustainable food futures. Solutions 7: 58-65.

EC, 2012. Guidelines on best practice to limit, mitigate or compensate soil sealing, European Commission, Publications Office of the European Union, Luxembourg.

Ewall, M., 2012. Legal tools for environmental equity vs. environmental justice. Sustainable Development Law & Policy 13: 4.

FAO, 2021. Indigenous peoples' food systems: insights on sustainability and resilience from the front line of climate change, Rome, Italy. Available at: https://doi.org/10.4060/cb5131en

FAO & ITPS, 2022. Urbanisation and soil sealing. Food and Agriculture Organization of the United Nations and Intergovernmental Technical Panel on Soils, Rome, Italy.

Gerritsen, A.L., Stuiver, M. and Termeer, C.J.A.M., 2013. Knowledge governance: an exploration of principles, impact, and barriers. Science and Public Policy 40: 604-615. https://doi.org/10.1093/scipol/sct012

Gruezmacher, M. and van Assche, K., 2022. Crafting strategies for sustainable local development – a community guide. In Planning. https://doi.org/10.17418/B.2021.9789493164178

Gunderson, L.H. and Holling, C.S., 2002. Panarchy: understanding transformations in human and natural systems. Island Press, Washington DC, USA.

Gupta, C., Campbell, D., Munden-Dixon, K., Sowerwine, J., Capps, S., Feenstra, G. and Van Soelen Kim, J., 2018. Food policy councils and local governments: creating effective collaboration for food systems change. Journal of Agriculture, Food Systems, and Community Development 8: 11-28. https://doi.org/10.5304/jafscd.2018.08B.006

Haaland, C. and Van den Bosch, C.K., 2015. Challenges and strategies for urban green-space planning in cities undergoing densification: A review. Urban forestry & urban greening 14: 760-771. https://doi.org/10.1016/j.ufug.2015.07.009

Hajer, M., 1993. Discourse coalitions and the institutionalization of practice. In: Fischer, F. and Forester, J. (eds.) The argumentative turn in policy analysis and planning. Duke University Press, Durham, NC, USA, pp. 43-76. https://doi.org/10.1215/9780822381815-003

Hartig, T., Mitchell, R., De Vries, S. and Frumkin, H., 2014. Nature and health. Annual review of public health 35: 207-228.

Hölscher, K., Wittmayer, J.M., Hirschnitz-Garbers, M., Olfert, A., Walther, J., Schiller, G. and Brunnow, B., 2021. Transforming science and society? Methodological lessons from and for transformation research. Research Evaluation 30: 73-89. https://doi.org/10.1093/reseval/rvaa034

Horlings, L.G., Roep, D., Mathijs, E. and Marsden, T., 2020. Exploring the transformative capacity of place-shaping practices. Sustainability Science 15: 353-362. https://doi.org/10.1007/s11625-020-00787-w

Howard, E., 1898. To-morrow: a peaceful path to real reform. Swan Sonnenschein, London, UK.

Hunter, R.F., Cleland, C., Cleary, A., Droomers, M., Wheeler, B.W., Sinnett, D., Nieuwenhuijsen, M.J. and Braubach, M., 2019. Environmental, health, wellbeing, social and equity effects of urban green space interventions: a meta-narrative evidence synthesis. Environment international 130: 104923.

IPCC, 2022. Climate Change 2022: Impacts, adaptation, and vulnerability contribution of Working Group II to the Sixth Assessment Report of the Intergovernmental Panel on Climate Change. Cambridge University Press, Cambridge, UK.

Jennings, V., Larson, L. and Yun, J., 2016. Advancing sustainability through urban green space: Cultural ecosystem services, equity, and social determinants of health. International Journal of Environmental Research and Public Health 13: 196.

Kagono, T., 1985. Strategic vs evolutionary management: a U.S. – Japan comparison of strategy and organization, Amsterdam: North-Holland.

Kaufman, J. and Bailkey, M., 2000. Farming inside cities: entrepreneurial urban agriculture in the United States. Lincoln Institute of Land Policy.

Kemp, R., Rip, A. and Schot, J., 2001. Constructing transition paths through the management of niches, Path dependence and creation. Lawrence Erlbaum, pp. 269-299.

Knapp, M., Bauer, A., Perkins, M. and Snell, T., 2013. Building community capital in social care: is there an economic case? Community Development Journal 48: 313-331.

Kruize, H., van der Vliet, N., Staatsen, B., Bell, R., Chiabai, A., Muiños, G., Higgins, S., Quiroga, S., Martinez-Juarez, P. and Aberg Yngwe, M., 2019. Urban green space: creating a triple win for environmental sustainability, health, and health equity through behavior change. International Journal of Environmental Research and Public Health 16: 4403.

Krznaric, R., 2020. The good ancestor. The Experiment Publishing, New York, NY, USA, 336 pp.

Lagendijk, A. and Cornford, J., 2000. Regional institutions and knowledge – tracking new forms of regional development policy. Geoforum 31: 209-218. https://doi.org/10.1016/S0016-7185(99)00031-7

Latour, B., 1987. Science in action: how to follow scientists and engineers through society. Harvard University Press, Cambridge, MA, USA.

Leclère, D., Obersteiner, M., Barrett, M., Butchart, S.H.M., Chaudhary, A., De Palma, A., DeClerck, F.A.J., Di Marco, M., Doelman, J.C., Dürauer, M., Freeman, R., Harfoot, M., Hasegawa, T., Hellweg, S., Hilbers, J.P., Hill, S.L.L., Humpenöder, F., Jennings, N., Krisztin, T., Mace, G.M., Ohashi, H., Popp, A., Purvis, A., Schipper, A.M., Tabeau, A., Valin, H., van Meijl, H., van Zeist, W.-J., Visconti, P., Alkemade, R., Almond, R., Bunting, G., Burgess, N.D., Cornell, S.E., Di Fulvio, F., Ferrier, S., Fritz, S., Fujimori, S., Grooten, M., Harwood, T., Havlík, P., Herrero, M., Hoskins, A.J., Jung, M., Kram, T., Lotze-Campen, H., Matsui, T., Meyer, C., Nel, D., Newbold, T., Schmidt-Traub, G., Stehfest, E., Strassburg, B.B.N., van Vuuren, D.P., Ware, C., Watson, J.E.M., Wu, W. and Young, L., 2020. Bending the curve of terrestrial biodiversity needs an integrated strategy. Nature 585: 551-556. https://doi.org/10.1038/s41586-020-2705-y

Lemstra, A.-B., Noordermeer, R., Slob, T.J., Gremmen, M., Bosch, C.v.d., Rees, A.v., Biesmeijer, K. and Westerink, J., 2020. Groene Cirkels Kaas en Bodemdaling. Available at: https://www.online-broadcast.com/kaasenbodemdaling.

Loorbach, D., Frantzeskaki, N. and Huffenreuter, R.L., 2015. Transition management: taking stock from governance experimentation. Journal of corporate citizenship: 48-66. https://doi.org/10.9774/GLEAF.4700.2015.ju.00008

Maas, J., Verheij, R.A., Groenewegen, P.P., De Vries, S. and Spreeuwenberg, P., 2006. Green space, urbanity, and health: how strong is the relation? Journal of Epidemiology & Community Health 60: 587-592. https://doi.org/10.1136/jech.2005.043125

Morseletto, P., 2020. Restorative and regenerative: exploring the concepts in the circular economy. Journal of Industrial Ecology 24: 763-773. https://doi.org/10.1111/jiec.12987

NL Greenlabel, 2021. Manifest Bouwen voor natuur, drie bouwstenen voor natuurlijke dorpen en steden. Available at: https://tinyurl.com/3pct9pfk.

Nonaka, I., 1994. A dynamic theory of organizational knowledge creation. Organization Science 5: 14-37. https://doi.org/10.1287/orsc.5.1.14

Nonaka, I. and Takeuchi, H., 1995. The knowledge-creating company: How Japanese companies create the dynamics of innovation. Oxford University Press, Oxford, UK.

Pahl-Wostl, C. and Hare, M., 2004. Processes of social learning in integrated resources management. Journal of Community & Applied Social Psychology 14: 193-206. https://doi.org/10.1002/casp.774

Pascucci, S., 2020. Circular food economies. Routledge handbook of sustainable and regenerative food systems. Routledge, London, UK, pp. 318-335.

Pauleit, S. and Duhme, F., 2000. Assessing the environmental performance of land cover types for urban planning. Landscape and Urban Planning 52: 1-20. https://doi.org/10.1016/S0169-2046(00)00109-2

Pineo, H., 2020. Towards healthy urbanism: inclusive, equitable, and sustainable (THRIVES) – an urban design and planning framework from theory to praxis. Cities & Health. https://doi.org/10.1080/23748834.2020.1769527

Prieto-Sandoval, V., Jaca, C. and Ormazabal, M., 2018. Towards a consensus on the circular economy. Journal of cleaner production 179: 605-615. https://doi.org/10.1016/j.jclepro.2017.12.224

Resilience Alliance, 2010. Assessing resilience in social-ecological systems: Workbook for practitioners. Version 2.0, Resilience Alliance.

Rip, A., 2000. Fashions, lock-ins and the heterogeneity of knowledge production. The future of knowledge production in the academy. Society for Research into Higher Education and Open University, Buckingham, UK, pp. 28-39.

Rip, A., 2002. Science for the 21st century. The future of science and the humanities. Amsterdam University Press, Amsterdam, the Netherlands, pp. 99-148.

Rip, A. and Kemp, R., 1998. Technological change. In: Rayner, S. and Malone, E.L. (eds.) Human choice and climate change. Battelle Press, Columbus, OH, USA, pp. 327-399.

Roggema, R., Tillie, N. and Keeffe, G., 2021. Nature-based urbanization: scan opportunities, determine directions and create inspiring ecologies. Land 10: 651. https://doi.org/10.3390/land10060651

Roosendaal, L., Herens, M., de Roo, N., Stuiver, M., Pittore, K., Soma, K. and Hetterscheid, B., 2020. City region food system governance: guiding principles and lessons learned from case studies around the world, Wageningen Centre for Development Innovation, Wageningen, the Netherlands.

Rotmans, J., 2014. Stad vanuit een Transitiebril. Available at: https://www.stadsleven.nu/2014/12/23/stad-vanuit-een-transitiebril-column-jan-rotmans/.

Rutten, J. and Stuiver, M., 2022. Groen en gemeenteraadsverkiezingen: aandacht voor groen in de partijprogramma's voor de gemeenteraadsverkiezingen van 2022. In: Wageningen Environmental Research (ed.), Wageningen.

Snep, R., 2014. Biodiversiteit op de brouwerij, Wageningen University & Research, Wageningen, the Netherlands.

Sonneveld, M.P.W., 2004. Impressions of interactions: land as a dynamic result of co-production between man and nature, Wageningen, the Netherlands.

Steenbreek, 2021. Verkoeling in eigen tuin: zes tips. Available at: https://steenbreek.nl/verkoeling-in-eigen-tuin/.

Stuiver, M., 2008. Regime Change and Storylines. A sociological analysis of manure practices in contemporary Dutch dairy farming. Wageningen University and Research, Wageningen, the Netherlands.

Stuiver, M. and Koffijberg, M., 2022a. Goed samenwerken met 'de werkvloer': De groene stad, hoe doe je dat? In: D.L. Wageningen Environmental Research, Wageningen, the Netherlands.

Stuiver, M. and Koffijberg, M., 2022b. Hoe ga je om met zorgen van bewoners over natuurinclusief? De groene stad, hoe doe je dat? Wageningen Environmental Research, Wageningen, the Netherlands.

Stuiver, M. and Koffijberg, M., 2022c. Hoe overtuig je de 'boardroom' ? De groene stad, hoe doe je dat? Wageningen Environmental Research, Wageningen, the Netherlands.

Stuiver, M. and Koffijberg, M., 2022d. Hoe versterk je als gemeente de regie op vergroening?! De groene stad, hoe doe je dat? Wageningen Environmental Research Wageningen, the Netherlands.

Stuiver, M. and Koffijberg, M., 2022e. Kan sociale woningbouw ook natuurinclusief en #hoedan? De groene stad, hoe doe je dat? Wageningen Environmental Research, Wageningen, the Netherlands.

Stuiver, M. and Koffijberg, M., 2022f. Natuurinclusief bouwen kan vandaag al beginnen: de groene stad, hoe doe je dat? Wageningen Environmental Research, Wageningen, the Netherlands.

Stuiver, M. and O'Hara, S., 2021. Food connects Washington DC in 2050 – a vision for urban food systems as the centerpieces of a circular economy. Sustainability 13: 7821. https://doi.org/10.3390/su13147821

Termeer, C.J., Dewulf, A. and Biesbroek, G.R., 2017. Transformational change: governance interventions for climate change adaptation from a continuous change perspective. Journal of Environmental Planning and Management 60: 558-576. https://doi.org/10.1080/09640568.2016.1168288

Timmermans, W., Lenzholzer, S., Voskamp, I., Struckman, L., Maagdenberg, G., Weppelman, I., Mashhoodi, B., Dill, S., Cortesão, J., De Haas, W., Van Hattum, T., Luo, S., van Rooij, S., Roosenschoon, O., Sterk, M., Stremke, S., Wamelink, W., Verzandvoort, S. and Harkema, T., 2022. De stad van 2120: natuurlijk! Project number: BO-60-003-004, Wageningen University & Research, Wageningen, the Netherlands. https://doi.org/10.18174/565411

Ulanowicz, R.E., 2021. Socio-ecological networks: a lens that focuses beyond physics. Frontiers in Ecology and Evolution 9. https://doi.org/10.3389/fevo.2021.643122

UN-Habitat III, 2017. The new urban agenda. United Nations, New York, NY, USA.

UN, 2015. Transforming our world: the 2030 Agenda for Sustainable Development. United Nations, New York, NY, USA.

Van 't Hooft, K., 2014. Endogenous development in practice, towards well-being of people and ecosystems. Available at: https://edepot.wur.nl/534539.

Van Hattum, T., Blauw, M., Jensen, M. and Bruin, K., 2016. Towards water smart cities: climate adaptation is a huge opportunity to improve the quality of life in cities, Wageningen University & Research, Wageningen, the Netherlands.

Voskamp, I.M., de Luca, C., Polo-Ballinas, M.B., Hulsman, H. and Brolsma, R., 2021. Nature-based solutions tools for planning urban climate adaptation: state of the art. Sustainability 13: 6381.

WCED, 1987. Our common future. Oxford University Press, New York, NY, USA.

Westerink, J., Pleijte, M., Schrijver, R., Dam, R.v., Krom, M.d. and Boer, T.d., 2021. Can a 'good farmer' be nature-inclusive? Shifting cultural norms in farming in The Netherlands. Journal of Rural Studies 88: 60-70. https://doi.org/10.1016/j.jrurstud.2021.10.011

Westley, F.R., Tjornbo, O., Schultz, L., Olsson, P., Folke, C., Crona, B. and Bodin, Ö., 2013. A theory of transformative agency in linked social-ecological systems. Ecology and Society 18. https://doi.org/10.5751/ES-05072-180327

Wiskerke, J.S., 2015. Urban food systems. In: De Zeeuw, H. and Drechzel, P. (eds.) Cities and agriculture. Routledge, London, UK, pp. 19-43.

Yudina, A., 2017. Garden city: supergreen buildings, urban skyscapes and the new planted space. Thames & Hudson, New York, NY, USA.

www.ingramcontent.com/pod-product-compliance
Lightning Source LLC
Chambersburg PA
CBHW061353240926
51846CB00007B/1396
9789086863839